Logic, Logic, and Logic

Logic, Logic, and Logic

George Boolos

With Introductions and Afterword by John P. Burgess

Edited by Richard Jeffrey

Harvard University Press
Cambridge, Massachusetts
London, England

Copyright © 1998 by the President and Fellows of Harvard College
All rights reserved
Printed in Canada
This book has been digitally reprinted. The content
remains identical to that of previous printings.

First Harvard University Press paperback edition, 1999

Library of Congress Cataloging-in-Publication Data

Boolos, George.
Logic, Logic, and Logic / George S. Boolos ; edited by Richard C. Jeffrey,
with introductions and afterword by John P. Burgess.
     p.   cm.
Includes bibliographical references and index.
ISBN 0-674-53766-1 (cloth)
ISBN 0-674-53767-X (pbk.)
1. Logic. I. Jeffrey, Richard C. II. Title.
BC51.B58 1998   97-37668
160–dc21

Contents

Editorial Preface

George Boolos (1940–1996) designed this book shortly before his untimely death. He provided the title, and a table of contents, but not the kind of introductory notes now common in such volumes, relating the selected papers to each other and to the larger literature. John Burgess has provided such introductions here, together with an afterword surveying George's work on provability logic.

Of the thirty items selected, one has not been published or scheduled for publication elsewhere, while five have been previously anthologized, four of these in William Demopoulos' collection *Frege's Philosophy of Mathematics*. For these five items the text here follows the reprint. A technical appendix accompanying one item has been omitted here.

In addition to the works mentioned in the text, the Bibliography for this volume includes George's complete bibliography, compiled by him.

There are two systematic departures from the form of the original articles: Throughout, we have used a uniform modern notation for logical symbols, and all quotation marks are double quotes (except in one article that is *about* quotation marks, where the usage of the original has been followed exactly).

Part I of the present collection contains papers on the related subjects of set theory and second-order logic, including articles on the logic of plurals. Part II contains contributions to Frege studies. Part III contains articles on mathematical induction, examples on the lengths of proofs, and articles relating to Gödel's incompleteness theorems. George identified the last two items in this part as "lighter papers." We have changed George's ordering in one instance, moving the one other item identified as a "lighter paper," a recent popular piece on Frege, from its original antepenultimate position in the collection to a position at the beginning of Part II, where it serves to supplement the introductory note.

One of the publisher's referees, besides affirming that introductory notes would be desirable even if these could not come from the George's hand, suggested that such notes should be divided into sections corresponding to the parts of the volume. We have followed this suggestion. The same referee suggested that such notes should be supplemented with an account of some of the more technical papers that were not included. The largest body of excluded material (nineteen out of the thirty-five papers not included, plus two books) is work on applications of modal logic to proof theory, the subject now known as provability logic. For this subject the single best account is George's book, *The Logic of Provability*. But an afterword about George's work in this subject has been supplied as suggested, at the end of this volume.

George was as modest about making claims for himself as he was generous in acknowledging the contributions of others, and in writing of his works in our introductory notes we have tried to be as matter-of-fact in tone as he would have been himself. This preface provides a more appropriate place to speak briefly of the importance of his contributions.

While his work on the subject is beyond the scope of the present collection, he has indisputably been one of the major players in provability logic, both as an author of technical papers, and as the author both of the first and of the latest books on the subject. His work on Frege forms the core of Demopoulos' anthology, and his unique and irreplaceable role as a leader in the new direction in Frege studies is universally recognized. His work on plural quantifiers is of interest to philosophers and logicians and linguists alike, and is justly celebrated; it has moreover been the starting point for ambitious further projects by others. If in these last decades the iterative conception of set, long scarcely known to philosophers, has attracted much attention, and if during the same period second-order logic, long viewed suspiciously, has regained its respectability, no single influence has done more to bring about these changes than that of his early papers on these topics.

In sum, in virtually every area in which he has worked, his contributions are widely and gratefully acknowledged. And yet the full magnitude of his contribution to logic only becomes apparent when his papers in diverse areas are brought together, as they are for the first time here.

J.P.B. and R.J.

Editor's Acknowledgments

George's widow, Sally Sedgwick, joins me in thanking the colleagues, students and friends of George who generously devoted many hours of their time to the production of this volume. First and foremost we owe thanks to Ulrich Meyer, who provided the necessary LaTeX typesetting mastery, got the nitty-gritty technical details in order, and supervised the proofreading.

The hands-on LaTeX typesetting was the work of Michael Fara, Delia Graff, T. J. Mather, Ulrich Meyer, and Cara Spencer, under the generalship of Ulrich Meyer, who also did the lion's share of the close work. With the help of Richard Heck, Sally Sedgwick drew up a list of proofreaders, and recruited them. The proofreaders were David Auerbach, Andrew Botterell, Ann Bumpus, John Burgess, Emily Carson, Michael Glanzberg, Warren Goldfarb, Delia Graff, Michael Hallett, Richard Heck, Tamara Horowitz, Darryl Jung, Penelope Maddy, Mathieu Marion, Paolo Mancosu, Vann McGee, James Page, Susan Russinoff, Lisa Sereno, Sanford Shieh, Jason Stanley, Gabriel Uzquiano, and Linda Wetzel. The index is the work of Ulrich Meyer, Michael Hallett, and Emily Carson. Richard Heck and Giovanni Sambin provided valuable comments on earlier versions of the afterword, but of course bear no responsibility for errors in the final product.

Sally Sedgwick bears much responsibility for the book's existence. My own role has been minor: George wanted me to be the editor, so I have put my name in that slot, and done some of the busy work, but the vast bulk of the editorial labor has been done by others, as noted above. My proudest accomplishment as editor has been the recruitment of John Burgess.

For all of us this has been a labor of love.

R.J.

I Studies on Set Theory and the Nature of Logic

Introduction

Set Theory

Georg Cantor, the founder of set theory, published neither explicit axioms nor an intuitive account of the conception of set he was assuming. Gottlob Frege, the founder of modern logic, did give explicit axioms in the logical system he proposed as a foundation for mathematics. Unfortunately, he assumed a naive conception of set, on which for any condition there is a set whose elements are all and only the things for which that condition holds; and such a conception leads to contradiction when applied to the condition "is a set that is not an element of itself." If we assume that there is a set whose elements are all and only the things for which this condition holds, inconsistency results when we ask whether the condition holds for that set itself, whether that set itself is an element of itself. The answer seems to be that it is if and only if it isn't. This paradox was noted by Bertrand Russell, after whom it is named, and by Ernst Zermelo.

Russell proposed a formal system of his own as a foundation for mathematics, and Zermelo proposed a list of axioms for set theory, which with additions and amendments due to Abraham Fraenkel is generally accepted by set theorists today. Though Zermelo at first only claimed for his axioms that they seem to yield the theorems of Cantor while avoiding the paradox of Russell, he later gave an intuitive account of the underlying conception of set he was assuming, which also is widely accepted among set theorists today. It is instructive to compare the Russellian and Zermelodic conceptions, which may also be called the *stratified* and *cumulative* conceptions. On both conceptions sets form a hierarchy, with some sets lying at higher levels than others. Sometimes the temporal metaphor of sets formed at later and later stages is used in place of the spatial metaphor of sets lying at higher and higher levels, in which case one speaks of the *iterative* conception.

On the stratified conception, the elements of any set must come from the level immediately below. At the bottom or level zero are whatever individuals or non-sets there are assumed to be; and for purposes of providing a framework for mathematics one must assume there are infinitely many

3

of them. Above these at level one lie sets whose elements come from level zero. Among these is the empty set ∅, for instance. Above these at level two lie sets whose elements come from level one. Among these is the unit set of the empty set {∅}, for instance. Above these at level three lie sets whose elements come from level two. Among these is the unit set of the unit set of the empty set {{∅}}, for instance. And so on. There is no top. On the cumulative conception, the elements of any set must come from levels below, and may come from any levels below. At the bottom or level zero are any individuals or non-sets there are assumed to be; though for purposes of providing a framework for mathematics one need not assume there are any. Above these at level one lie sets whose elements come from level zero. Above these at level two lie sets whose elements include items from level one, but may also include items from level zero. Above these at level three lie sets whose elements include items from level two, but may also include items from levels one or zero. And so on. Above all finite levels at level omega lie sets whose elements include items from arbitrarily high finite levels, and may contain elements from any finite level. Among these is for instance the set {∅, {∅}, {{∅}}, . . .}. Above these at level omega plus one lie sets whose elements include items from level omega, but may also include items from finite levels. And so on. There is no top.

Article 1 had as its first aim to provide philosophers with an accessible account of the cumulative or iterative conception. For in the early 1970s this conception, though familiar to set theorists, was little known to philosophers, who tended to assume that the only intuitive conception of set was the inconsistent naive one, and that the generally accepted axioms were merely an *ad hoc* list. The article had as a second aim to argue that while the axioms of what is known as *Zermelo set theory* or Z are indeed implicit in the iterative conception of set, the remaining two axioms of the set theory generally accepted today are not implied by that conception, however attractive or plausible they may be on other grounds. One of these axioms is called *replacement,* and since its addition was proposed by Fraenkel, the result of adding it Zermelo set theory Z is called *Zermelo–Fraenkel set theory* or ZF. The other is the *axiom of choice* or AC, Z+AC and ZF+AC being called ZC and ZFC respectively. Z is, though different in form from Russell's system, of roughly equal strength with it; ZF is stronger. Z is probably sufficient for mainstream core mathematics; ZFC is certainly needed for higher set theory.

Many other commentators have taken a more expansive view, considering it to be part of the iterative conception that the hierarchy of sets is supposed to be as "high" as possible, with as many levels of sets as there possibly could be, and as "wide" as possible, with as many sets at a given level as there possibly could be given what sets there are at lower levels, and considering replacement to be in some sense implicit in the demand for "height," and choice in the demand for "width." One reason given in the article for nonetheless regarding choice and replacement as additions to, rather than implications of, the iterative conception, is that one can write down explicit axioms for a theory of levels or stages—this is done in the article—corresponding closely to what is explicitly said in informal expositions of the iterative conception, which axioms do in a strict logical sense imply the axioms of Z but do *not* so imply the remaining axioms of ZFC. The status of choice and replacement, and more generally of set theory, is a question revisited in three later articles in this part.

Article 6 presents an improved list of axioms for the theory of levels or stages, deriving from work of Dana Scott (discussed further in Article 24 of Part III). It contrasts these with a "limitation of size" conception of sets, deriving from work on Frege (reported in Article 11 of Part II), represented by an axiom that may be called the *small extensions principle,* according to which for any condition there is a set of all and only the things for which the conditions holds *provided there are not too many such things.* It is argued that some axioms of ZFC follow from the theory of levels or stages and some from the theory of small extensions, but that there is no single, unified conception of set from which all the axioms follow.

Article 7 is an introduction to a posthumously published lecture of Kurt Gödel. Introductions should not themselves require introductions, lest an infinite regress develop. But it may be said here that some of the "basic theorems on foundations of mathematics" whose "implications" are discussed by Gödel are relevant to the concerns of this part, and that while the aim of the article is to introduce Gödel's views rather than to dispute them, occasion does arise to take issue with Gödel's view that all the axioms of ZFC "force themselves upon us as true," arguing that even if they force themselves upon us as *corresponding to some natural and intuitive conception,* which is as may be, they do not force themselves upon as *corresponding to anything real.*

Article 8 answers the question of its title, "Must we believe in set theory?" in the way one might expect from what has been said about some of the earlier papers: "Not all of it!" One argument for an answer of an unrestricted "Yes!" is the "phenomenological" one that given a certain picture of sets the usual axioms of set theory force themselves upon us. To this the response was indicated already in the preceding article. Another argument is the "pragmatic" one that mathematics, for which set theory provides the accepted framework, is indispensable for science. To this the response is made that in fact nothing like the whole of ZFC is needed for the development of scientifically applicable mathematics, Z being more than enough for such a purpose.

The article also takes note of a consideration advanced by David Lewis, that given the comparative historical records of success and failure in mathematics and philosophy, it would be comically immodest for philosophers to claim to have disproved accepted mathematics. It also takes note of a not unrelated argument, deriving ultimately from Rudolf Carnap, claiming that so long as we acquiesce in accepted mathematical criteria for answering questions in a branch of mathematics, then there is enough agreement about criteria for evaluating answers to give our question a meaning; whereas things are quite otherwise if we begin appealing to extra-scientific philosophical criteria. The article agrees that the Ludovician considerations have force against finitists and nominalists who reject any set theory at all, but insists that the Carnapian argument does not force us to accept *every* axiom of generally accepted set theory. (Actually, the Carnapian argument is called "rubbish.")

The article ends expressing regret in a footnote that there has not been time to discuss explicitly the views of Solomon Feferman and Penelope Maddy. Feferman has been a sympathetic commentator on a restrictive view known as *predicativism,* which accepts much less than Z, and Maddy on an expansive view known as *cabalism,* which accepts much more than ZFC—of which views a bit more will be said below. It is indeed to be regretted that a symposium involving the author and such representatives of more restrictive and more expansive views never took place.[1]

Second-Order Logic

We assume the reader has some familiarity with *first-order* logic, the logic of all modern textbooks. It has two levels, *sentential* logic, or the logic of the truth-functional *connectives* of constant truth, constant falsehood, negation, conjunction, disjunction, and the conditional and biconditional ($\top, \bot, \neg, \wedge, \vee, \rightarrow, \leftrightarrow$), and *predicate* logic, or the logic of the universal and existential *quantifiers* ($\forall, \exists$). A convenient formulation takes $\bot, \rightarrow, \forall$ as the only primitive logical particles, with everything else considered a definitional abbreviation. (For instance, $\neg A$ abbreviates $A \rightarrow \bot$, $\exists x\, Ux$ abbreviates $\neg\forall x \neg Ux$.)

---

[1] To mention some other work of the author related to the material in this section, Gödel proves that if ZFC is consistent then so is ZFC+CH, where CH is Cantor's continuum hypothesis, which was up to Gödel's time the most famous unproved conjecture in set theory. The iterative conception of set heuristically underlies this proof of Gödel's, which proceeds by producing a *constructible* hierarchy of sets *within* the full cumulative hierarchy. The constructible hierarchy in effect allows at a given level not *arbitrary* sets of elements from lower levels, but only ones that are appropriately *definable.* The author's earliest work, in the articles (Boolos and Putnam, 1968) and (Boolos, 1970a), was an examination under a hand lens of the finer structure of Gödel's constructible hierarchy (which Ronald B. Jensen was to examine under an electron microscope, with very surprising results).

The theory of first-order logic, or first-order meta-logic, has two sides. The syntactic or proof-theoretic side gives a characterization of the logical theoremhood of a formula in terms of the existence of a derivation constituting a proof of that formula. Here a derivation is a sequence of formulas or steps, each of which is either an axiom from a specified list or follows from earlier formulas in the sequence by a rule from a specified list, and a derivation constitutes a proof of the formula that is its last step.

The semantic or model-theoretic side gives a characterization of the logical validity of a formula in terms of the non-existence of an interpretation constituting a counter-model for the formula. Here an interpretation consists of non-empty set, to serve as the universe of the interpretation, or domain over which variables range, plus an assignment to each relation-symbol in the language of some relation on that universe, to serve as the interpretation of that symbol; and an interpretation constitutes a model (respectively, counter-model) for a formula if that formula comes out true (respectively, false) under that interpretation. The Gödel completeness theorem, which says that theoremhood coincides with validity, connects the two sides of the theory.

Now second-order logic allows not only first-order quantifications, exemplified by $\forall x(Ux)$ or $\exists x(Ux)$ or $\forall x \exists y(Rxy \land Sxy)$ but also second-order quantifications, exemplified by $\forall U(Ux)$ or $\exists U(Ux)$ or $\forall R \exists S(Rxy \land Sxy)$. Here quantifiers like $\forall U$ with one-place U are called *monadic,* and quantifiers like $\forall R$ with two-place R are called *dyadic.* There is a standard extension of the proof theory for first-order logic to second-order logic. The notion of derivation is changed only by the addition of new axioms, most importantly the scheme of *comprehension,* according to which for any formula φ, the following counts as an axiom:

$$\exists U \forall x(Ux \leftrightarrow \varphi(x))$$

There is also a standard extension of the model theory for first-order logic to second-order logic. The notion of interpretation is unchanged, but there is an added clause in the definition of what it is for a formula to be true in an interpretation, a clause to handle second-order quantifications. Its effect is as if one read $\exists U$ as "there is a subset U of the universe" and Ux as "x is an element of U." Unfortunately, no completeness theorem holds, and second-order logic is in some other respects less tractable that first-order logic. Against this negative feature of lesser tractability must be set the positive feature of greater expressiveness, best seen by a few examples. As a first example, the relation $x = y$ of *identity* between first-order entities, which has to be taken as undefined when first-order logic is considered in isolation, becomes definable as follows:

$$\forall x \forall y(x = y \leftrightarrow \forall U(Ux \leftrightarrow Uy))$$

It may be mentioned that the corresponding relation $U \equiv V$ of *coextensiveness* between second-order entities need not be taken as undefined either, since it is definable, and using only first-order quantification at that, as follows:

$$\forall U \forall V (U \equiv V \leftrightarrow \forall x(Ux \leftrightarrow Vx))$$

As a second example, the relationship $U \cong V$ of *equinumerosity* is definable only by using (dyadic) second-order quantification, as follows:

$$\exists R(\forall x(Ux \rightarrow \exists y(Vy \wedge \forall z(Rxz \leftrightarrow z = y))) \wedge \forall y(Vy \rightarrow \exists x(Ux \wedge \\ \forall z(Rzy \leftrightarrow z = x))))$$

This says that there is a function from the x such that Ux to the y such that Vy that is both *one-to-one* or *injective*, meaning that distinct arguments x and x' are never assigned by the function the same value y, and also *onto* or *surjective*, meaning that every such y is the value of the function for some such x as argument. This means intuitively that there are just as many x such that Ux as there are y such that Vy.

As a third example, consider the following three conditions, which may or may not hold for a given R:

Reflexivity $\forall x(Rxx)$
Symmetry $\forall x \forall y(Rxy \rightarrow Ryz)$
Transitivity $\forall x \forall y \forall z((Rxy \wedge Ryz) \rightarrow Rxz)$

Correspondingly there are three operations associating a given R with the strongest S implied by R that is reflexive or symmetric or transitive, the reflexive or symmetric or transitive *closure* ρR or σR or τR of R. These are definable as follows:

$$\forall x \forall y(\rho Rxy \leftrightarrow (x = y \vee Rxy)$$
$$\forall x \forall y(\sigma Rxy \leftrightarrow (Ryx \vee Rxy)$$
$$\forall x \forall y(\tau Rxy \leftrightarrow \forall U((\forall x'(Rxx' \rightarrow Ux') \wedge \forall z \forall z'((Uz \wedge Rzz') \rightarrow \\ Uz')) \rightarrow Uy))$$

Note that while the first two are definable using only first-order quantification, the third is definable only using (monadic) second-order quantification. To give the stock example, if Rxy amounts to "x is a parent of y," then τRxy amounts to "x is an ancestor of y." This example explains the alternate name *ancestral* of R for τR. To give another example, if Rxy amounts to "natural number x immediately precedes natural number y," then τRxy amounts to "natural number x is less than natural number y." This example begins to indicate the importance of the notion of ancestral.

Article 3 had as its first aim to provide philosophers with an account of second-order logic, and to argue for the genuinely *logical* character of such second-order notions as the ancestral, and of such second-order results as its transitivity. For as of the early 1970s second-order logic was little known to many philosophers, and regarded as not really *logic* by many others. Another aim was to argue for the claim that first-order but not second-order logic can be applied to set theory, a point argued at greater length in another article.

Article 2, heretofore unpublished, was officially a reply to a paper by Charles Parsons, "Sets and classes," (Parsons, 1983c), and unofficially also a reply to a paper by D. A. Martin of the same title, still unpublished. The question with which the paper is largely concerned is this. Does it make sense to use second-order quantifiers $\forall U$, $\exists U$ in a situation where the first-order quantifiers $\forall x$, $\exists x$ range over entities that do not form a set, as is the case if they are taken to range over all sets?

Russell held that it does not make sense even to use first-order quantifiers to range over all sets. For he held that it makes sense to use quantifiers "for all such-and-such" or "for some such-and-such" only when the such-and-such form a set; and there is no set of all sets. In formulating his own system, whose underlying intuitive conception was described above, Russell accordingly had no variables ranging over all sets, but rather had separate styles of variables ranging over each level of sets. In its most illiberal form, Russell's doctrine leads to predicativism, which does not even permit quantification over all sets at any one level, but requires that the levels be ramified into a hierarchy of orders, whose intricacies need not concern us here. In a much liberalized form, Russell's doctrine is compatible with the acceptance even of ZFC, subject to the reservation that one is never taken to be *unambiguously* quantifying over *absolutely* all sets, rather than just those up to some very high level in the hierarchy. Parsons in his paper defended something like a much liberalized version of Russell's doctrine.

Zermelo held that it makes sense to use first-order quantifiers to range over all sets, and even to use second-order quantifiers when first-order quantifiers are so used. Indeed, he himself so used second-order quantifiers in stating one of his set-theoretic axioms, the following axiom of *separation:*

$$\forall U \forall x \exists y \forall z (z \in y \leftrightarrow (z \in x \land U z))$$

He read $\forall U$ as "for every definite property U," and read $U z$ as "z has the definite property U," with "definite" meaning "not vague." The later systems known in the literature as the Bernays–Gödel and Morse–Kelley systems retain this second-order axiom, but read $\forall U$ as "for every class U," and $U z$ as "z is a member of the class U." They add explicit axioms for "classes," which in effect make "classes" into just one additional layer added above the whole hierarchy of sets. Martin, a cabalist or member of the group of prominent set theorists who call themselves the "cabal," adumbrated in his paper a novel theory of "classes" in which these are something more than just a top layer to the hierarchy of sets.

Fraenkel proposed as an amendment to Zermelo to work in a purely first-order language. He proposed substituting for the second-order axiom of separation an axiom *scheme* of first-order separation. Here an axiom scheme is a rule according to which all formulas of a certain form are to count as axioms. The axiom scheme of first-order separation is the rule according to which for any first-order formula φ the following counts as an axiom:

$$\forall x \exists y \forall z (z \in y \leftrightarrow (z \in x \land \varphi(z)))$$

Fraenkel thus in effect took first-order but not second-order logic to be applicable to set theory. Such a position is explicitly defended in Article 2.

An old observation of Georg Kreisel should be mentioned here. Kreisel noted that though it is common to use first-order quantifiers ranging over all sets, the standard definition of logical validity considers only interpretations in which the first-order quantifiers range only over some *set* of things. He went on to note that there is really no problem here, because by the completeness theorem, validity as standardly defined coincides with theoremhood as standardly defined, and it is clear that the axioms and rules involved in the standard definition of theoremhood are acceptable even when the first-order quantifiers range over all sets; but then he went on further to note that no such solution is available in the case of second-order logic, where there is no completeness theorem.[2]

Plural Quantification

Articles 4 and 5 are closely related, both being concerned with *plural* quantification, "some things, the *U*s, are such that ...," as contrasted with *singular* quantification, "some thing x is such that ..." Quantificational constructions in natural language that resist formalization in first-order terms were a recognized topic in the literature of linguistics and logic, so it is somewhat surprising that plural quantification had prior to these papers received little or no attention. That it is not formalizable in terms of first-order singular quantification is one of the points argued in these articles.

---

[2] Again to mention related work, articles (Boolos, 1970b) and (Boolos, 1973) concern the Löwenheim–Skolem and Beth theorems, two fundamental results about the tractability of first-order logic, both worthy to be placed alongside the Gödel completeness theorem, and both sometimes cited for their philosophical bearings. For second-order logic, the Löwenheim–Skolem theorem fails; but the Beth theorem, which is a substantial result at the first-order level, becomes a trivial truism at the second-order level. This perhaps shows that it is not *in all respects* the case that first-order logic is always more tractable than second-order logic.

Plural quantification provides an alternative interpretation of second-order logic. Recall that on a conventional interpretation, the first-order entities are elements of some set of things, the universe of the interpretation, and the second-order entities are subsets of that universe. $\exists U$ and Ux are read "there is a subset U of the universe" and "x is an element of the set U." On the alternative interpretation, the first-order entities are some things, which need not form a set, and there are no second-order entities. For second-order quantification is not interpreted as quantification over a different sort of thing, but rather as a different sort of quantification, plural rather than singular, over the same things. $\exists U$ and Ux are read "there are some things, the Us" and "x is one of the Us." The availability of this new interpretation makes reconsideration of some positions taken in earlier articles needful. For one thing, it is now not so clear that second-order logic cannot be applied to set theory. For another thing, it is now not so clear that the standard definition of validity in terms of models whose universes are sets is appropriate (as per Kreisel's remarks recalled above).

Some note should be taken here of a limitation, noted by the nominalist Hartry Field, especially since optimism about the potential role of plural quantification recurs in some later articles, especially in Part II. The limitation is that plurals provide a reading of monadic second-order quantification, such as could replace quantification over one-place relations or collections of single elements of the universe, but not a reading of dyadic second-order quantification, such as could replace quantification over two-place relations or collections of ordered pairs of elements of the universe. This is a significant limitation in the general case, but not in the case of set theory, where the ordered pair of any two elements of the universe is itself already an element of the universe.

Some note should also be taken here of another interpretation of second-order quantification that has sometimes been proposed: the theory of parts and wholes known as *mereology* (Stanisław Leśniewski) or the *calculus of individuals* (Nelson Goodman). Suffice it to say here that this theory does supply another surrogate for monadic second-order logic, provided the things in the universe considered are non-overlapping, no two having any parts in common. Especially interesting in such a case is the combination of mereology and plural quantification, which provides a surrogate for monadic third-order logic. The power of this combination only becomes clear for the first time in the book of David Lewis, *Parts of Classes* (Lewis, 1991). Indeed it only becomes *fully* clear in the co-authored appendix thereto, where A. P. Hazen points out that monadic third-order logic provides a surrogate for dyadic second-order logic.[3]

---

[3] Again to mention related work, the author's interest in natural language was evinced as early as article (Boolos, 1969). Topics pertaining to quantification in natural language were pursued in two squibs in *Linguistic Inquiry*, articles (Boolos, 1981a) and (Boolos, 1984c), the latter being concerned with plural quantification.

1

The Iterative Conception of Set

A set, according to Cantor, is "any collection. . . into a whole of definite, well-distinguished objects . . . of our intuition or thought."[1] Cantor also defined a set as a "many, which can be thought of as one, i.e., a totality of definite elements that can be combined into a whole by a law."[2] One might object to the first definition on the grounds that it uses the concepts of *collection* and *whole*, which are notions no better understood than that of *set*, that there ought to be sets of objects that are not objects of our thought, that "intuition" is a term laden with a theory of knowledge that no one should believe, that *any* object is "definite," that there should be sets of ill-distinguished objects, such as waves and trains, etc., etc. And one might object to the second on the grounds that "a many" is ungrammatical, that if something is "a many" it should hardly be thought of as one, that *totality* is as obscure as *set*, that it is far from clear how laws can combine anything into a whole, that there ought to be other combinations into a whole than those effected by "laws," etc., etc. But it cannot be denied that Cantor's definitions could be used by a person to identify and gain some understanding of the sort of object of which Cantor wished to treat. Moreover, they do suggest—although, it must be conceded, only very faintly—two important characteristics of sets: that a set is "determined" by its elements in the sense that sets with exactly the same elements are identical, and that, in a sense, the clarification of which is one of the principal objects of the theory whose rationale we shall give, the elements of a set are "prior to" it.

It is not to be presumed that the concepts of *set* and *member of* can be

Reprinted with the kind permission of the editors from *The Journal of Philosophy* 68 (1971): 215–232.

[1] "Unter einer 'Menge' verstehen wir jede Zusammenfassung M von bestimmten wohlunterschiedenen Objekten m unserer Anschauung oder unseres Denkens (welche die 'Elemente' von M genannt werden) zu einem Ganzen"(Cantor, 1932), p. 282.

[2] " . . . jedes Viele, welches sich als Eines denken lässt, d.h. jeden Inbegriff bestimmter Elemente, welcher durch ein Gesetz zu einem Ganzen verbunden werden kann"(Cantor, 1932), p. 204.

explained or defined by means of notions that are simpler or conceptually more basic. However, as a theory about sets might itself provide the sort of elucidation about sets and membership that good definitions might be hoped to offer, there is no reason for such a theory to begin with, or even contain, a definition of "set." That we are unable to give informative definitions of *not* or *for some* does not and should not prevent the development of quantificational logic, which provides us with significant information about these concepts.

I. Naive set theory

Here is an idea about sets that might occur to us quite naturally, and is perhaps suggested by Cantor's definition of a set as a totality of definite elements that can be combined into a whole *by a law*.

By the law of excluded middle, any (one-place) predicate in any language either applies to a given object or does not. So, it would seem, to any predicate there correspond two sorts of thing: the sort of thing to which the predicate applies (of which it is true) and the sort of thing to which it does not apply. So, it would seem, for any predicate there is a set of all and only those things to which it applies (as well as a set of just those things to which it does not apply). Any set whose members are exactly the things to which the predicate applies—by the axiom of extensionality, there cannot be two such sets—is called the *extension* of the predicate. Our thought might therefore be put: "Any predicate has an extension." We shall call this proposition, together with the argument for it, the *naive conception of set*.

The argument has great force. How could there *not* be a collection, or set, of just those things to which any given predicate applied? Isn't anything to which a predicate applies similar to all other things to which it applies in precisely the respect that it applies to them; and how could there fail to be a set of all things similar to one another in this respect? Wouldn't it be extremely implausible to say, of any particular predicate one might consider, that there weren't two kinds of thing it determined, namely, a kind of thing of which it is true, and a kind of thing of which it is not true? And why should one not take these kinds of things to be sets? Aren't kinds sets? If not, what is the difference?

Let us denote by "$\mathcal{K}$" a certain standardly formalized first-order language, whose variables range over all sets and individuals (= non-sets), and whose nonlogical constants are a one-place predicate letter "S" abbreviating "is a set," and a two-place predicate letter "$\in$", abbreviating "is a member of." Which sentences of this language, together with their consequences, do we believe state truths about sets? Otherwise put, which formulas of $\mathcal{K}$ should we take as axioms of a set theory on the strength of our beliefs about sets?

If the naive conception of set is correct, there should (at least) be a set of just those things to which φ applies, if φ is a formula of $\mathcal{K}$. So (the universal closure of) $\ulcorner \exists y(Sy \wedge \forall x(x \in y \leftrightarrow \varphi)) \urcorner$ should express a truth about sets (if no occurrence of "y" in φ is free).

We call the theory whose axioms are the axiom of extensionality (to which we later recur), i.e., the sentence

$$\forall x \forall y(Sx \wedge Sy \wedge \forall z(z \in x \leftrightarrow z \in y) \rightarrow x = y)$$

and all formulas $\ulcorner \exists y(Sy \wedge \forall x(x \in y \leftrightarrow \varphi)) \urcorner$ (where "y" does not occur free in φ) *naive set theory*.

Some of the axioms of naive set theory are the formulas

$$\exists y(Sy \wedge \forall x(x \in y \leftrightarrow x \neq x))$$

$$\exists y(Sy \wedge \forall x(x \in y \leftrightarrow (x = z \vee x = w)))$$

$$\exists y(Sy \wedge \forall x(x \in y \leftrightarrow \exists w(x \in w \wedge w \in z)))$$

$$\exists y(Sy \wedge \forall x(x \in y \leftrightarrow (Sx \wedge x = x)))$$

The first of these formulas states that there is a set that contains no members. By the axiom of extensionality, there can be at most one such set. The second states that there is a set whose sole members are z and w; the third, that there is a set whose members are just the members of members of z.

The last, which states that there is a set that contains all sets whatsoever, is rather anomalous; for if there is a set that contains all sets, a universal set, that set contains itself, and perhaps the mind ought to boggle at the idea of something's *containing* itself. Nevertheless, naive set theory is simple to state, elegant, initially quite credible, and natural in that it articulates a view about sets that might occur to one quite naturally.

Alas, it is inconsistent.

Proof of the inconsistency of naive set theory
(Russell's paradox)

No set can contain all and only those sets which do not contain themselves. For if any such set existed, if it contained itself, then, as it contains *only* those sets which do not contain themselves, it would not contain itself; but if it did not contain itself, then, as it contains *all* those sets which do not contain themselves, it would contain itself. Thus any such set would have to contain itself if and only if it did not contain itself. Consequently, there is no set that contains all and only those sets which do not contain themselves.

This argument, which uses no axioms of naive set theory, or any other set theory, shows that the sentence

$$\neg\exists y(Sy \wedge \forall x(x \in y \leftrightarrow (Sx \wedge \neg x \in x)))$$

is *logically valid* and, hence, is a theorem of any theory that is expressed in $\mathcal{K}$. But one of the axioms and, hence, one of the theorems, of naive set theory is the sentence

$$\exists y(Sy \wedge \forall x(x \in y \leftrightarrow (Sx \wedge \neg x \in x)))$$

Naive set theory is therefore inconsistent.

II. The iterative conception of set

Faced with the inconsistency of naive set theory, one might come to believe that any decision to adopt a system of axioms about sets would be *arbitrary* in that no explanation could be given why the particular system adopted had any greater claim to describe what we conceive sets and the membership relation to be like than some other system, perhaps incompatible with the one chosen. One might think that no answer could be given to the question: why adopt *this* particular system rather than that or this other one? One might suppose that any apparently consistent theory of sets would have to be unnatural in some way or fragmentary, and that, if consistent, its consistency would be due to certain provisions that were laid down for the express purpose of avoiding the paradoxes that show naive set theory inconsistent, but that lack any independent motivation.

One might imagine all this; but there is another view of sets: the *iterative conception of set*, as it is sometimes called, which often strikes people as entirely natural, free from artificiality, not at all ad hoc, and one they might perhaps have formulated themselves.

It is, perhaps, no more natural a conception than the naive conception, and certainly not quite so simple to describe. On the other hand, it is, as far as we know, consistent: not only are the sets whose existence would lead to contradiction not assumed to exist in the axioms of the theories that express the iterative conception, but the more than fifty years of experience that practicing set theorists have had with this conception have yielded a good understanding of what can and what cannot be proved in these theories, and at present there just is no suspicion at all that they are inconsistent.[3]

---

[3]The conception is well known among logicians; a rather different version of it is sketched in (Shoenfield, 1967), ch. 10. I learned of it principally from Putnam, Kripke, and Donald Martin. Authors of set-theory texts either omit it or relegate it to back pages; philosophers, in the main, seem to be unaware of it, or of the preeminence of ZF, which may be said to embody it. It is due primarily to Zermelo and Russell.

The standard, first-order theory that expresses the iterative conception of set as fully as a first-order theory in the language $\mathcal{L}$ of set theory[4] can, is known as *Zermelo–Fraenkel set theory*, or "ZF" for short. There are other theories besides ZF that embody the iterative conception: one of them, Zermelo set theory, or "Z", which will occupy us shortly, is a *subsystem* of ZF in the sense that any theorem of Z is also a theorem of ZF; two others, von Neumann–Bernays–Gödel set theory and Morse–Kelley set theory, are supersystems (or extensions) of ZF, but they are most commonly formulated as second-order theories.

Other theories of sets, incompatible with ZF, have been proposed.[5] These theories appear to lack a motivation that is independent of the paradoxes in the following sense: they are not, as Russell has written, "such as even the cleverest logician would have thought of if he had not known of the contradictions."[6] A final and satisfying resolution to the set-theoretical paradoxes cannot be embodied in a theory that blocks their derivation by artificial technical restrictions on the set of axioms that are imposed *only because* paradox would otherwise ensue; these other theories survive only through such artificial devices. ZF alone (together with its extensions and subsystems) is not only a consistent (apparently) but also an independently motivated theory of sets: there is, so to speak, a "thought behind it" about the nature of sets which might have been put forth even if, impossibly, naive set theory had been consistent. The thought, moreover, can be described in a rough, but informative way without first stating the theory the thought is behind.

In order to see why a conception of set other than the naive conception might be desired even if the naive conception were consistent, let us take another look at naive set theory and the anomalousness of its axiom, "$\exists y(Sy \wedge \forall x(x \in y \leftrightarrow (Sx \wedge x = x)))$."

According to this axiom there is a set that contains all sets, and therefore there is a set that contains itself. It is important to realize how odd the idea of something's containing itself is. Of course a set can and must *include* itself (as a subset). But *contain* itself? Whatever tenuous hold on the concepts of *set* and *member* were given one by Cantor's definitions of "set" and one's ordinary understanding of "element," "set," "collection," etc. is altogether lost if one is to suppose that some sets are members of themselves. The idea is paradoxical not in the sense that it is contradictory to suppose that some set is a member of itself, for, after all, "$\exists x(Sx \wedge x \in x)$" is obviously consistent, but that if one understands "$\in$" as meaning "is a

---

[4] $\mathcal{L}$ contains (countably many) variables, ranging over (pure) sets, "$=$", and "$\in$", which is its sole nonlogical constant.

[5] For example, Quine's systems NF and ML.

[6] (Russell, 1959), p. 80.

member of," it is very, very peculiar to suppose it true. For when one is told that a set is a collection into a whole of definite elements of our thought, one thinks: Here are some things. Now we bind them up into a whole.[7] *Now* we have a set. We don't suppose that what we come up with after combining some elements into a whole could have been one of the very things we combined (not, at least, if we are combining two or more elements).

If $\exists x(Sx \wedge x \in x)$, then $\exists x \exists y(Sx \wedge Sy \wedge x \in y \wedge y \in x)$. The supposition that there are sets x and y each of which belongs to the other is almost as strange as the supposition that some set is a self-member. There is of course an infinite sequence of such cyclical pathologies: $\exists x \exists y \exists z(Sx \wedge Sy \wedge Sz \wedge x \in y \wedge y \in z \wedge z \in x)$, etc. Only slightly less pathological are the suppositions that there is an ungrounded set,[8] or that there is an infinite sequence of sets $x_0, x_1, x_2, \ldots$, each term of which belongs to the previous one.

There does not seem to be any argument that is guaranteed to persuade someone who really does not see the peculiarity of a set's belonging to itself, or to one of its members, etc., that these states of affairs are peculiar. But it is in part the sense of their oddity that has led set-theorists to favor conceptions of set, such as the iterative conception, according to which what they find odd does not occur.

We describe this conception now. Our description will have three parts. The first is a rough statement of the idea. It contains such expressions as "stage," "is formed at," "earlier than," "keep on going," which must be exorcised from any formal theory of sets. From the rough description it sounds as if sets were continually being created, which is not the case. In the second part, we present an axiomatic theory which partially formalizes the idea roughly stated in the first part. For reference, let us call this theory the *stage theory*. The third part consists in a derivation from the stage theory of the axioms of a theory of sets. These axioms are formulas of $\mathcal{L}$, the language of set theory, and contain none of the metaphorical expressions which are employed in the rough statement and of which abbreviations are found in the language in which the stage theory is expressed.

Here is the idea, roughly stated:

A set is any collection that is formed at some stage of the following process: Begin with individuals (if there are any). An individual is an object that is not a set; individuals do not contain members. At stage zero (we count from zero instead of one) form all possible collections of individuals. If there are no individuals, only one collection, the null set, which contains no

---

[7]We put a "lasso" around them, in a figure of Kripke's.

[8]x is *ungrounded* if x belongs to some set z, each of whose members has a member in common with z.

members, is formed at this 0th stage. If there is only one individual, two sets are formed: the null set and the set containing just that one individual. If there are two individuals, four sets are formed; and in general, if there are n individuals, 2^n sets are formed. Perhaps there are infinitely many individuals. Still, we assume that one of the collections formed at stage zero is the collection of all individuals, however many of them there may be.

At stage one, form all possible collections of individuals and sets formed at stage zero. If there are any individuals, at stage one some sets are formed that contain both individuals and sets formed at stage zero. Of course some sets are formed that contain only sets formed at stage zero. At stage two, form all possible collections of individuals, sets formed at stage zero, and sets formed at stage one. At stage three, form all possible collections of individuals and sets formed at stages zero, one, and two. At stage four, form all possible collections of individuals and sets formed at stages zero, one, two, and three. Keep on going in this way, at each stage forming all possible collections of individuals and sets formed at earlier stages.

Immediately after all of stages zero, one, two, three, ..., there is a stage; call it stage omega. At stage omega, form all possible collections of individuals formed at stages zero, one, two, ... One of these collections will be the set of all sets formed at stages zero, one, two, ...

After stage omega there is a stage omega plus one. At stage omega plus one form all possible collections of individuals and sets formed at stages zero, one, two, ..., and omega. At stage omega plus two form all possible collections of individuals and sets formed at stages zero, one, two, ..., omega, and omega plus one. At stage omega plus three form all possible collections of individuals and sets formed at earlier stages. Keep on going in this way.

Immediately after all of stages zero, one, two, ..., omega, omega plus one, omega plus two, ..., there is a stage, call it stage omega plus omega (or omega times two). At stage omega plus omega form all possible collections of individuals and sets formed at earlier stages. At stage omega plus omega plus one ... ...

... omega plus omega plus omega (or omega times three) ...

... (omega times four) ...

... omega times omega ... ...

Keep on going in this way ...

According to this description, sets are formed over and over again: in fact, according to it, a set is formed at every stage later than that at which it is first formed. We could continue to say this if we liked; instead we shall say that a set is formed only once, namely, at the earliest stage at which, on our old way of speaking, it would have been said to be formed.

That is a rough statement of the iterative conception of set. According to this conception, no set belongs to itself, and hence there is no set of all sets; for every set is formed at some earliest stage, and has as members only individuals or sets formed at still earlier stages. Moreover, there are not two sets x and y, each of which belongs to the other. For if y belonged to x, y would have had to be formed at an earlier stage than the earliest stage at which x was formed, and if x belonged to y, x would have had to be formed at an earlier stage than the earliest stage at which y was formed. So x would have had to be formed at an earlier stage than the earliest stage at which it was formed, which is impossible. Similarly, there are no sets x, y, and z such that x belongs to y, y to z, and z to x. And in general, there are no sets $x_0, x_1, x_2, \ldots, x_n$ such that x_0 belongs to x_1, x_1 to $x_2, \ldots, x_{n-1}$ to x_n, and x_n to x_0. Furthermore it would appear that there is no sequence of sets $x_0, x_1, x_2, x_3, \ldots$, such that x_1 belongs to x_0, x_2 belongs to x_1, x_3 belongs to x_2, and so forth. Thus, if sets are as the iterative conception has them, the anomalous situations do not arise in which sets belong to themselves or to others that in turn belong to them.

The sets of which ZF in its usual formulation speaks ("quantifies over") are not all the sets there are, if we assume that there are some individuals, but only those which are formed at some stage under the assumption that there are no individuals. These sets are called *pure* sets. All members of a pure set are pure sets, and any set, all of whose members are pure, is itself pure. It may not be obvious that any pure sets are ever formed, but the set Λ, which contains no members at all, is pure, and is formed at stage 0. $\{\Lambda\}$ and $\{\{\Lambda\}\}$ are also both pure and are formed at stages 1 and 2, respectively. There are many others. From now on, we shall use the word "set" to mean "pure set."

Let us now try to state a theory, the stage theory, that precisely expresses much, but not all, of the content of the iterative conception. We shall use a language, $\mathcal{J}$, in which there are two sorts of variables: variables "x", "y", "z", "w", $\ldots$, which range over sets, and variables "r", "s", "t", which range over stages. In addition to the predicate letters "$\in$" and "$=$" of $\mathcal{L}$, $\mathcal{J}$ also contains two new two-place predicate letters "E", read "is earlier than," and "F", read "is formed at." The rules of formation of $\mathcal{J}$ are perfectly standard.

Here are some axioms governing the sequence of stages:

(I) $\forall s \neg sEs$ (No stage is earlier than itself.)

(II) $\forall r \forall s \forall t((rEs \land sEt) \rightarrow rEt)$ (*Earlier than* is transitive.)

(III) $\forall s \forall t(sEt \lor s = t \lor tEs)$ (*Earlier than* is connected.)

(IV) $\exists s \forall t(t \neq s \rightarrow sEt)$ (There is an earliest stage.)

(V) $\forall s \exists t(sEt \land \forall r(rEt \rightarrow (rEs \lor r = s)))$ (*Immediately* after any stage there is another.)

Here are some axioms describing when sets and their members are formed:

(VI) $\exists s(\exists t\, tEs \land \forall t(tEs \rightarrow \exists r(tEr \land rEs)))$ (There is a stage, not the earliest one, which is not immediately after any one stage. In the rough description, stage omega was such a stage.)

(VII) $\forall x \exists s(xFs \land \forall t(xFt \rightarrow t = s))$ (Every set is formed at some unique stage.)

(VIII) $\forall x \forall y \forall s \forall t((y \in x \land xFs \land yFt) \rightarrow tEs)$ (Every member of a set is formed *before*, i.e., at an earlier stage than, the set.)

(IX) $\forall x \forall s \forall t(xFs \land tEs \rightarrow \exists y \exists r(y \in x \land yFr \land (t = r \lor tEr)))$ (If a set is formed at a stage, then, at or after any earlier stage, at least one of its members is formed. So it never happens that all the members of a set are formed before some stage, but the set is not formed at that stage, if it has not been formed already.)

We may capture part of the content of the idea that at any stage every *possible* collection (or set) of sets formed at earlier stages is formed (if it has not yet been formed) by taking as axioms all formulas $\ulcorner \forall s \exists y \forall x(x \in y \leftrightarrow (\chi \land \exists t(tEs \land xFt)))\urcorner$, where χ is a formula of the language $\mathscr{J}$ in which no occurrence of "y" is free. Any such axiom will say that for any stage there is a set of just those sets to which χ applies that are formed before that stage. Let us call these axioms *specification axioms*.

There is still one important feature contained in our rough description that has not yet been expressed in the stage theory: the analogy between the way sets are *inductively generated* by the procedure described in the rough statement and the way the natural numbers $0, 1, 2, \ldots$ are inductively generated from 0 by the repeated application of the successor operation. One way to characterize this feature is to assert a suitable induction principle concerning sets and stages; for, as Frege, Dedekind, Peano, and others have enabled us to see, the content of the idea that objects of a certain kind are inductively generated in a certain way is just the proposition than an appropriate induction principle holds of those objects.

The principle of mathematical induction, the induction principle governing the natural numbers, has two forms, which are interderivable on certain assumptions about the natural numbers. The first version of the principle is the statement

$$\forall P[(P0 \land \forall n[Pn \rightarrow PSn]) \rightarrow \forall n\, Pn]$$

which may be read, "If 0 has a property and if whenever a natural number
has the property its successor does, then every natural number has the
property." The second version is the statement

$$\forall P[\forall n(\forall m[m < n \to Pm] \to Pn) \to \forall n\, Pn]$$

It may be read, "*If each natural number has a property provided that all
smaller natural numbers have it, then every natural number has the property.*"

The induction principle about sets and stages that we should like to assert
is modeled after the second form of the principle of mathematical induction.
Let us say that a stage s *is covered by* a predicate if the predicate applies to
every set formed at s. Our analogue for sets and stages of the second form
of mathematical induction says that *if each stage is covered by a predicate
provided that all earlier stages are covered by it, then every stage is covered
by the predicate.* The full force of this assertion can be expressed only with
a second-order quantifier. However, we can capture some of its content by
taking as axioms all formulas

$$\ulcorner \forall s(\forall t(tEs \to \forall x(xFt \to \theta)) \to \forall x(xFs \to \chi)) \to \forall s\forall x(xFs \to \chi)\urcorner$$

where χ is a formula of $\mathcal{J}$ that contains no occurrences of "t" and θ is just
like χ except for containing a free occurrence of "t" wherever χ contains a
free occurrence of "s". [Observe that "$\forall x(xFs \to \chi)$" says that χ applies
to every set formed at stage s and, hence, that s is covered by χ.] We call
these axioms *induction axioms.*

III. Zermelo set theory

We complete the description of the iterative conception of set by showing
how to derive the axioms of a theory of sets from the stage theory. The
axioms we derive speak only about sets and membership: they are formulas
of $\mathcal{L}$.

The axiom of the null set: $\exists y \forall x \neg x \in y$. (*There is a set with no
members.*)

Derivation. Let $\chi = $ "$x = x$". Then

$$\forall s \exists y \forall x(x \in y \leftrightarrow (x = x \land \exists t(tEs \land xFt)))$$

is a specification axiom, according to which, for any stage, there is a set
of all sets formed at earlier stages. As there is an earliest stage, stage 0,
before which no sets are formed, there is a set that contains no members.
Note that, by axiom (IX) of the stage theory, any set with no members

is formed at stage 0; for if it were formed later, it would have to have a member (that was formed at or after stage 0).

The axiom of pairs: $\forall z \forall w \exists y \forall x (x \in y \leftrightarrow (x = z \lor x = w))$. (*For any sets v and w, not necessarily distinct, there is a set whose sole members are z and w.*)

Derivation. Let $\chi = $ "$(x = z \lor x = w)$". Then

$$\forall s \exists y \forall x (x \in y \leftrightarrow ((x = z \lor x = w) \land \exists t(t\mathrm{E}s \land x\mathrm{F}t)))$$

is a specification axiom, according to which, for any stage, there is a set of all sets formed at earlier stages that are identical with either z or w. Any set is formed at some stage. Let r be the stage at which z is formed; s, the stage at which w is formed. Let t be a stage later than both r and s. Then there is a set of all sets formed at stages earlier than t that are identical with z or w. So there is a set containing just z and w.

The axiom of unions: $\forall z \exists y \forall x (x \in y \leftrightarrow \exists w(x \in w \land w \in z))$. (*For any set z, there is a set whose members are just the members of members of z.*)

Derivation. "$\forall s \exists y \forall x (x \in y \leftrightarrow (\exists w(x \in w \land w \in z) \land \exists t(t\mathrm{E}s \land x\mathrm{F}t)))$" is a specification axiom, according to which, for any stage, there is a set of all members of members of z formed at earlier stages. Let s be the stage at which z is formed. Every member of z is formed before s, and hence every member of a member of z is also formed before s. Thus there is a set of all members of members of z.

The power-set axiom: $\forall z \exists y \forall x (x \in y \leftrightarrow \forall w(w \in x \to w \in z))$. (*For any set z, there is a set whose members are just the subsets of z.*)

Derivation. "$\forall s \exists y \forall x (x \in y \leftrightarrow (\forall w(w \in x \to w \in z) \land \exists t(t\mathrm{E}s \land x\mathrm{F}t)))$" is a specification axiom, according to which, for any stage, there is a set of all subsets of z formed at earlier stages. Let t be the stage at which z is formed and let s be the next later stage. If x is a subset of z, then x is formed before s. For otherwise, by axiom (IX), there would be a member of x that was formed at or after t and, hence, that was not a member of z. So there is a set of all subsets of z formed before s, and hence a set of all subsets of z.

The axiom of infinity:

$$\exists y(\exists x(x \in y \land \forall z \neg z \in x) \land \forall x(x \in y \to \exists z(z \in y \land \forall w(w \in z \leftrightarrow$$
$$(w \in x \lor w = x)))))$$

(*Call a set null if it has no members. Call z a successor of x if the members of z are just those of x and x itself. Then there is a set which contains a null set and which contains a successor of any set it contains.*)

Derivation. Let us first observe that every set x has a successor. For let y be a set containing just x and x (axiom of pairs), and let w be a set containing just x and y (axiom of pairs again), and let z contain just the members of members of w (axiom of unions). Then z is a successor of x, for its members are just x and x's members. Next, note that if z is a successor of x, x is formed at r, and t is the next stage after r, then z is formed at t. For every member of z is formed before t. So z is formed at or before t, by axiom (IX). But x, which is in z, is formed at r. So z cannot be formed at or before r. So z cannot be formed before t. Now, by axiom (VI), there is a stage s, not the earliest one, which is not immediately after any stage. "$\forall s \exists y \forall x (x \in y \leftrightarrow (x = x \wedge \exists t(t\mathrm{E}s \wedge x\mathrm{F}t)))$" is a specification axiom, according to which, for any stage, there is a set of all sets formed at earlier stages. So there is a set y of all sets formed before s. y thus contains all sets formed at stage 0, and hence contains a null set. And if y contains x, y contains all successors of x (and there are some), for all these are formed at stages immediately after stages before s and, hence, at stages themselves before s.

Axioms of separation (Aussonderungsaxioms): *All formulas*

$$\ulcorner \forall z \exists y \forall x (x \in y \leftrightarrow (x \in z \wedge \varphi)) \urcorner$$

where φ is a formula of $\mathcal{L}$ in which no occurrence of "y" is free.

Derivation. If φ is a formula of $\mathcal{L}$ in which no occurrence of "y" is free, then $\ulcorner \forall s \exists y \forall x (x \in y \leftrightarrow ((x \in z \wedge \varphi) \wedge \exists t(t\mathrm{E}s \wedge x\mathrm{F}t))) \urcorner$ is a specification axiom, which we may read, "for any stage s, there is a set of all sets formed at earlier stages, which belong to z and to which φ applies." Let s be the stage at which z is formed. All members of z are formed before s. So, for any z, there is a set of just those members of z to which φ applies, which we would write, $\ulcorner \forall z \exists y \forall x (x \in y \leftrightarrow (x \in z \wedge \varphi)) \urcorner$. A formal derivation of an Aussonderungsaxiom would use the specification axiom described and axioms (VII) and (VIII) of the stage theory.

Axioms of regularity: *All formulas*

$$\ulcorner \exists x \varphi \rightarrow \exists x (\varphi \wedge \forall y (y \in x \rightarrow \neg \psi)) \urcorner$$

where φ does not contain "y" at all and ψ is just like φ except for containing an occurrence of "y" wherever φ contains a free occurrence of "x".

Derivation. The idea: Suppose φ applies to some set x'. x' is formed at some stage. That stage is therefore not covered by $\ulcorner \neg \varphi \urcorner$. By an induction axiom, there is then a stage s not covered by $\ulcorner \neg \varphi \urcorner$, although all stages earlier than s are covered by $\ulcorner \neg \varphi \urcorner$. Since s is not covered by $\ulcorner \neg \varphi \urcorner$, there is an x, formed at s, to which $\ulcorner \neg \varphi \urcorner$ does not apply, i.e., to which φ applies.

If y is in x, however, y is formed before s, and hence the stage at which it is formed is covered by $\ulcorner \neg \varphi \urcorner$. So $\ulcorner \neg \varphi \urcorner$ applies to y (which is what $\ulcorner \neg \psi \urcorner$ says).

For a formal derivation, contrapose, reletter, and simplify the induction axiom

$$\ulcorner \forall s (\forall t (tEs \to \forall x (xFt \to \neg \varphi)) \to \forall x (xFs \to \neg \varphi)) \to \\ \forall s \forall x (xFs \to \neg \varphi) \urcorner$$

so as to obtain

$$\ulcorner \exists s \exists x (xFs \wedge \varphi) \to \exists s \exists x (xFs \wedge \varphi \wedge \forall y \forall t (tEs \wedge yFt \to \neg \psi)) \urcorner$$

Assume $\ulcorner \exists x \varphi \urcorner$. Use axiom (VII) and modus ponens to obtain

$$\ulcorner \exists s \exists x (xFs \wedge \varphi \wedge \forall y \forall t (tEs \wedge yFt \to \neg \psi)) \urcorner$$

Use axioms (VII) and (VIII) to obtain $\ulcorner \exists x (\varphi \wedge \forall y (y \in x \to \neg \psi)) \urcorner$ from this.

The axioms of regularity (partially) express the analogue for sets of the version of mathematical induction called the *least-number principle:* if there is a number that has a property, then there is a least number with that property. The analogue itself has been called the *principle of set theoretical induction.*[9] Here is an application of set-theoretical induction.

Theorem *No set belongs to itself.*

Proof. Suppose that some set belongs to itself, i.e., that $\exists x \, x \in x$.

$$\exists x \, x \in x \to \exists x (x \in x \wedge \forall y (y \in x \to \neg y \in y))$$

is an axiom of regularity. By modus ponens, then, some set x belongs to itself though no member of x (not even x) belongs to itself. This is a contradiction. ∎

The axioms whose derivations we have given are those statements which are often taken as axioms of ZF and which are deducible from all (sufficiently strong[10]) theories that can fairly be called formalizations of the iterative conception, as roughly described. (The axiom of extensionality has a special status, which we discuss below.) Other axioms than those we have given could have been taken as axioms of the stage theory. For example, we could have fairly taken as an axiom a statement asserting the

---

[9]By Tarski, among others.

[10] "Sufficiently strong" may here be taken to mean "at least as strong as the stage theory."

existence of a stage, not immediately later than any stage, but later than some stage that is itself neither the earliest stage nor immediately later than any stage. Such an axiom would have enabled us to deduce a stronger axiom of infinity than the one whose derivation we have given, but this stronger statement is not commonly taken as an axiom of ZF. We could also have derived other statements from the stage theory, such as the statement that no set belongs to any of its members, but this statement is never taken as an axiom of ZF. We do not believe that the axioms of replacement or choice can be inferred from the iterative conception.

One of the axioms of regularity,

$$\forall z(\exists x\, x \in z \to \exists x(x \in z \land \forall y(y \in x \to \neg y \in z)))$$

is sometimes called *the* axiom of regularity; in the presence of other axioms of ZF, all the other axioms of regularity follow from it. The name "Zermelo set theory" is perhaps most commonly given to the theory whose axioms are "$\forall x \forall y(\forall z(z \in x \leftrightarrow z \in y) \to x = y)$," i.e., the axiom of extensionality, and the axioms of the null set, pairs, and unions, the power-set axiom, the axiom of infinity, all the Aussonderungsaxioms, and the axiom of regularity.[11] With the exception of the axiom of extensionality, then, all the axioms of Zermelo set theory follow from the stage theory.

IV. Zermelo–Fraenkel set theory

The axioms of replacement. ZF is the theory whose axioms are those of Zermelo set theory and all axioms of replacement.[12] A formula of $\mathcal{L}$ is an axiom of replacement if it is the translation into $\mathcal{L}$ of the result "substituting" a formula of $\mathcal{L}$ for "F" in

$$F \text{ is a function} \to \forall z \exists y \forall x(x \in y \leftrightarrow \exists w(w \in z \land F(w) = x))$$

There is an extension of the stage theory from which the axioms of replacement could have been derived. We could have taken as axioms all instances (that can be expressed in $\mathcal{J}$) of a principle which may be put, "If each set is correlated with at least one stage (no matter how), then for any set z there is a stage s such that for each member w of z, s is later than some stage with which w is correlated." This *bounding* or *cofinality* principle is an attractive further thought about the interrelation of sets and stages, but it does seem to us to be a *further* thought, and not one that can be said to have been meant in the rough description of the iterative

---

[11](Zermelo, 1908) took as axioms versions of the axioms of extensionality, the null set, pairs (and unit set), unions, the power-set axiom, the axiom of infinity, the Aussonderungsaxioms, and the axiom of choice.

[12]Sometimes the axiom of choice is also considered one of the axioms of ZF.

conception. For that there are exactly ω_1 stages does not seem to be excluded by anything said in the rough description; it would seem that R_{ω_1} (see below) is a model for any statement of $\mathcal{L}$ that can (fairly) be said to have been implied by the rough description, and not all of the axioms of replacement hold in R_{ω_1}.[13] Thus the axioms of replacement do not seem to us to follow from the iterative conception.

Adding the axioms of replacement to those of Zermelo set theory enables us to define a sequence of sets, $\{R_\alpha\}$, with which the stages of the stage theory may be identified. Suppose we put $R_0 = $ the null set; $R_{\alpha+1} = R_\alpha \cup$ the power set of R_α, and $R_\lambda = \cup_{\beta<\lambda} R_\beta$ (λ a limit ordinal)—axioms of replacement ensure that the operation R is well-defined—and say that s is a stage if $\exists \alpha\, s = R_\alpha$, that x is formed at s if x is subset but not a member of s, and that s is earlier than t if, for some α, β, $s = R_\alpha$, $t = R_\beta$, and $\alpha < \beta$. Then we can prove as theorems of ZF not only the translations into the language of set theory of the axioms of the stage theory, but also those of all those stronger axioms asserting the existence of stages further and further "out" that might have been suggested by the rough description (and those of the instances of the bounding principle which are expressible in $\mathcal{J}$ as well). ZF thus enables us to describe and assert the full first-order content of the iterative conception within the language of set theory.

Although they are not derived from the iterative conception, the reason for adopting the axioms of replacement is quite simple: they have many desirable consequences and (apparently) no undesirable ones. In addition to theorems about the iterative conception, the consequences of replacement include a satisfactory if not ideal[14] theory of infinite numbers, and a highly desirable result that justifies inductive definitions on well-founded relations.

The axiom of extensionality. The axiom of extensionality enjoys a special epistemological status shared by none of the other axioms of ZF. Were someone to deny another of the axioms of ZF, we would be rather more inclined to suppose, on the basis of his denial alone, that he believed that axiom false than we would if he denied the axiom of extensionality. Although "there are unmarried bachelors" and "there are no bachelors" are equally preposterous things to say, if someone were to say the former, he would far more invite the suspicion that he did not mean what he said than someone who said the latter. Similarly, if someone were to say, "there are distinct sets with the same members," he would thereby justify us in thinking his usage nonstandard far more than someone who asserted the denial of some other axiom. Because of this difference, one might be tempted to

---

[13]Worse yet, R_{δ_1} would also seem to be such a model. (δ_1 is the first nonrecursive ordinal.)

[14]An ideal theory would decide the continuum hypothesis, at least.

call the axiom of extensionality "analytic," true by virtue of the meanings of the words contained in it, but not to consider the other axioms analytic.

It has been persuasively argued, by Quine and others, however, that until we have an acceptable explanation of how a sentence (or what it says) can be true in virtue of meanings, we should refrain from calling *anything* analytic. It seems probable, nevertheless, that whatever justification for accepting the axiom of extensionality there may be, it is more likely to resemble the justification for accepting most of the classical examples of analytic sentences, such as "all bachelors are unmarried" or "siblings have siblings" than is the justification for accepting the other axioms of set theory. That the concepts of *set* and *being a member of* obey the axiom of extensionality is a far more central feature of our use of them than is the fact that they obey any other axiom. A theory that denied, or even failed to affirm, some of the other axioms of ZF might still be called a set theory, albeit a deviant or fragmentary one. But a theory that did not affirm that the objects with which it dealt were identical if they had the same members would only by charity be called a theory of *sets* alone.

The axiom of choice. One form of the axiom of choice, sometimes called the "multiplicative axiom," is the statement, "For any x, if x is a set of nonempty disjoint sets (two sets are disjoint if nothing is a member of both), then there is a set, called a *choice set* for x, that contains exactly one member of each of the members of x."

It seems that, unfortunately, the iterative conception is neutral with respect to the axiom of choice. It is easy to show that, since, as is now known, neither the axiom of choice nor its negation is a theorem of ZF, neither the axiom nor its negation can be derived from the stage theory. Of course the stage theory, which is supposed to formalize the rough description, could be extended so as to decide the axiom. But it seems that no additional axiom, which would decide choice, can be inferred from the rough description without the assumption of the axiom of choice itself, or some equally uncertain principle, in the inference. The difficulty with the axiom of choice is that the decision whether to regard the rough description as implying a principle about sets and stages from which the axiom could be derived is as difficult a decision, because essentially the same decision, as the decision whether to accept the axiom.

Suppose that we tried to derive the axiom by arguing in this manner: Let x be a set of nonempty disjoint sets. x is formed at some stage s. The members of members of x are formed at earlier stages than s. Hence, at s, if not earlier, there is a set formed that contains exactly one member of each member of x. But to assert this is to beg the question. How do we know that such a choice set *is* formed? If a choice set is formed, it is indeed formed at or before s. But how do we know that one is formed at all? To argue that at s we can choose one member from each member of x and so form a choice set for x is also to beg the question: "we *can't* choose" one member from each member of x if there is no choice set for x.

To say this is not to say that the axiom of choice is not both obvious and indispensable. It is only to say that the justification for its acceptance is not to be found in the iterative conception of set.

2

Reply to Charles Parsons' "Sets and Classes"

I want to begin by discussing the considerations which lead Professor Parsons to a conception of set theory on which "the language of set theory is systematically ambiguous." I shall concern myself only with those of his views that are discussed in his present paper, and not with those that are set forth in his other, recent articles on truth, ontology, and the paradoxes. Parsons mentions a "general maxim in set theory that any set theory we can formulate can plausibly be extended by assuming that there is a *set* that is a (standard) model of it," from which it follows that we could produce no discourse in the language which could be interpreted as true if and only if the quantifiers range over absolutely all sets. Now this maxim has an obvious problematical feature: it appears not to be consistently incorporable into set theory. Both the second incompleteness theorem and the well-foundedness of "∈" prevent a consistent set theory from implying the existence of a standard model of itself. Presumably, though, a "general maxim in set theory" is only a rule or license or whatever about what is permissible in doing set theory and not a statement about sets and sets alone. What I find disturbing is what this maxim suggests to Parsons about the inscrutability of reference. Suppose that you and I are "discoursing," and that our "discourse" consists of enunciations in the language of ZF:

> You: Every set has a power set!
> Me: No fooling! Sets are the same if they have the same members!!
> You: Wow!! No set contains all sets!!!

Etc. If discourses are finite, the reflection principle implies that the quantifiers in any such discourse can be reinterpreted to range over the members of some one fixed set, *salva veritate*; the general maxim is needed only

This paper was written in 1974 and is here published for the first time. Charles Parsons' article can be found in (Parsons, 1983c).

for "infinite discourses." But *salva veritate* is not synonymous with *salvis omnibus*; that truth is preserved does not mean that anything whatsoever is. In particular the speakers may have been talking about all sets, and reinterpreting what they say in such a way that it is not about all sets is changing the meaning of what was said if not the truth value; to report our speakers as having spoken only about the members of some set would be to misrepresent what they said. (Notice, by the way, that no dubious—or even doubted—notion of meaning is in question here; the references (ranges) of the quantifiers are being misrepresented.) Parsons asks whether an interpretation of such a discourse that makes the quantifiers range over a set would be incorrect. My inclination is to answer that it need not be incorrect as far as the truth-values of the statements that were made go; but it would be an incorrect account of what might be called the referential content of the discourse. If our speakers were speaking about all sets, it would be incorrect "in respect of" what was said to report them as talking about the members of a standard model of ZF, or even the members of a natural model of ZF. If, as many believe, the axiom of constructibility is false, and Gödel's L is not Cantor's paradise, then to suppose our speakers to be speaking about constructible sets alone if they were speaking about *all* sets is to make a mistaken supposition, even if the truth-values of what was said are unaffected. An interesting question is whether there are reasons for adopting a "relativist" view according to which there is a fact of the matter whether or not the range of the quantifiers in such a discourse is countable, but not whether or not the range is this or that natural model of ZF.

Let me belabor the point. Suppose that there were three interpreters of our discourse. According to one, the quantifiers in our discourse ranged over the members of some suitably large $R(\alpha)$, according to the second, over the members of some $R(\beta)$ that happened to be an elementary extension of $R(\alpha)$, and according to the third, over (absolutely) all sets. Does Parsons wish to maintain that both of the interpretations of the discourse offered by interpreters 1 and 2 could be correct (with respect to content)? If so, would an interpretation according to which the quantifiers range over members of a countable model be incorrect? And if not, why should we not suppose that an interpretation according to which the quantifiers ranged over the members of some set was incorrect after all?

Doubtless one's set-theoretic utterances can always be reinterpreted so that they are not about all sets but only about all members of some set. Doubtless too nothing can forestall claims to the effect that one was really talking only about the members of some set or force a hardened skeptic to cease claiming that no one can know that some particular quantification was really over all sets. And doubtless too not all philosophers will stop concluding from their inability to find an effective answer to a certain kind of skepticism that there is no fact of the relevant matter. But that truth-preserving reinterpretation of a discourse is always possible does not mean that it is ever obligatory. And the temptation to reinterpret is seldom particularly strong: if one learns of some startling new consequence of the existence of a measurable cardinal, one is not in the least tempted to think that the discoverer is speaking only about the members of some set that is a model of ZF or even that he may not be quantifying over absolutely as many sets as there really are. Professor Martin's remark that set theorists have made it clear that the range of variables in set theory can not be a *set* in some more inclusive concept of set seems to me to be dead right. That much *is* clear, as is the extraordinary perversity of thinking, "Well, of course, not a set in any of those (set many) concepts to which Martin was referring."

Parsons considers the case of a person who, knowing no set theory other than ZF, comes to accept the existence of (strongly) inaccessible cardinals as at least plausible. Part of a commonly given argument for the existence of inaccessibles goes something like this: "Every number that is the sum or product of fewer than omega (two, zero) numbers that are themselves smaller than omega (two, zero) is smaller than omega (two, zero). No other finite numbers have this property (the product of zero zeros is one). It's not plausible, is it, that zero, two, and omega should be the only numbers with this property? And any such number larger than omega will be strongly inaccessible." Suppose that Parsons' ZF-ite comes to believe that there may well be inaccessibles because of some such argument as this. Parsons asks whether he has been persuaded of the possible truth of something left open by his theory "as it was" or has had his conception of set changed, and says, "We all know of arguments for there being 'no fact of the matter' about such a question." I feel strongly tempted to say: (1) the ZF-ite has been persuaded of something left open by ZF; (2) his conception of set has probably been changed only in that he now believes something *about the very same sets* that he did not hitherto believe about them; his views about what sets are like will have changed, but he need not, as far as I can see, be holding views about new sets on that account alone. "However that may be," continues Parsons, "it is hard to see how your understanding of the quantifiers of set theory could not at least be taken to be *vague*, so that reading them as ranging over sets of rank less than the first strong inaccessible would be an otherwise correct way of making them precise." I find this statement particularly puzzling. Isn't the person who is convinced by this argument convinced precisely that one of the objects he was quantifying over was an inaccessible cardinal? Is the idea this: at t_0 the ZF-ite quantifies over only sets of rank less than that of the least inaccessible cardinal (though he doesn't know this); later, at t_1, he quantifies over a larger domain which contains all the sets he quantified over at t_0 together with an inaccessible; it is as the result of hearing the common argument for inaccessibles (or some other one) that he comes at t_1 to quantify over the larger domain? But why should we believe that this account of the matter is the correct one rather than the simpler one: that at t_0 and t_1 he quantified over the same sets and at t_1 believed something about those sets (viz., that one of them was an inaccessible) that he did not believe at t_0? It seems to me that what's (a little bit) less vague at t_1 is not his understanding of the quantifiers of set theory, but his views about the way sets are, about the nature of the set-theoretic universe. Why does Parsons speak of an *otherwise* correct way of making [the quantifiers] more precise? Would reading them as ranging over the sets belonging (say) to the minimal model also be an otherwise correct way of making them more precise? Remember that our ZF-ite knows no set theory beyond ZF. Perhaps it will be replied that there is no fact of the matter about what the range of the quantifiers is, that any truth-value preserving way of reading the quantifiers is "as correct as" any other. Well, maybe there *is* no fact of the matter. But that there is none does not seem to me to have been established by the considerations offered by Parsons about inaccessible cardinals or the general maxim in set theory and I suppose I'm still not persuaded of the existence of a systematic ambiguity in the use of the language of set theory.

I now want to discuss Parsons' claim, endorsed by Martin, that the use of schematic letters in inscriptions that follow assertion-signs such as ("Theorem 7.19") in treatises on set theory either involves the authors of those treatises in unintended and possibly unwanted quantification over classes or involves them in asserting that each of a certain infinite set of sentences is true. Parsons and Martin agree—and I agree with them—that authors of set theory texts do not intend merely to indicate that each of a certain infinite set of sentences is provable in ZF (as it might be). As Parsons says, "the authors of the book want their readers to learn not just facts about provability in a certain formal system, but also the facts about *sets* expressed by such formulae when [understood] in the natural way." Now I am by no means sure about the following line of argument, but I am not persuaded that set-theorists' use of schematic letters is as embarrassing or guilty as Parsons and Martin think it is. What is not clear to me is that in writing "Theorem 7.19 $A \subseteq \mathrm{On} \to \mathrm{Ord}(\cup(A))$" they are making any single statement or assertion at all. To be sure, by writing this they wish to indicate their acceptance of each of the propositions expressed by some instance of the schema following "7.19." They may well also want their readers to accept those propositions too. Another way they might

have achieved the same end might have been to have written down the
conjunction of the first hundred instances of 7.19, perhaps with a diagram
illustrating the structural similarity of those conjuncts and then written
down "& ... " This would have had a number of disadvantages, the least of
which is imprecision. But it is not clear to me that all that they ought to
have wanted to do could not have been done in this manner. Parsons says
that "the actual statements made by the material mode versions of such
theorems" are ambiguous.

I suppose I'm not clear why Parsons thinks that there are any such things
as the actual statements made by the material-mode versions of such theorems or at least why anyone who writes down "Theorem 7.19 ..." must
be taken to be making any such statement. Parsons notes that if one takes
the theorems as literally about classes, then a formalization of their language would involve quantification over classes; if one takes them as claims
that every statement of a certain form is true, then a formalization would
require a truth-predicate. Undeniably. But what I'm not certain about
is whether formalization is in this instance formalization of an (informally
made) assertion. Why must we take "$A \subseteq \text{On} \to \text{Ord}(\cup(A))$" as expressing
any generalization at all, whether about classes or truth? What is wrong
with saying that by writing what they write they indicate their acceptance
of each of an infinite class of propositions? Why must we suppose that the
only way they can do this is by writing something which must be taken
to express a general assertion of some sort? In *Set Theory and Its Logic*,[1]
Quine writes, "When I state a definition in terms of 'x', 'y', 'Fx', etc., I
mean it for all choices of sentences (of whatever theory we may be working
in) in the position of 'Fx'." Presumably the same would go for theorems
as for definitions. What does Quine mean by "I mean it for all choices of
sentences"? I don't think he means, "I mean that all instances are true,"
but rather something like "I mean to be taken as asserting—or to have
committed myself to—each instance." Why can one not mean, or indicate acceptance of, each of infinitely many assertions without asserting a
generalization about those assertions?

A three-quarters serious suggestion that I should like to make is that we
interpret authors of set theory texts as having irremediably bad handwriting, and that we take inscriptions in those texts of so-called "schemata"
or "metatheorems" such as "$A \subseteq \text{On} \to \text{Ord}(\cup(A))$" to be badly produced
tokens of an infinite paragraph type. The type would of course consist of
infinitely many sentences together with their periods (there is no need to
introduce infinitary languages), and each of the sentences would express
only what could be expressed by a particular sentence of the language of

[1](Quine, 1969).

ZF. This somewhat far-fetched way of looking at the matter enables us both to agree with Martin in saying that the set theorist seemingly succeeds in simultaneously asserting each of the propositions expressed by the instances of the schema and to avoid taking set theorists as quantifying over classes or using a truth or satisfaction-predicate.

Let me try to be a little bit less fanciful. Obviously, no human being, not even a set theorist, can actually write down infinitely many sentences. It is because of such things as the shortness of human life, the atomic character of matter, the impossibility of traveling faster than the speed of light, etc. that no one can inscribe so many sentences. There is, however, a conventional way of signalling or indicating that one wishes to be committed to each proposition expressed by a sentence that is an instance of some one schema, and that is to write down the schema itself after the words "Theorem" or "Meta-theorem." In doing so, it seems to me, one need not be taken as even hinting that one is either talking about classes or utilizing a truth-predicate, and we need not take anyone who so signals a commitment to infinitely many propositions as having *asserted that* he is so committed. We can take him as having asserted, or at least committed himself to, all of those infinitely many propositions in virtue of having done something other than making a general assertion. And since we can take him as having done this, why can we not regard that schema as expressing precisely what would have been expressed by that infinitely long paragraph, whose production we believe to be physically impossible?

What now of classes? Parsons says that "it seems that a perspective is always possible according to which your classes are really sets." Martin's tentative suggestion is that classes are given all at once, by the properties that determine which objects are members of them. On Martin's theory, no class is a set, but some classes will be members of others. One possible reaction to the introduction of classes—collections that are not sets—is this: "Wait a minute! I thought that set theory was supposed to be a theory about all, 'absolutely' all, the collections that there were and that 'set' was synonymous with 'collection'." If it is replied that Cantor, who ought to have known, asserted that a set was a collection *into a whole* of definite, well-differentiated objects of our intuition or thought, and thereby left open the possibility that there were collections which were not collections into a whole, and that Cantor later introduced the term "inconsistent multiplicity" which presumably applied to the collection of all sets, among many others, then we may ask why that particular inconsistent multiplicity— call it *V*—to which all sets belong is not itself a quite definite and very well differentiated object of our thought, and then why there is not a *set* containing (say) the null set and *V*? Unsatisfactory as ZF may be as far as its inability to discuss truth and satisfaction for its own language goes, it is no worse in that respect than any other presently known formalized theory. Kelley–Morse set theory cannot be the last word in set theories: the considerations that prompt its adoption prompt the adoption of a still stronger theory. And of a yet stronger one. If one admits that there are proper classes at all, oughtn't one to take seriously the possibility of an iteratively generated hierarchy of collection-theoretic universes in which the sets which ZF recognizes play the role of the ground-floor objects? I can't believe that any such view of the nature of "∈" can possibly be correct. Are the reasons for which one believes in classes really strong enough to make one believe in the possibility of such a hierarchy?

My own view is a "nominalist" or a strict "settist" one: there are no classes—or at least we do not as yet have sufficiently strong reason to believe that there are—and the only belongees that there are are the sets of which ZF (possibly ZF with individuals) treats. I could, I suppose, be brought to believe in classes if a theory such as the one sketched by Martin toward the end of his paper could be developed in a satisfactory way. I'm not sure whether such a theory could be satisfactory without disposing of the Epimenides paradox. How is this to be done? There are at least two points to be made about the Epimenides: the first is that an acceptable resolution must deal not just with "I am false" but also with "I am false or truth-valueless," i.e., with "I am not true." The second is that there are versions of the Epimenides which use only modus ponens, conditionalization, substitution of equals for equals, and Tarski's convention (T) for the derivation of a contradiction or an obvious falsehood. One such is due to Curry: it involves considering the sentence named "Sam," which is the sentence "If Sam is true, then 2+2 = 5." Principles of logic dealing with negation are not needed for producing certain semantic paradoxes, and rejecting excluded middle alone does not suffice to block these. I suppose that what worries me about Martin's outline of a theory is that I don't have even a hazy idea how allowing certain sentences to have indeterminate truth-values will enable us to get round the various versions of the Epimenides or the theorem on the indefinability of truth. Perhaps at some point in the discussion we might discuss how the outline is to be filled in.

3

On Second-Order Logic

I shall discuss some of the relations between second-order logic, first-order logic, and set theory. I am interested in two quasi-terminological questions, viz., the extent to which second-order logic is (or is to be counted as) logic, and the extent to which it is set theory. It is of little significance whether second-order logic may bear the (honorific) label "logic" or must bear "set theory." What matter, of course, are the reasons that can be given on either side. It seems to be commonly supposed that the arguments of Quine and others for not regarding second- (and higher-) order logic as logic are decisive, and it is against this view that I want to argue here. I shall be concerned mainly with Quine's critique of second-order logic and with some of the reasons that can be offered in support of applying neither, one, or both of the terms "logic" and "set theory" to second-order logic.[1]

The first of Quine's animadversions upon second-order logic that I shall discuss is to be found in the section of his *Philosophy of Logic*[2] called "Set Theory in Sheep's Clothing." Much of this section is devoted to dispelling two confusions which we can easily agree with Quine in deploring: that of supposing that "$(\exists F)$" and "(F)" say that some (all) predicates (i.e., predicate-expressions) are thus and so, and that of supposing that quan-

Reprinted with the kind permission of the editors from *The Journal of Philosophy* 72 (1975): 509–527.

I am grateful to Richard Cartwright, Oswaldo Chateaubriand, Fred Katz, and James Thomson for helpful criticism.

[1]My motive in taking up this issue is that there is a way of associating a truth of second-order logic with each truth of arithmetic; this association can plausibly be regarded as a "reduction" of arithmetic to set theory. It is described in ch. 18 of (Boolos and Jeffrey, 1974). I am inclined to think that the existence of this association is the heart of the best case that can be made for logicism and that unless second-order logic has *some* claim to be regarded as logic, logicism must be considered to have failed totally. I see the reasons offered in this paper on behalf of this claim as part of a partial vindication of the logicist thesis. I don't believe we yet have an assessment that is as just as it could be of the extent to which Frege, Dedekind, and Russell succeeded in showing logic to be the ground of mathematical truth.

[2](Quine, 1970); parenthetical page references to Quine are to this book.

tification over attributes has relevant ontological advantages over quantification over sets. What I wish to dispute is his assertion that the use of predicate letters as quantifiable variables is to be deplored, even when the values of those variables are sets, on the ground that predicates are not *names* of their extensions. Quine writes, "Predicates have attributes as their 'intensions' or meanings (or would if there were attributes) and they have sets as their extensions; but they are names of neither. Variables eligible for quantification therefore do not belong in predicate positions. They belong in name positions" (67).

Let us grant that predicates are not names. Why must we then suppose, as the "therefore" in Quine's sentence would indicate we must, that variables eligible for quantification do not belong in predicate positions? Quine earlier (66–67) gives this argument:

> Consider first some ordinary quantifications: "$(\exists x)(x$ walks)," "$(x)(x$ walks)," "$(\exists x)(x$ is prime)." The open sentence after the quantifier shows "x" in a position where a name could stand; a name of a walker, for instance, or of a prime number. The quantifications do not mean that names walk or are prime; what are said to walk or to be prime are things that could be named *by* names in those positions. To put the predicate letter "F" in a quantifier, then, is to treat predicate positions suddenly as name positions, and hence to treat predicates as names of entities of some sort. The quantifier "$(\exists F)$" or "(F)" says not that some or all predicates are thus and so, but that some or all entities of the sort named by predicates are thus and so.

If Quine had argued:

> Consider some extraordinary quantifications: "$(\exists F)($Aristotle $F)$," "$(F)($Aristotle $F)$," "$(\exists F)(17 \, F)$." The open sentence after the quantifier shows "F" in a position where a predicate could stand; a predicate with an extension in which Aristotle, for instance, or 17 might be. The quantifications do not mean that Aristotle or 17 are in predicates; what Aristotle or 17 are said to be in are things that could be had *by* predicates in those positions. To put the variable "x" in a quantifier, then, is to treat name positions suddenly as predicate positions, and hence to treat names as predicates with extensions of some sort. The quantifier "$(\exists x)$" or "(x)" says not that some or all names are thus and so, but that some or all extensions of the sort had by names are thus and so.

we should have wanted to say that the last two statements were false and
did not follow from what preceded them. It seems to me that the same
ought to be said about the argument Quine actually gives.

To put "*F*" in a quantifier may be to treat "*F*" as having a *range* but it
need not be to treat predicate positions as name positions nor to treat predicates as names of entities of any sort. Quine seems to suppose that because
a variable of the more ordinary sort, an individual variable, always occurs
in positions where a name but not a predicate could occur, the same must
hold for every sort of variable. We may grant that the ordinary quantifications mean what Quine says they mean. But we are not thereby committed
to any paraphrase containing "name" (or any of its cognates) that purports
to give the meaning of our extraordinary quantifications. Perhaps someone
might suppose that variables must always *name* the objects in their range,
albeit only "indefinitely" or "temporarily." However, we have no reason
not to think that there might be a sort of variable, a predicate variable,
that ranges over the objects in its range (these will be extensions) but does
not *name* them "indefinitely" or any other way; rather, predicate variables
will *have* them "indefinitely," as (constant) predicates have their extensions
"definitely." Such variables would not be names of any sort, not even "indefinite" ones, but would have a range containing those objects (extensions)
which could be had by predicates in predicate positions.

It may be that a suggestion is lurking that an adequate referential account
of the truth conditions of sentences cannot be given unless it is supposed
that all variables act as names that (indefinitely) name the objects in their
range. But this is not the case. Although variables must have a range
containing suitable objects, it need not be that variables of every sort indefinitely name the objects in their ranges. "$(\exists F)$" does not have to be
taken as saying that some entities of the sort named by predicates are thus
and so; it can be taken to say that some of the entities (extensions) had by
predicates contain thus and such. So some variables eligible for quantification might well belong in predicate positions and not in name positions.
And taking "Fx" to be true if and only if that which "x" names is in
the extension of "F" in no way commits us to supposing that "F" names
anything at all.

In the same section of *Philosophy of Logic* Quine has some advice for
the logician who wants to admit sets as values of quantifiable variables and
also wants distinctive variables for sets. The logician should not, Quine
says, write "Fx" and thereupon quantify on "F", but should instead write
"$x \in \alpha$" and then, if he wishes, quantify on "α". The advantage of the new
notation is thought to be its greater explicitness about the set-theoretic
presuppositions of second-order logic. There is an important distinction
between first- and second-order logic with regard to those presuppositions,

which may be part of the reason Quine insists on regarding "*F*", "*G*", etc. in first-order formulas as schematic letters and not quantifiable variables. In order to give a theory of truth for a first-order language which is materially adequate (in Tarski's sense) and in which such laws of truth as "The existential quantification of a true sentence is true" can be proved, it is not necessary to assume that the predicates of the language have extensions, although it does appear to be necessary to make this assumption in order to give such a theory for a second-order language.

There are reasons for not taking Quine's advice, however. One is that the notation Quine recommends abandoning represents certain aspects of logical form in a most striking way.[3] Another, and more important, reason is that the usual conventions about the use of special variables like "α" guarantee that rewriting second-order formulas in Quine's way can result in the loss of validity or implication. For example, "$\exists F \forall x\, Fx$" is valid, but "$\exists \alpha \forall x\, x \in \alpha$" is not; and "$x = z$" is implied by "$\forall Y(Yx \to Yz)$" but not by "$\forall \alpha(x \in \alpha \to z \in \alpha)$."

Quine disparages second-order logic in two further ways: reading him, one gets the sense of a culpable involvement with Russell's paradox and of a lack of forthrightness about its existential commitments. "This hypothesis itself viz., '$(\exists y)(x)(x \in y \equiv Fx)$' falls dangerously out of sight in the so-called higher-order predicate calculus. It becomes '$(\exists G)(x)(Gx \equiv Fx)$', and thus evidently follows from the genuinely logical triviality '$(x)(Fx \equiv Fx)$' by an elementary logical inference. Set theory's staggering existential assumptions are cunningly hidden now in the tacit shift from schematic predicate letter to quantifiable set variable" (68). Quine, of course, does not assert that higher-order predicate calculi are inconsistent. But even if they are consistent,[4] the validity of "$\exists X \forall x(Xx \leftrightarrow \neg x \in x)$," which certainly looks contradictory, would at any rate seem to demonstrate that their existence assumptions must be regarded as "vast." A problem now arises: although "$\exists X \exists x\, Xx$" and "$\exists X \forall x\, Xx$" are also valid, "$\exists X \exists x \exists y(Xx \wedge Xy \wedge x \neq y)$" is not valid; it would thus seem that, despite its affinities with set theory and its vast commitments, second-order logic is not committed to the existence of even a two-membered set. Both of these difficulties, it seems to me, can be resolved by examining the notion of validity in second-order logic. This

---

[3]For instance, writing out the definition of the ancestral aR_*b in this notation:

$$\forall F(\forall x(aRx \to Fx) \wedge \forall x \forall y(Fx \wedge xRy \to Fy) \to Fb)$$

shows it to be obtained from an ordinary first-order formula by prefixing a universal quantifier, and suggests an interesting question: Is there an *existential* quantification of a first-order formula that is a satisfactory definition of the ancestral? (The answer is no.)

[4]Gentzen showed that the problem of their consistency had a very easy positive solution. See (Gentzen, 1936).

examination seems to show a certain surprising weakness in second-order logic.

When is a sentence valid in second-order logic? When it is true under all its interpretations. When does it follow from others? When it is true under all its interpretations under which all the others are true. What, then, is an interpretation of a second-order sentence? If we are considering "standard" second-order logic in which second-order quantifiers are regarded as ranging over *all* subsets of, or relations on, the range of the first-order quantifiers,[5] we may answer: exactly the same sort of thing an interpretation of a first-order sentence is, viz., an ordered pair of a nonempty set D and an assignment of a function to each nonlogical constant in the sentence. The domain of the function is the set of all n-tuples of members of D if the constant is of degree n, and the range is a subset of D if the constant is a function constant and a subset of $\{T, F\}$ if it is a predicate constant. [Names (sentence letters) are function (predicate) constants of degree 0; functions from the set of all 0-tuples of members of D into an arbitrary set E are of course members of E.] We need not explicitly mention separate ranges for the second-order variables that may occur in the sentence. An existentially quantified sentence $\exists\alpha F(\alpha)$ is then true under an interpretation I just in case $F(\beta)$ is true under some interpretation J that differs from I (if at all) only in what it assigns to the constant β, which is presumed not to occur in $\exists\alpha F(\alpha)$ and presumed to be of the same logical type[6] as the variable α. The other clauses in the definition of *truth in an interpretation* are exactly as you would suppose them to be. Notice that in this account no mention is made of what sort (individual, sentential, function, or predicate) of variable α is; α may be any sort of variable at all. Notice also that, if only individual variables are allowed, the account is just a paraphrase of one standard definition of *truth in an interpretation*. The definition changes neither the conditions under which a first-order sentence is true in an interpretation nor the account of what an interpretation is, but merely extends in the obvious way the account given in (say) Mates's *Elementary Logic*[7] or Jeffrey's *Formal Logic*[8] to cover the new sorts of quantified sentences that arise in second-order logic. Quine has stressed the discontinuities between first- and second-order logic so emphatically and for so long that the obvious and striking continuities may be forgotten. In Mates's book, for example, nineteen laws of validity are stated, of which all but one (the compactness theorem) hold for second-order logic. Thus

---

[5]Only "standard" or "full" second-order logic is considered in this paper.

[6]Two symbols are of the same logical type if they are of the same degree and are either both predicate symbols or both function symbols.

[7](Mates, 1972).

[8](Jeffrey, 1981).

there is a standard account of the concepts of validity and consequence for first-order sentences, and there is an obvious, straightforward, non-ad hoc way of extending that account to second-order sentences.[9]

We can now see what is shown by the validity of

$$\exists X \forall x (Xx \leftrightarrow \neg x \in x).$$

First of all, the sentence *is* valid: given any I, we can always find a suitable J in which "$\forall x (Bx \leftrightarrow \neg x \in x)$" is true by assigning to "B" the set of all objects in the domain I that do not bear to themselves the relation that I assigns to "$\in$". Since the domain of I is a set, one of the axioms of set theory (an *Aussonderungsaxiom*) guarantees that there will always be such a subset of the domain. But without a guarantee that there is a set of all sets, we cannot conclude from the validity of "$\exists X \forall x (Xx \leftrightarrow \neg x \in x)$" that there is a set of all non-self-membered sets. And we have guarantees galore that there is no set of all sets. We do, of course, land in trouble if we suppose that "x" ranges over all sets, that "X" ranges over all sets of objects over which "x" ranges, and that "$\in$" has its usual meaning; for then "$\exists X \forall x (Xx \leftrightarrow \neg x \in x)$" would be false. But that it would then be false does not show it to be invalid; for there is no interpretation whose domain contains all sets.

Our difficulty is thus circumvented, but at some cost. We must insist that we mean what we say when we say that a second-order sentence is valid if true under all its interpretations, and that an interpretation is an ordered pair of a *set* and an assignment of functions to constants.

There is thus a limitation on the uses of second-order logic to which first-order logic is not subject. Examples such as "$\exists X \forall x (Xx \leftrightarrow \neg x \in x)$" and "$\exists X \forall x \, Xx$," both valid, seem to show that it is impermissible to use the notation of second-order logic in the formalization of discourse about certain sorts of objects, such as sets or ordinals, in case there is no *set* to which all the objects of that sort belong. This restriction does not apply, as it appears, to first-order logic: ZF (Zermelo–Fraenkel set theory) is couched in the notation of first-order logic, and the quantifiers in the sentences expressing the theorems of the theory are presumed to range over all sets, even though (if ZF is right) there is no set to which all sets belong. In the case of "$\exists X \forall x \, Xx$," we cannot assume, for example, that the quantifier "$\forall x$" ranges over all ordinals, for then "$\exists X \forall x \, Xx$" would be true iff there were a set to which all ordinals belong, and there is no such set. Nor can we assume that it ranges over all the sets that there are, for it would then be true iff there were a set of all sets. Thus if we wish (as we do)

---

[9]In Part IV of (Quine, 1972), Quine extends the notion of validity to first-order sentences with identity and discusses higher-order logic at length, but does not describe the extension of the notion of validity to second-order logic.

to maintain that both sentences are true (because valid) and also wish to preserve the standard account of the conditions under which sentences are true, we cannot suppose that all sets belong to the range of "$\forall x$" in either, or that all ordinals belong to the range of "$\forall x$" in "$\exists X \forall x \, Xx$." There is of course a step from supposing that the quantifier "$\forall x$" in "$\exists X \forall x \, Xx$" may not be assumed to range over all sets to supposing that all members of the range of first-order quantifiers in second-order sentences used to formalize a certain discourse must be contained in some one set (which depends upon the discourse), and there might be ways of not taking it. But all the difficulties do appear to have the same source, and seem to point to the impermissibility of second-order discourse about all sets, all ordinals, etc.

(We have been assuming all along that ZF is correct and that *sets* are the only "set-like" objects there are, the only objects to which membership is borne. If, however as certain extensions of ZF assert, there are also certain *classes*, which are not sets, but which sets may be members of, then of course we are free to interpret "$\exists X \forall x \, Xx$" as saying that there is a class to which all sets belong and thus to suppose that "$\forall x$" ranges over all sets in "$\exists X \forall x \, Xx$." But even if classes do exist, there is again a distinction between first- and second-order notation that is significantly like the distinction just described: we may use the former but not the latter to discuss *all members of the counterdomain* (the right field) *of* "$\in$". One of the lessons of Russell's paradox is that if we read "Xx" as "(OBJECT) X bears R to (object) x," then the range of first-order quantifiers in second-but not first-order sentences may not contain all OBJECTS.)

There is a similar, but less significant, restriction on the use of the notation of first-order logic. One who uses it to formalize some discourse is committed (in the absence of special announcements to the contrary) to the non-emptiness of the ontology of the discourse and also to the presence in the ontology of references of any names that occur in the formalization. The use of names in formalization can be avoided, however as Quine has pointed out, and various formulations of first-order logic exist in which the empty domain is permitted. But there is a striking difference between the commitment to non-emptiness of an ontology and the commitment to sethood: we believe that our own ontology is non-empty, but not that it forms a set! The contradictions appear, therefore, to teach us not that second-order logic may be inconsistent (as Quine perhaps intimates), but that it seems impossible that any "universal characteristic" should be couched in the notation of second-order logic.

What now of the existence assumptions of second- and higher-order logic, which Quine calls both "vast" and "staggering"? *Set theory* (ZF) certainly makes staggering existence *claims*, such as that there is an infinite cardinal number κ that is the κth infinite cardinal number (and hence that there is a set with that many members). Quine maintains that higher-order logic involves "outright assumption of sets the way [set theory] does."[10] Of course there are differences between set theory and higher-order logic: all set theories agree that there is a set containing at least two objects, but, as noted, "$\exists X \forall x \exists y (Xx \wedge Xy \wedge x \neq y)$" is not valid, for it is false in all one-membered interpretations. Let us try to see what the ways are in which second-order logic involves assuming the existence of sets.

First of all, "in second-order logic one quantifies over sets." There are certain (second-order) sentences of any given language that will be classified by second-order logic as logical truths (i.e., as valid), even though they assert, under any interpretation of the language whose domain forms a set, the existence of certain sorts of sub*sets* of the domain. (The sort depends upon the interpretation.) "$\exists X \forall x (Xx \leftrightarrow \neg x \in x)$" and "$\exists X \forall x (Xx \leftrightarrow x = x)$" are two examples. Thus, unless there exist sets of the right sorts, these sentences will be false under certain interpretations.

Now one may be of the opinion that no sentence ought to be considered as a truth of *logic* if, no matter how it is interpreted, it asserts that there are *sets* of certain sorts. Similarly, one might hold that the truth of "$\exists f \forall x \, Rf(x)x$" ought not to *follow* from that of "$\forall x \exists y \, Ryx$" (even if the axiom of choice is true), or one might think that it is not *as a matter of logic* that there is a set with certain closure properties if Smith is not an ancestor of Jones (i.e., not a parent, not a grandparent, etc.).

The view that logic in "topic-neutral" is often adduced in support of this opinion: the idea is that the special sciences, such as astronomy, field theory, or set theory, have their own special subject matters, such as heavenly bodies, fields, or sets, but that logic is not about any sort of thing in particular, and, therefore, that it is no more in the province of logic to make assertions to the effect that sets of such-and-such sorts exist than to make claims about the existence of various types of planets. The subject matter of a particular science, what the science is about, is supposed to be determined by the range of the quantifiers in statements that formulate the assertions of the science; logic, however, is not supposed to have any special subject matter: there is neither any sort of thing that may not be quantified over, nor any sort that must be quantified over.

I know of no perfectly effective reply to this view. But, in the first place, one should perhaps be suspicious of the identification of subject matter and range. (Is elementary arithmetic really not *about* addition, but only *about* numbers?) And then it might be said that logic is not so "topic-neutral" as it is often made out to be: it can easily be said to be about the notions of negation, conjunction, identity, and the notions expressed

---

[10](Quine, 1969), p. 258.

by "all" and "some," among others (even though these notions are almost never quantified over). In the second place, unlike *planet* or *field*, the notions of *set, class, property, concept,* and *relation,* etc. *have* often been considered to be distinctively logical notions, probably for some such very simple reason as that anything whatsoever may belong to a set, have a property, or bear a relation. That some set- or relation-existence assertions are counted as logical truths in second- or higher-order systems does not, it seems to me, suffice to disqualify them as systems of logic, as a system would be disqualified if it classified as a truth of logic the existence of a planet with at least two satellites. Part 3 of the *Begriffsschrift,* for example, where the definition of the ancestral was first given, is as much a part of a treatise on logic as are the first two parts; the first occurrence of a second-order quantifier in the *Begriffsschrift* no more disqualifies it from that point on as a work on logic than does the earlier use of the identity sign or the negation sign. Poincaré's wisecrack, "La logique n'est plus stérile. Elle engendre la contradiction," was cruel, perhaps, but not unfair. And many of us first learned about the ancestral and other matters from a work not unreasonably entitled *Mathematical Logic.*

Another way in which second- but not first-order logic involves existential and other sorts of set-theoretic assumptions is this: via Gödelization and because of the completeness theorem, elementary arithmetic ("Z") is a suitable background for the development of a significant theory of validity of first-order formulas. A notion of "validity," coextensive with the usual one (truth of the universal closure in all interpretations), can be defined in the language of Z via Gödelization, and the validity of each valid formula (and no others) can then be proved in the theory, as can many general laws of validity. Moreover, the invalidity of many invalid sentences can also be demonstrated. In contrast, not only is there no hope of proving the validity of each valid second-order sentence in elementary arithmetic, the notion of second-order validity cannot even be *defined* in the language of *second-order* arithmetic. We can effectively associate with each first-order sentence a statement of arithmetic of a particularly simple form that is true if and only if the first-order sentence is valid, but no such association is even remotely possible for second-order sentences.[11] Worse, for many highly problematical statements of set theory (such as the continuum hypothesis) there exist second-order sentences that are valid if and only if those statements are true. Thus the metatheory of second-order logic is hopelessly set-theoretic, and the notion of second-order validity possesses many if not all of the epistemic debilities of the notion of set-theoretic truth.

---

[11]There is a precise sense in which the set of valid second-order sentences is *staggeringly* undecidable: it is not definable in nth-order arithmetic, for any n. Its "Löwenheim number" is also staggeringly high.

On the other hand, although it is not hard to have some sympathy for the view that no notion of validity should be so extravagantly distant from the notion of proof, we should not forget that validity of a first-order sentence is just truth in all its interpretations. (The equation of first-order validity with provability effected by the completeness theorem would be miraculous if it weren't so familiar.) And, as we shall see below, there are notions of (first-order) logical theory which, unlike *validity*, can be adequately treated of only in a background theory that is stronger than elementary arithmetic.

While comparing set theory and second-order logic, we ought to remark in passing that the definability in set theory of the notion of second-order validity at once guarantees both the nonexistence of a reduction of the notion of set-theoretical truth to that of second-order validity and the existence of a reduction in the opposite direction: no effective—indeed no set-theoretically definable—function that assigns formulas of second-order logic to sentences of set theory assigns second-order logical truths to all and only the truths of set theory (otherwise set-theoretical truth would be set-theoretically definable). However, the function that assigns to each formula of second-order logic the sentence of set theory that asserts that the formula *is* a second-order logical truth reduces second-order validity to set-theoretical truth. Thus each of the notions in the series ⟨first-order validity, first-order arithmetical truth, second-order arithmetical truth, second-order validity, set-theoretical truth⟩ can always be reduced via effective functions to later ones but never to earlier ones; the notions are thus in order of increasing strength of one certain sort.

Quine writes (66) that "the logic capable of encompassing [the reduction of mathematics to logic] was logic inclusive of set theory." If second-order logic is "inclusive of set theory," it would seem to have to count as valid some nontrivial theorems of set theory, and if, among those counted as valid, there were some to the effect that certain kinds of sets existed, second-order logic might seem to involve excessive ontological commitments in yet another way. And it may easily seem that second-order logic involves such commitments. For "$\exists X \forall x \neg Xx$" and "$\exists X \forall y (Xy \leftrightarrow y \subseteq x)$" are both valid and might be thought to assert that the null and power sets exist, just as all set theories say.

It seems, however that there is a serious difficulty in supposing that *any* second-order sentence asserts, for example, that there is a set with no members; it seems that no second-order sentence asserts the same thing as any theorem of set theory, and hence that not even the smallest fragment of set theory is, in this sense, included in second-order logic.

Consider the question "What does '$\forall x\, x = x$' assert?" One may answer, "Why, that everything is identical with itself." But if one answers thus, one must realize that one's answer has a determinate sense only if the reference (range) of "everything" is fixed. A more cautious answer might be "Why, that everything in the domain (whatever the domain may be) is identical with itself." If the natural numbers are in question "$\forall x \exists y\, y < x$" is false; if the rationals, true. (It seems to me that the ordinary Peano–Russell notation is less than ideal in not representing in a sufficiently vivid way the partial dependence of truth-value upon domain. In some ways it would be nicer if each quantifier were required to wear a subscript that indicated its range. It seems that the design of standard notation is influenced by the archaic view that logic is about some one fixed domain of *objects* or *individuals*, and that a logical truth is a sentence that is true no matter what relations on the domain are assigned to the predicate letters in the sentence.)

Thus the correct answer to the question, "What does '$\exists X \forall x \neg X x$' assert?" would seem to be something like "That depends upon what the domain is supposed to be (and also upon how that domain is 'given' or 'described'). But, whatever the domain may be, '$\exists X \forall x \neg X x$' will assert that there is a subset of the domain to which none of its members belong."

It should now appear that no valid second-order sentence can assert the same thing as any theorem of set theory. For a second-order sentence, whether valid or not, asserts something only with respect to an interpretation, whose domain may not be taken to contain all sets. But if the sentence were to assert what any particular set-theoretic statement asserts, its domain, it would seem, would have to contain all sets. "$\exists X \forall x\, X x$" is valid, but does not assert that there is a universal set, which, if ZF is correct, is false; rather, it asserts that there is a subset of the domain (whichever set that may be) to which everything in the domain belongs. The quantifiers in the first-order sentences that express the assertions of ZF range over objects that do not together constitute a set. We have argued that the ranges of the variables in second-order sentences must be sets. If so, it is hard to see how any second-order sentence could express or assert what any theorem of ZF does, or that second-order logic counts as valid some significant theorems of set theory.

There is a clear sense, however, in which second-order logic can at least be said to be committed to the assertion that an empty set exists. For since the empty set is a subset of the domain of every interpretation whatsoever and is the only set to which no members of any domain belong, "$\exists X \forall x \neg X x$" may be taken to assert the existence of the empty set independently of any interpretation, and second-order logic may thus be regarded as committed to its existence too. Moreover, higher- and higher-order logics will be committed in the same way to more and more sets.[12] In the case of second-order

---

[12] I owe this point to Oswaldo Chateaubriand.

logic, though, the commitment is exceedingly modest; the null set is the only set to whose existence second-order logic can be said to be committed.

One sense, already noted, in which the use of second- but not first-order logic commits one to the existence of sets in this: If L_1 is the first-order fragment of an interpreted second-order language L_2 whose domain D contains no sets, then there are many logical truths of L_1 that claim the existence of objects in D with certain properties, but there are none that claim the existence of subsets of D; however, among the logical truths of L_2 there are many such: for each predicate of L_2 with one free individual variable, there is a logical truth of L_2 that asserts the existence of a subset of D that is the extension of the predicate.

We have already seen definitions of validity and consequence for second-order sentences which bring out the obvious continuity of second- with first-order logic: validity and consequences are, as always, truth in all appropriate interpretations; the definition of an interpretation remains unchanged, as does the account of the conditions under which a first-order sentence is true in an interpretation. The account needs only to be *supplemented* with new clauses for the new sorts of sentence that arise in second-order logic. The supplementation may be given in separate clauses for each new sort of quantifier, which will be perfectly analogous to those for individual quantifiers. It may also be given in a general account of the conditions under which a sentence beginning with a quantifier is true in an interpretation, which applies uniformly to all sorts of quantifiers, and of which the clauses for sentences beginning with individual quantifiers are special cases. The existence of such a definition provides *a* strong reason for reckoning second-order logic as logic. We come now to a second virtue of second-order logic, the well-known superiority of its "expressive" capacity.

If we conjoin the first two "Peano postulates," replace constants by variables, and existentially close, we obtain

$$\exists z \exists S (\forall x\, z \neq S(x) \wedge \forall x \forall y (S(x) = S(y) \rightarrow x = y))$$

a sentence true in just those interpretations whose domains are (Dedekind) infinite. If we do the same for the induction postulate, we obtain

$$\exists z \exists S \forall X (X z \wedge \forall x (X x \rightarrow X S(x)) \rightarrow \forall x\, X x)$$

which is true in just those interpretations with countable domains. Thus the notions of infinity and countability can be characterized (or "expressed") by second-order sentences, though not by first-order sentences (as the compactness and Skolem–Löwenheim theorems show). Although first-order logic's expressive capacity is occasionally quite surprising, there are many interesting notions such as *well-ordering, progression, ancestral,* and *identity* that cannot be characterized in first-order logic (first-order logic without "$=$" in the case of *identity*!), but that can be characterized in second-. And the second-order characterizations of notions like these offer a way of regarding as inconsistent certain apparently inconsistent (infinite) sets of statements, each of whose finite subsets is consistent—a way that is not available in (compact) first-order logic. Four examples of such sets are { "Smith is an ancestor of Jones," "Smith is not a parent of Jones," "Smith is not a grand-parent of Jones," ... }, { "It is not the case that there are infinitely many stars," "There are at least two stars," "There are at least three stars," ... }, { "R is a well-ordering," "$a_1 R a_0$," "$a_2 R a_1$," "$a_3 R a_2$," ... }, and of course, { "x is a natural number," "x is not zero," "x is not the successor of zero," ... }.[13]

Compare these four sets with { "Not: there are at least three stars," "Not: there are no stars," "Not: there is exactly one star," "Not: there are exactly two stars"} and { "R is a linear ordering," "$a_0 R a_1$," "$a_1 R a_2$," "Not: $a_0 R a_2$"}. There is a translation into the notation of first-order logic under which the latter two sets of statements are *formally* inconsistent. Moreover, the translation, together with an explanation of the conditions under which the translations are true in interpretations, provides an important part of the explanation of the inconsistency of the two sets. One would have hoped that the same sort of thing might be possible for the four former sets. It seems impossible, on reflection, that all the statements in any one of these four sets should be true; it also seems that the reasons for this impossibility would have to be of the same character as those which explain the inconsistency of the latter two sets, the kind of reason it has always been the business of logic to give. That the logic taught in standard courses demonstrably cannot represent the inconsistency of our four sets of sentences shows not that they are consistent after all, but that not all (logical) inconsistencies are representable by means of that logic. One may suspect that the second-order account of these inconsistencies is not the "correct" account and that perhaps some sort of infinitary logic might more accurately reflect the logical form of the sentences in question; in any event, second-order logic does not muff these cases altogether. In addition, then, to there being a "straightforward" extension of the definitions of *valid sentence* and *consequence of* from first- to second-order logic, another reason for regarding second-order logic as logic is that there are notions of a palpably logical character (*ancestral, identity*), which can be defined in second-order logic (but not first-) and which figure critically in inferences whose validity second-order logic (but not first-) can represent.

Let us turn now to the failure of the completeness theorem for second-

---

[13](Tarski, 1960), p. 410.

order logic, which can hardly be regarded as one of second-order logic's happier features. The existence of a sound and complete axiomatic proof procedure and the effectiveness of the notion of proof guarantee that the set of valid sentences of first-order logic is effectively generable; Church's theorem shows that it is not effectively decidable. There are decidable fragments of first-order logic, e.g., monadic logic with identity, but decidability vanishes if even a single two-place letter is allowed in quantified sentences. However, in a 1919 paper called "Untersuchungen über die Axiome des Klassenkalküls ..."[14] Skolem showed that the class of monadic second-order sentences, in which only individual and one-place predicate variables and constants may occur, is also decidable.

Discussing the contrast between classical first-order quantification theory and an extension of it containing "branching" quantifiers, Quine writes,

> ...there is reason, and better reason, to feel that our previous conception of quantification ...is not capriciously narrow. On the contrary, it determines an integrated domain of logical theory with bold and significant boundaries, designate it as we may. One manifestation of these boundaries is the following. The logic of quantification in its unsupplemented form admits of complete proof procedures for validity (90).

The extension is then noted not to admit of complete proof procedures.

> A remarkable concurrence of diverse definitions of logical truth ...suggested to us that the logic of quantification as classically bounded is a solid and significant unity. Our present reflections on branching quantification further confirm this impression. It is at the limits of the classical logic of quantification, then, that I would continue to draw the line between logic and mathematics (91).

Completeness cannot by itself be a sufficient reason for regarding the line between first- and second-order logic as the line between logic and mathematics. We have seen, first, that monadic logic differs from full first-order logic on the score of *decidability*, every bit as significant a property as *completeness*; we have further seen that this difference persists into second-order logic; and we have discussed at length the fact that we can extend to second-order sentences the definition of *truth in an interpretation* without change in the notation of an *interpretation*. How, then, can the *semi*-effectiveness of the set of first-order logical truths be thought to provide much of a reason for distinguishing logic from mathematics? Why *completeness* rather

----

[14]Reprinted in (Skolem, 1970), pp. 67–101, and esp. pp. 93–101.

than *decidability* or *interpretation*? Of course there is a big difference between second- and first-order logic; there are many. There are also big
differences among various fragments of first-order logic, between second-
and third-order logic, and between second-order logic and set theory.

Quine does not state that the completeness theorem by itself provides
sufficient reason for drawing the line, however. Another reason, or more
of the reason, is given by what he calls the "remarkable concurrence of
diverse definitions of logical truth." One of these diverse definitions is the
usual one: a sentence (or "schema," in Quine's terminology) is a logical
truth if it is satisfied by every model, i.e., if it is true under all its interpretations. The other is that a sentence of a reasonably rich language is a
logical truth if truths alone come of it by substitution of (open) sentences
for its simple component sentences. The languages in question are interpreted languages (otherwise the notion of truth of a sentence of a language,
used in the definition, would be incomprehensible), and their grammar has
been "standardized," i.e., put into the notation of the first-order predicate
calculus, *without* function signs or identity. As usual, "reasonably rich" has
to do with arithmetic. For Quine's purposes, a language may be taken to
be reasonably rich if its ontology contains all natural numbers (or an isomorphic copy) and its ideology contains a one-place predicate letter true of
the natural numbers (or their copies) and two three-place predicate letters
representing the sum and product operations.

By appealing to a generalization of Löwenheim's theorem that is due to
Hilbert and Bernays—any satisfiable schema is satisfied by a model whose
domain is the set of natural numbers and whose predicates are assigned relations on natural numbers *that can be defined in arithmetic*—Quine proves
a result he calls remarkable: a schema is provable (in some standard system)
if and only if it is valid (true in all its interpretations), if and only if every
substitution instance of it in any given reasonably rich object language is
true. Dually, a schema is irrefutable if and only if it is satisfiable (true in
at least one interpretation), if and only if some substitution instance of it
in the object language is true. (The equivalence of validity and provability,
and of satisfiability and irrefutability, is guaranteed by the completeness
theorem.)

For the purposes of this theorem, Quine cannot count the identity sign
as a logical symbol: "$\exists x \exists y \neg x = y$" is a schema and also a sentence whose
only substitution instance is itself (if "$=$" counts as a logical symbol), which
is true (since there exist at least two objects in the domain of the object
language), but which is not a logical truth according to the usual definition,
for it is false in all one-membered interpretations.

A second minor point about the definition is that it just does not work if the object language is not reasonably rich.[15] But the language of arithmetic, interpreted in the usual way, is certainly reasonably rich, or becomes so when "+" and "·" are supplanted by three-place predicate letters.

The theorem may be remarkable, but it is not, I think, remarkably remarkable. A distinction can be drawn between two kinds of completeness theorem that can be proved about systems of logic: between weak and strong completeness theorems. A weak completeness theorem shows that a sentence is provable whenever it is valid; a strong theorem, that a sentence is provable *from a set* of sentences whenever it is a logical consequence of the set. Most of the usual proofs of the weak completeness of systems of first-order logic can be expanded quite easily to proofs of the strong completeness of those systems. The strong completeness of first-order logic can be expressed: a set of sentences is satisfiable if it lacks a refutation. (A refutation of a set of sentences is a proof of the negation of a conjunction of members of the set.)

It seems to me that the concurrence of the two accounts of the concept of logical truth cannot be called remarkably remarkable if their extensions to the relation of logical consequence do not concur. If there is a reasonably rich language and a set of sentences in that language which is satisfiable according to the usual account but which cannot be turned into a set of truths by (simultaneous, uniform) substitution of open sentences of the language, then the interest of the alternative definition of logical truth is somewhat diminished, for it is a definition that cannot be extended to kindred logical relations in the correct manner. And, as it happens, there is a satisfiable set of sentences of a reasonably rich language with this property. Proof is given in the appendix.

The compactness theorem might be thought to provide a way out of the difficulty. Since a set is satisfiable if and only if all its finite subsets are satisfiable, we might propose to define satisfiability by saying that a set is satisfiable just in case every conjunction of its members has a true substitution instance. So there turn out to be three accounts of satisfiability of sets of sentences, the account just mentioned, truth in some one model, and irrefutability.

But this concurrence is not in the least remarkable. The strong completeness theorem is remarkable; and the Löwenheim–Hilbert–Bernays theorem is remarkable. The concurrence of the two definitions of validity of single sentences—truth in all interpretations and truth of all instances—is remarkable too, *because both definitions have some antecedent plausibility as correct explications of a pre-theoretical notion of logical validity* ("truth regardless of what the nonlogical words mean"). The definition of satisfia-

---

[15]See (Hinman, Kim, and Stich, 1968).

bility of a set as "truth of some instance of each conjunction of schemata in the set" has no such plausibility as an account of satisfiability. It even sounds wrong.

One ought then to be wary of the claims that the concurrence of diverse definitions of logical truth is remarkable and that this concurrence suggests that classical quantificational logic is a "solid and significant unity." One of the definitions is a definition of logical truth only in virtue of a remarkable theorem about first-order logic; another cannot be generalized properly. Does classical quantificational logic then fail to be a significant and solid unity? Certainly not.

Appendix

We consider two first-order languages (without "="), L and M, whose predicate letters are F, Z, S, P, T, and G. The variables of both languages range over the natural numbers, and both specify that F is true of all natural numbers, that Z is true of zero alone, and that S, P, and T are predicate letters for successor, sum, and product, respectively. L specifies that G is true of all natural numbers. L is a reasonably rich language. Let A be the set of Gödel number of truths of L. A is not definable in L. Finally, M specifies that G is true of all and only the members of A.

Let B be the set of truths of M. B is satisfiable. But B cannot be turned into a set of truths of L by substitution of open sentences of L for the predicate letters F, Z, S, P, T, and G. For, if it could, A would be recursive in the extensions in L of the open sentences substituted for Z, S, and G, and hence A would be definable in L; for the extensions would certainly be definable in L, and *definable in L* is closed under *recursive in*.

Let "$E(\vartheta)$" abbreviate "the extension in L of the open sentence substituted for ϑ," The reason that A would be recursive in $E(Z)$, $E(S)$, $E(G)$ is that, for each natural number n, $\exists x_0 x_1 \ldots x_{n-1} x_n (Z x_0 \wedge S x_0 x_1 \wedge \ldots \wedge S x_{n-1} x_n)$ is in B; if $n \in A$, then $\forall x_0 x_1 \ldots x_{n-1} x_n (Z x_0 \wedge S x_0 x_1 \wedge \ldots \wedge S x_{n-1} x_n \rightarrow G x_n)$ is in B; and if $n \notin A$, then $\forall x_0 x_1 \ldots x_{n-1} x_n (Z x_0 \wedge S x_0 x_1 \wedge \ldots \wedge S x_{n-1} x_n \rightarrow \neg G x_n)$ is in B. Then, to determine whether $n \in A$, we may use "oracles" for $E(Z)$ and $E(S)$ to find an $(n+1)$-tuple $a_0, a_1, \ldots, a_{n-1}, a_n$ of natural numbers such that a_0 is in $E(Z)$ and the n pairs $a_0, a_1, \ldots,$ and a_{n-1}, a_n are in $E(S)$, and then use an oracle for $E(G)$ to determine whether a_n is in $E(G)$. a_n is in $E(G)$ iff $n \in A$. The procedure is recursive in $E(Z), E(S), E(G)$.

We have thus shown that B is a satisfiable set of sentences of the reasonably rich language L which cannot be turned into a set of truths by (simultaneous, uniform) substitution of open sentences of L for the predicate letters of L which occur in the sentences in B.

4

To Be is to Be a Value of a Variable (or to Be Some Values of Some Variables)

Are quantification and cross-reference in English well represented by the devices of standard logic, i.e., variables x, y, z, ..., the quantifiers $\forall$ and $\exists$, the usual propositional connectives, and the equals sign? It's my impression that many philosophers and logicians think that—on the whole—they are. In fact, I suspect that the following view of the relation between logic and quantificational and referential features of natural language is fairly widely held:

No one (the view begins) can think that the propositional calculus contains all there is to logic. Because of the presence in natural language of quantificational words like "all" and "some" and words used extensively in cross-reference, like "it," "that," and "who," there is a vast variety of forms of inference whose validity cannot be adequately treated without the introduction of variables and quantifiers, or other devices to do the same work. Thus everyone will concede that the predicate calculus is at least a part of logic.

Indispensable to cross-reference, lacking distinctive content, and pervading thought and discourse, *identity* is without question a logical concept. Adding it to the predicate calculus significantly increases the number and variety of inferences susceptible of adequate logical treatment.

And now (the view continues), once identity is added to the predicate

Reprinted with the kind permission of the editors from the *The Journal of Philosophy* 81 (1984): 430–449.

I am grateful to Richard Cartwright, Helen Cartwright, James Higginbotham, Judith Thomson, and the editors of the *Journal of Philosophy* for helpful comments, criticism, and discussion. Helen Cartwright's valuable unpublished Ph.D. dissertation, "Classes, Quantities, and Non-singular Reference" deals at length with many of the issues with which the present paper is concerned.

calculus, there would not appear to be all that many valid inferences whose validity has to do with cross-reference, quantification, and generalization which cannot be treated in a satisfactory way by means of the resulting system. It may be granted that there are certain valid inferences, involving so-called "analytic" connections, which cannot be handled in the predicate calculus with identity. But the validity of these inferences has nothing to do with quantification in natural language, and it may thus be doubted whether a logic that does nothing to explain their validity is thereby deficient.

In any event (the view concludes), the variety of inferences that cannot be dealt with by first-order logic (with identity) is by no means as great or as interesting as the variety that can be handled by the predicate calculus, even without identity, but not by the propositional calculus.

It is the conclusion of this view that I want to take exception to. (At one time I thought the whole view was probably true.) It seems to me that we really do not know whether there is much or little in the province of logic that the first-order predicate calculus with identity cannot treat. In the first part of this paper I shall present and discuss some data which suggest that there may be rather more than might be supposed, that there may be an interesting variety both of quantificational and referential constructions in natural language that cannot be represented in standard logical notation and of valid inferences for whose validity these constructions are responsible. Whether quantification and cross-reference in English are well represented by standard logic seems to me to be an open question, at present.

Several kinds of constructions, sentences, and inferences that cannot be symbolized in first-order logic are known. Perhaps the best known of these involve numerical quantifiers such as "more," "most," and "as many," e.g., the inference

> Most democrats are left-of-center.
> Most democrats dislike Reagan.
> Therefore, some who are left-of-center dislike Reagan.

Another is the construction "For every A there is a B," which, although it might appear to be symbolizable in first-order notation, cannot be so represented, for it is synonymous with "There are at least as many Bs as As."[1] The construction is not of recent date; it is exemplified in a couplet from 1583 by one T. Watson:[2]

> For every pleasure that in love is found,
> A thousand woes and more therein abound.

[1]Cf. (Boolos, 1981a).
[2]See the entry for "for" in the Oxford English Dictionary.

Jaakko Hintikka has offered a number of examples of sentences that cannot, he claims, be represented in first-order logic.[3] One of these is:

> Some relative of each villager and some relative of each townsman hate each other.

There appears to be a consensus regarding this sentence, viz., that if it is O.K., then it can be symbolized in standard first-order logic as follows:

$$\forall x \forall y \exists z \exists w (Vx \wedge Ty \rightarrow Rzx \wedge Rwy \wedge Hzw \wedge Hwz \wedge z \neq w)$$

I find this sentence marginally acceptable at best and not acceptable if not symbolizable as above.

Jon Barwise has offered "The richer the country, the more powerful is one of its officials" as another example of a sentence that cannot be symbolized in first-order logic.[4] However, since the sentence seems to me, at any rate, to mean "Whenever x is a richer country than y, then x has (at least) one official who is more powerful than *any* official of y," it also seems to me to have a first-order symbolization:

$$\forall x \forall y ([Cx \wedge Cy \wedge xRy] \rightarrow \exists w [wOx \wedge \forall z (zOy \rightarrow wPz)])$$

Are there better examples?

Perhaps the best-known example of a sentence whose quantificational structure cannot be captured by means of first-order logic is the Geach–Kaplan sentence, cited by W. V. Quine in *Methods of Logic*[5] and *The Roots of Reference*[6]:

(A)   Some critics admire only one another.

(A) is supposed to mean that there is a collection of critics, each of whose members admires no one not in the collection, and none of whose members admires himself. If the domain of discourse is taken to consist of the critics and Axy to mean "x admires y," then (A) can be symbolized by means of the *second*-order sentence:

(B) $\exists X (\exists x\, Xx \wedge \forall x \forall y [Xx \wedge Axy \rightarrow x \neq y \wedge Xy])$

And since (B) is not equivalent to any first-order sentence, (A) cannot be correctly symbolized in first-order logic.

---

[3] (Hintikka, 1974).
[4] (Barwise, 1979).
[5] (Quine, 1982).
[6] (Quine, 1982), p. 293, where "people" is substituted for "critics" in the example.

The proof, due to David Kaplan, that (B) has no first-order equivalent is simple and exhibits an important technique in showing nonfirstorderizability: Substitute the formula $(x = 0 \lor x = y + 1)$ for Axy in (B), and observe that the result:

(C) $\exists X (\exists x \, Xx \land \forall x \forall y [Xx \land (x = 0 \lor x = y + 1) \to x \neq y \land Xy])$

is a sentence that is true in all nonstandard models of arithmetic but false in the standard model.[7]

I must confess to a certain ambivalence regarding the Geach–Kaplan sentence. Although it usually strikes me as a quite acceptable sentence of English, it doesn't invariably do so. (The "only" seems to want to precede the "admires" but the intended meaning of the sentence forces it to stay put.) I find that if the predicates in the example are changed in what one might have supposed to be an inessential way matters are improved slightly:

> Some computers communicate only with one another.
> Some Bostonians speak only to one another.
> Some critics are admired only by one another.

I don't have any idea why replacing the transitive verb "admires" by a verb or verb phrase taking an accompanying prepositional phrase helps matters, but it does seem to me to do so.

I turn now from this brief survey of known examples of sentences not representable in first-order logic to examination of some other nonfirstorderizable sentences. Like the Geach–Kaplan sentence but unlike the sentences involving "most," these sentences *look* as if they "ought to be" symbolizable in first-order logic. They contain plural forms such as "are" and "them," and it is in large measure because they contain these forms that they cannot be represented in first-order logic.

Consider the following sentence, which, however, contains no plurals and which can be symbolized in first-order logic:

(D)   There is a horse that is faster than Zev and also faster than the sire of any horse that is slower than it.

---

[7] To see that (C) is true in any nonstandard model, take as X the set of all nonstandard elements of the model. X is nonempty, does not contain 0, hence contains only successors, and contains the immediate predecessor of any of its members. To see that it is false in the standard model, suppose that there is some suitable set X of natural numbers. X must be nonempty: if its least member x is 0, let $y = 0$; otherwise $x = y + 1$ for some y. Since x is least, y is not in X, and "Xy" is false. The nonfirstorderizability of "For every A there is a B" can be established in a similar way: Select variables x and y not found in any presumed first-order equivalent, substitute $[(1) < x + 5 \land \neg \exists y \, 3 \cdot y = (1)]$ for $A(1)$, substitute $[(1) < x + 5 \land \exists y \, 3 \cdot y = (1)]$ for $B(1)$, and existentially quantify the result with respect to x; the result would be true in all nonstandard models but false in the standard model.

Quantifying over horses, and using 0, s, $>$, and $<$ for "Zev," "the sire of," "is faster than," and "is slower than," respectively, we may symbolize (D) in first-order logic:

(E) $\exists x(x > 0 \wedge \forall y[y < x \rightarrow x > s(y)])$

Sentence (F), however, cannot be symbolized in first-order logic:

(F)   There are some horses that are faster than Zev and also faster than the sire of any horse that is slower than them.

(F) differs from (D) only in that some occurrences in (D) of the words "is," "a," "horse," and "it" have been replaced by occurrences of their plural forms "are," "some," "horses," and "them." The content of (F) is given slightly more explicitly in:

(G)   There are some horses that are all faster than Zev and also faster than the sire of any horse that is slower than all of them.

I take it that (F) and its variant (G) can be paraphrased: there is a nonempty collection (class, totality) X of horses, such that all members of X are faster than Zev and such that, whenever any horse is slower than all members of X, then all members of X are faster than the sire of that horse.[8] (F) and (G) can be symbolized by means of the second-order sentence (domain and denotations as above):

(H) $\exists X(\exists x\, Xx \wedge \forall x(Xx \rightarrow x > 0) \wedge \forall y[\forall x(Xx \rightarrow y < x) \rightarrow \forall x(Xx \rightarrow x > s(y))])$

(H) is equivalent to no first-order sentence; for it is false in the standard model of arithmetic (under the obvious reinterpretation) but true in any nonstandard model, since the set of nonstandard elements of the model will always be a suitable value for X. Thus (F) cannot be symbolized in first-order logic.[9]

(F) is not an especially pretty sentence. It is hard to understand, awkward, and contrived. But ugly or not, it is a perfectly grammatical sentence of English, which has, as far as I can see, the meaning given above and no other. Moreover, such faults as it has appear to be fully shared by (D).

[8] Zev won the Kentucky Derby in 1923.

[9] Cf. (Boolos, 1984c). In an important unpublished manuscript entitled "Plural Quantification," Lauri Carlson has given "If some numbers all are natural numbers, one of them is the smallest of them," as an example of a sentence that cannot be symbolized in the first-order predicate calculus. I have heard it claimed that this is not a proper sentence of English. Perhaps it is not, but "If there are some numbers all of which are natural numbers, then there is one of them that is smaller than all the others," surely is. I am grateful to Irene Heim for calling this reference to my attention.

Another example, shorter and perhaps more intelligible:

(I)   There are some gunslingers each of whom has shot the right foot of
at least one of the others.

(I) may be rendered in second-order logic:

(J) $\exists X(\exists x\, Xx \wedge \forall x[Xx \rightarrow \exists y(Xy \wedge y \neq x \wedge Bxy)])$

(Here we quantify over gunslingers and use B for "has shot the right foot
of.") By substituting $x = y + 1$ for Bxy, we may easily see that (J) is
equivalent to no first-order sentence. (Alternatively, we may note that
if we negate (J), substitute $y \leq x$ for Bxy, and make some elementary
transformations, we obtain

$$\forall X(\exists x\, Xx \rightarrow \exists x[Xx \wedge \forall y(Xy \wedge y \leq x \rightarrow y = x)])$$

a formula that expresses the least-number principle, which is one version of
the principle of mathematical induction.)

When used as a demonstrative pronoun, "that" is marked for number,
as singular, but when used as a relative pronoun, as in (F), it is unmarked
for number, i.e., can be used in either the singular or plural. "Who,"
"whom," and "whose," however, are unmarked for number when used either
as relative or as interrogative pronouns. "Which" is also unmarked for
number as a relative pronoun, but "which ones," when it can be used,
is strongly preferred to "which" as an interrogative plural form; it may
well be that interrogative "which," like demonstrative "that," is marked as
singular.

It is the plural forms in (F) and (I), as well as the unmarkedness of
"that" and "whom," that are responsible for the nonfirstorderizability of
these sentences. And by taking a cue from the well-known second-order
definitions of "x is a standard natural number" and "x is an ancestor of
y," we can use plurals to define these notions in English (in terms of "zero"
and "successor of" and in terms of "parent of," respectively):

(K)   If there are some numbers of which the successor of any one of them
is also one, then if zero is one of them, x is one of them.

(L)   If there are some persons of whom each parent of any one of them is
also one, then if each parent of y is one of them, x is one of them;
and someone is a parent of y.

There are some comments on (K) and (L) to be made: (a) "which" and
"whom" are used in these sentences as we have noticed they can be used,
in the plural. (b) Instead of saying "of which the successor of any one of

them is also one," one could as well say "of which the successor of any one of them is also one of them": at least one "them" is needed to cross-refer to the "witnessing" values of "which": this "them" is sometimes called a *resumptive* pronoun, and appears to be needed to capture the force of $\forall y(Xy \rightarrow Xs(y))$, with its two occurrences of X. (c) Like (F) and (I), (K) and (L) cannot be given correct first-order symbolizations, and thus the following (valid) inference cannot be represented in first-order logic:

> If there are some persons of whom each parent of any one of
> them is also one, then if each parent of Yolanda is one of them,
> Xavier is one of them; and someone is a parent of Yolanda.
> Every parent of someone red is blue.
> Every parent of someone blue is red.
> Yolanda is blue.
> Therefore, Xavier is either red or blue.

(To see that this is a valid inference, consider the persons who are either red or blue. By the second and third premisses, every parent of any one of these persons is also one of them; and since Yolanda is blue, each of her parents is red, hence red or blue, and hence one of these persons. Thus Xavier is also one of them and thus either red or blue.) (d) The "there are"s in the antecedents of course express universal quantification, as does the "there is" in "If there is a logician present, he should leave." (e) Like (F), (K) and (L) are somewhat ungainly, in part because of the resumptive "them" they contain, but principally because of the complexity of the thoughts they express. However, they seem to be perfectly acceptable vehicles for the expressions of those very thoughts. And although they are indeed contrived—they have been contrived *to take advantage of referential devices that are available in English*—the fact that they are so hardly begins to bear on the question whether they are ungrammatical, unintelligible, or in some other way unacceptable.

The suggestion that it is the complexity of the thoughts expressed in (K) and (L) that is responsible for their ungainliness rather than the presence of any construction not properly a part of English draws support from the ease and naturalness with which "x is identical to y" may be defined in the same style: if there are some things of which x is one, then y is one of them too. (Or: it is not the case that there are some things of which x is one, but of which y is not one.)

Another example, of a different sort, is:

(M)   Each of the numbers in the sequence $1, 2, 4, 8, \ldots$ is greater
       than the sum of all the numbers in the sequence that precedes it.

(M) states something true, which, using a mixture of logical and arithmetical notation, we can express as follows:

(N) $\forall x \forall y (Px \wedge y = \Sigma \{z : Pz \wedge x > z\} \rightarrow x > y)$

In (N), Σ is a sign for a function from sets of objects in a domain to objects in that domain and attaches to a variable and a formula to form a term in which that variable is bound. Signs for such functions are simply not part of the primitive vocabulary of first-order logic, although on occasion mention of functions of this type can be paraphrased away (e.g. "the least of the numbers z such that ... Z ... "). No one function sign of the ordinary sort can do full justice to "the sum of the numbers z such that ... z ... ," as can be seen by considering:

(O)   Although every power of 2 is 1 greater than the sum of all the powers of 2 that are smaller than it, not every power of 3 is 1 greater than the sum of all the powers of 3 that are smaller than it.

We certainly cannot symbolize (O) as:

$$\forall x \forall y (Px \wedge y = f(x) \rightarrow x = y + 1) \wedge \neg \forall x \forall y (Qx \wedge y = f(x) \rightarrow$$
$$x = y + 1)$$

and were we to try to improve matters by changing the second occurrence of f to an occurrence of (say) g, we should fail to depict the recurrence of the semantic primitive "the sum of ... " in the second conjunct of (O). Nor could any ordinary function sign express the dependencies that may obtain between predicates contained in " ... z ... " and those found in the surrounding context.

A short and sweet example of the same type is:

No number is the sum of all numbers.

The last example for the moment of a sentence whose meanings cannot all be captured in first-order logic is one that is again found in Quine's *Methods of Logic*—but not, this time, in the final part of the book, "Glimpses Beyond." It is the sentence (P):

(P)   Some of Fiorecchio's men entered the building unaccompanied by anyone else.

On Quine's analysis of this sentence, it can be represented as $\exists x (Fx \wedge Ex \wedge \forall y [Axy \rightarrow Fy])$, where Fx, Ex, and Axy mean "x was one of Fiorecchio's men," "x entered the building," and "x was accompanied by y."[10] Quine

---

[10] (Quine, 1982), p. 197. Quine uses K, F, and H instead of F, E, and A, respectively.

states that "x was unaccompanied by anyone else" clearly has the intended meaning "Anyone accompanying x was one of Fiorecchio's men."

Quine's is certainly one reading this sentence bears: there are some Fiorecchians each of whom entered the building unaccompanied by anyone who wasn't a Fiorecchian. But since (P) appears, at times, to mean something like:

> There were some men, see.
> They were all Fiorecchio's men.
> They entered the building.
> And they weren't accompanied by anyone else.

it can also be understood to mean: there are some Fiorecchians each of whom entered the building unaccompanied by anyone who wasn't one of *them*. On this stronger reading, there is no asymmetry between the predicates "x was one of Fiorecchio's men" and "x entered the building," "else" means "not one of them," and the whole can be symbolized by:

$$\exists X(\exists x\, Xx \wedge \forall x(Xx \to Fx) \wedge \forall x(Xx \to Ex) \wedge$$
$$\forall x \forall y(Xx \wedge Axy \to Xy))$$

whose nonfirstorderizability can be seen in the usual way, by substituting $x > 0$ for both Fx and Ex and $x = y + 1$ for Axy.

It is because of these examples that I think that the question whether the first-order predicate calculus with identity adequately represents quantification, generalization, and cross-reference in natural language ought to be regarded as a question that hasn't yet been settled.

Changing the subject somewhat, I now want to look at a number of sentences whose most *natural* representations are given by second-order formulas, but second-order formulas that turn out to be equivalent to first-order formulas.

The sentence:

(Q)   There are some monuments in Italy of which no one tourist has seen all.

might appear to require a second-order formula for its correct symbolization, e.g.,

(R) $\exists X(\exists x\, Xx \wedge \forall x[Xx \to Mx] \wedge \neg\exists y[Ty \wedge \forall x(Xx \wedge Syx)])$

Of course, (Q) can be paraphrased:

(S)   No tourist has seen all the monuments in Italy.

and this can be symbolized in first-order logic as:

(T) $\exists x\, Mx \wedge \neg\exists y[Ty \wedge \forall x(Mx \to Syx)]$

which is equivalent to (R).[11] But just as $\neg\neg p$ can sometimes be a better symbolization than p of "It's not the case that John didn't go," e.g., if p were used to symbolize "John went," so (R) captures more of the quantificational structure of (Q) than does the equivalent (T). (Q) might appear to say that there is a (nonempty) collection of monuments in Italy and no tourist has seen every member of this collection; (S) doesn't begin to hint at collections of monuments. Nevertheless, (Q) and (S) say the same thing, if any two sentences do, and (R) and (T) are, predictably enough, equivalent.

Another example of the same "collapsing" phenomenon:

(U) Mozart composed a number of works, and every tolerable opera with an Italian libretto is one of them.

has the second-order symbolization:

(V) $\exists X(\exists x\, Xx \wedge \forall x(Xx \to Mx) \wedge \forall x(Tx \to Xx))$

But as (U) says what (W) says:

(W) Mozart composed a number of works, and every tolerable opera with an Italian libretto is a work that Mozart composed.

so (V) is equivalent to the first-order

(X) $\exists x\, Mx \wedge \forall x(Tx \to Mx)$

The construction "Every ... is one of them" bears watching; suffice it for now to observe that it is a perfectly ordinary English phrase.

Collapses can also occur unexpectedly. (Through a publisher's error) the sentence:

(Y) Some critics admire one another and no one else.

meaning (approximately), "There is a collection of critics, each of whom admires all and only the *other* members of the collection," and possessing the second-order symbolization:

(Z) $\exists X(\exists x\exists y[Xx \wedge Xy \wedge x \neq y] \wedge \forall x[Xx \to \forall y(Axy \leftrightarrow \{Xy \wedge y \neq x\})])$

was claimed in the first American printing of the third edition of *Methods of Logic*[12] to be a sentence incapable of first-order representation. But although (Z) might appear to be susceptible to the same kind of treatment

---

[11]I take it that since (S) implies that there are some monuments in Italy, but does not imply that there are tourists, the conjunct $\exists x\, Mx$ is indispensable.

[12](Quine, 1972), pp. 238–239.

given out above, it was in fact observed by Kaplan to be equivalent to the
first-order formula:

(a) $\exists z(\exists y\, Azy \wedge \forall x[(z = x \vee Azx) \to \forall y(Axy \leftrightarrow \{(z = y \vee Azy)\wedge$
      $y \neq x\})])$

Consider now the sentence (b):

(b)   There are some sets that are such that no one of them is a
      member of itself and also such that every set that is not a
      member of itself is one of them. (Alternatively, There are some
      sets, no one of which is a member of itself, and of which every
      set that is not a member of itself is one.)

By quantifying over sets and abbreviating "is a member of" by $\in$, we may
use a second-order formula to symbolize (b):

(c) $\exists X(\exists x\, Xx \wedge \forall x[Xx \to \neg x \in x] \wedge \forall x[\neg x \in x \to Xx])$

(c) is obviously equivalent to (d):

(d) $\exists X(\exists x\, Xx \wedge \forall x[Xx \leftrightarrow \neg x \in x])$

Let us notice that (d) immediately implies $\exists x \neg x \in x$. Conversely, if $\exists x \neg x \in$
x holds, then there is at least one set in the totality X of sets that are not
members of themselves, and X witnesses the truth of (d). Thus (d) turns
out to be equivalent to $\exists x \neg x \in x$, the symbolization of an obvious truth
concerning sets.

(The worry over Russell's paradox which the reader may be experiencing
at this point may be dispelled by the observation that logical equivalence
is a model-theoretic notion, the "sets" just referred to may be taken to be
elements of the domain of an arbitrary model, and the "totalities," subsets
of the domain of the model.)

In view of the near-vacuity of (b) and the fact that instances of the
second-order comprehension schema $\exists X \forall x[Xx \leftrightarrow A(x)]$, including (e):

(e) $\exists X \forall x[Xx \leftrightarrow \neg x \in x]$

are logically valid under the standard semantics for second-order logic, the
collapse of (d) is not at all surprising. The rendering (d) of (b) is considerably more faithful to the semantic structure of (b) than is $\exists x \neg x \in x$,
however, and (b) is more nearly synonymous with (d) than with $\exists x \neg x \in x$.

But can we use (c) or (d) to represent (b) at all? May we use second-order
formulas like (c), (d), or (e) to make assertions about *all sets*?

Let's consider (e), which is slightly simpler than (c) or (d). (e) would
appear to say that there is a totality or collection X containing all and

only those sets x which are not members of themselves. Are we not here one the brink of a well-known abyss? Does not acceptance of the valid (e), understood as quantifying over all sets (with $\in$ taken to have its usual meaning), commit us to the existence of a set whose members are all and only those sets which are not members of themselves?

There are a number of ways out of this difficulty. One way, which I no longer favor, is to regard it as illegitimate to use a second-order formula when the objects over which the individual variables in the formula range do not form a set (just as it is illegitimate to use a first-order formula when there are *no* objects over which they range).[13] This stipulation keeps all instances of the comprehension principle as logical truths; it also enables one always to read the formula Xx as meaning that x is a *member* of the *set* X.

The principal drawback of this way out is that there are certain assertions about sets that we wish to make, which certainly cannot be made by means of a first-order formula—perhaps to claim that there is a "totality" or "collection" containing all and only the sets that do not contain themselves is to attempt to make one of these assertions—but which, it appears, could be expressed by means of a second-order formula if only it were permissible so to express them. To declare it illegitimate to use second-order formulas in discourse about all sets deprives second-order logic of its utility in an area in which it might have been expected to be of considerable value.

For example, the principle of set-theoretic induction and the separation (Aussonderung) principle virtually cry out for second-order formulation, as:

(f) $\forall X(\exists x\, Xx \rightarrow \exists x[Xx \wedge \forall y(y \in x \rightarrow \neg Xy)])$

and

(g) $\forall X \forall z \exists y \forall x(x \in y \leftrightarrow [x \in z \wedge Xx])$

respectively. It is, I think, clear that our decision to rest content with a set theory formulated in the first-order predicate calculus with identity, in which (f) and (g) are not even well-formed, must be regarded as a compromise, as falling short of saying all that we might hope to say. Whatever our reasons for adopting Zermelo–Fraenkel set theory in its usual formulation may be, we accept this theory because we accept a stronger theory consisting of a *finite* number of principles, among them some for whose complete expression second-order formulas are required.[14] We ought to be able to formulate a theory that reflects our beliefs.

---

[13]I took this view in Article 3 in this volume.

[14]Cf. the remarks about "full expression" and "part of the content" of various notions in Article 1 above.

We of course also wish to *maintain* such second-claims as are made by e.g., $\exists X \forall x [Xx \leftrightarrow \neg x \in x]$; if we are to utilize second-order logic in discourse about all sets, these comprehension principles must remain among the asserted statements. Nor do we want to take the second-order variables as ranging over some set-like objects, sometimes called "classes," which have members, but are not themselves members of other sets, supposedly because they are "too big" to be sets. Set theory is supposed to be a theory about *all* set-like objects.

How then can we legitimately claim that such (closed) formulas as $\exists X \forall x$ $[Xx \leftrightarrow \neg x \in x]$, (f), and (g) express truths, without introducing classes (set-like non-sets) into set theory and without assuming that the individual variables do not in fact range over all the sets there really are?

There is a simple answer. Abandon, if one ever had it, the idea that use of plural forms must always be understood to commit one to the existence of sets (or "classes," "collections," or "totalities") of those things to which the corresponding singular forms apply. The idea is untenable in general in any event: There are some sets of which every set that is not a member of itself is one, but there is no set of which every set that is not a member of itself is a member, as the reader, understanding English and knowing some set theory, is doubtless prepared to agree. Then, using the plural forms that are available in one's mother's tongue, translate the formulas into that tongue and see that the resulting English (or whatever) sentences express true statements. The sentences that arise in this way will lack the trenchancy of memorable aphorisms, but they will be proper sentences of English which, with a modicum of difficulty, can be understood and seen to say something true.

Applying this suggestion to:

(h) $\neg \exists X (\exists x\, Xx \wedge \forall x[Xx \rightarrow (x \in x \vee \exists y[y \in x \wedge Xy \wedge y \neq x])])$

which is equivalent to (f), we might obtain:

(i)   It is not the case that there are some sets each of which either contains itself or contains at least one of the others.

From Aussonderung we might perhaps get:

(j)   It is not the case that there are some sets that are such that it is not the case that for any set z there is a set y such that for any set x, x is a member of y if and only if x is a member of z and also one of them.

or, far more perspicuously,

(k) $\neg$ there are some sets such that $\neg \forall z \exists y \forall x[x \in y \leftrightarrow$
            $(x \in z \wedge x$ is one of them$)]$

(k) is of course neither an English sentence nor a well-formed formula of any reputable formalism—for that matter neither is (j), which contains the (non-English) variables x, y, and z—but is readily understood by anyone who understands both English and the first-order language of set theory. It would be somewhat laborious to produce a fully Englished version of (g), but the labor involved would be mainly due to the sequence $\forall z \exists y \forall x$ of *first-order* quantifiers that (g) contains. (j) and (k) are actually not quite right; properly they have the meaning:

(1) $\neg \exists X (\exists x\, Xx \wedge \neg \forall z \exists y \forall x [x \in y \leftrightarrow (x \in z \wedge Xx)])$

whereas the full Aussonderung principle omits the nonemptiness condition $\exists x\, Xx$; to get the full content in English of Aussonderung, however, we need only conjoin "and there is a set with no members" to (j) and $\exists y \forall x \neg x \in y$ to (k). This observation calls to our attention two small matters connected with plurals which must be taken up sooner or later.

Suppose that there is exactly one Cheerio in the bowl before me. Is it true to say that there are some Cheerios in the bowl? My view is no, not really, I guess not, but say what you like, it doesn't matter very much. Throughout this paper I have made the customary logician's assumption, which eliminates needless verbiage, that the use of plural forms does not commit one to the existence of two or more things of the kind in question.

On the side of literalness, however, I have assumed that use of such phrases as "some gunslingers" in "There are some gunslingers each of whom has either shot his own right foot or shot the right foot of at least one of the others" does commit one to—as one might say—a *nonempty* class of gunslingers, but not to a class containing two or more of them. Thus I suppose the sentence to be true in case there is exactly one gunslinger, who has shot his own right foot, but to be false if there aren't any gunslingers. It is this second assumption that is responsible for the ubiquitous $\exists x\, Xx$ in the formulas above.

Translation will be difficult from any logical formalism into a language such as English, which lacks a large set of devices for expressing cross-reference. And since plural pronouns like "them," although sometimes used as English analogues of second-order variables, much more frequently do the work of individual variables, translation from a second-order formalism containing infinitely many variables of both sorts into idiomatic, flowing, and easily understood English will be impossible nearly all of the time. My present point is that, in the cases of interest to us, the things we would like to say can be said, if not with Austinian or Austenian grace.

It is, moreover, clear that if English were augmented with various subscripted pronouns, such as "it$_x$," "that$_x$," "it$_y$," ..., "them$_X$," "that$_X$," "them$_Y$," ..., then any second-order formula[15] whose individual variables are understood to range over all sets could be translated into the augmented language, as follows: Translate Vv as "it$_v$ is one of them$_V$," $v \in v'$ as "it$_v$ is a member of it$_{v'}$," $v = v'$ as "it$_v$ is identical with it$_{v'}$," $\wedge$ as "and," $\neg$ as "not," and, where F^* is the translation of F, translate $\exists v F$ as "there is a set that$_v$ is such that F^*."

The clause for formulas $\exists V F$ is not quite so straightforward, because of the difficulty about nonemptiness mentioned above. It runs as follows: Let F^* be the translation of F, and let F^{**} be the translation of the result of substituting an occurrence of $\neg v = v$ for each occurrence of Vv in F. Then translate $\exists V F$ as "either there are some sets that$_V$ are such that F^*, or F^{**}."

For example, $(Xx \leftrightarrow \neg x \in x)$ comes out as "It$_x$ is one of them$_X$ iff it$_x$ is not a member of itself"; $\forall x(Xx \leftrightarrow \neg x \in x)$ as "Every set is such that it is one of them$_X$ iff it is not a member of itself"; and $\exists X \forall x(Xx \leftrightarrow \neg x \in x)$ as "Either there are some sets that are such that every set is one of them iff it is not a member of itself or every set is a member of itself." (We have, of course, improved the translations as we went along.)

I want to emphasize that the addition to English of operators "it$_{()}$," "that$_{()}$," "them$_{()}$," etc. or variables "x," "X," "y," etc. is not contemplated here. The "x" of "it$_x$" is not a variable but an index, analogous to "latter" in "the latter," or "seventeen" in "party of the seventeenth part"; "X" and "x" in "them$_X$" and "it$_x$" no more have *ranges* or *domains* than does "17" in "x_{17}." We could just as well have translated the language of second-order set theory into English augmented with pronouns such as "it$_{17}$," "them$_{1879}$," etc. or an elaboration of the "former"/"latter" usage. Note also that such augmentation will be needed for the translation into English of the language of *first*-order set theory as well.

Charles Parsons has pointed out to me that although second-order existential quantifiers can be rendered in the same manner we have described, it is curious that there appears to be no nonartificial way to translate second-order *universal* quantifiers, that the translation of $\forall X$ must be given indirectly, via its equivalence with $\neg \exists X \neg$. Because our translation "manual" relies so heavily on the phrases "there is a [singular count noun] that is such that ... it ..." and "there are some [plural count noun] that are such that ... they ...," the logical grammar of the construction these phrases exemplify is worth looking at.

Of course, in ordinary speech, the construction "that is/are such that ... it/they ..." is almost certain to be eliminable: the content of a sentence

---

[15]We assume that no quantifier in any formula occurs vacuously or in the scope of another quantifier with the same variable; every formula is equivalent to some formula satisfying this condition.

containing it can nearly always be conveyed in a much shorter sentence. But the difference between the two "that"s bears notice. The second one, following "such," is a "that" like the one found in oblique contexts and may be—as Donald Davidson has suggested that the "that" of indirect discourse is—a kind of demonstrative, used on an occasion to point to a subsequent utterance of an (open) sentence; the first "that," following the count noun and more frequently elided than the second, is no demonstrative, but a relative pronoun used to bind the "it" or "they" in the open sentence after "such that." Thus the first but not the second "that" works rather like the variable immediately following an ∃, binding occurrences of that same variable in a subsequent open formula. Whether the preceding count noun is singular or plural appears to make no difference to the quantificational role of the first "that"; as we have observed, "that" is not marked for number and can serve to bind either "it" or "they."

Whether any such second-order formula of the sort we have been considering can be translated into intelligible *un*augmented English is not an interesting question, and I shall leave it unanswered. Since English augmented in the manner I have described is intelligible to any native speaker who understands the term of art "party of the seventeenth part," I shall assume that devices like "it$_x$" and "them$_X$" are available in the language we use.

I take it, then, that there is a coherent and intelligible way of interpreting such second-order formulas as (e), (f), and (g) even when the first-order variables in these formulas are construed as ranging over all sets or set-like objects there are. The interpretation is given by translating them into the language we speak; the translations of (e), (f), and (g) are sentences we understand; and we can see that they express statements that we regard as true: after all, we do think it false that there are some sets each of which either contains itself or contains one of the others, and, once we cut through the verbiage, we do find it trivial that there are some sets none of which is a member of itself and of which each set that is not a member of itself is one. It cannot seriously be maintained that we do not *understand* these statements (unless of course we really *don't* understand them, as we wouldn't if, e.g., we knew nothing at all about set theory) or that any lack of clarity that attaches to them has anything to do with the plural forms found in the sentences expressing them. The language in which we think and speak provides the constructions and turns of phrase by means of which the meanings of these formulas may be explained in a completely intelligible way.

It may be suggested that sentences like (i) are intelligible, but only because we antecedently understand statements about collections, totalities, or sets, and that these sentences are to be analyzed as claims about the

existence of certain collections, etc. Thus "There are some gunslingers ..."
is to be analyzed as the claim that there is a collection of gunslingers ... [16]
The suggestion may arise from the thought that any precise and adequate
semantics for natural language must be interpretable in set theory (with
individuals). How else, one may wonder, is one to give an account of the
semantics of plurals?

One should not confuse the question whether certain sentences of our language containing plurals are intelligible with the question whether one can
give a semantic theory for those sentences. In view of the work of Tarski, it
should not automatically be expected that we can give an adequate semantics for English—whatever that might be—in English. Nothing whatever
about the intelligibility of those sentences would follow from the fact that
a systematic semantics for them cannot be given in set theory. After all,
the semantics of the language of ZF itself cannot be given in ZF.

In any event, as we have noticed, there are certain sentences that cannot
be analyzed as expressing statements about collections in the manner suggested, e.g., "There are some sets that are self-identical, and every set that
is not a member of itself is one of them." That sentence says something
trivially true; but the sentence "There is a collection of sets that are self-
identical, and every set that is not a member of itself is a member of this
collection," which is supposed to make its meaning explicit, says something
false.

I want now to consider the claim that a sentence of English like "There
are some sets of which every set that is not a member of itself is one" is
actually false, on the ground that this sentence *does* entail the existence
of an overly large set, one that contains all sets that are not members of
themselves.

The claim that this sentence entails the existence of this large set strikes
me as most implausible: there may be a set containing all trucks, but that
there is certainly doesn't seem to *follow* from the truth of "There are some
trucks of which every truck is one." Moreover, and more importantly, the
claim conflicts with a strong intuition, which I for one am loath to abandon,
about the meaning of English sentences of the form "There are some *A*s of
which every *B* is one," viz. that any sentence of this form means the same
thing as the corresponding sentence of the form "There are some *A*s and
every *B* is an *A*." If so, the sentence of the previous paragraph is simply

[16]In a similar vein, Lauri Carlson writes, "I take such observations as a sufficient
motivation for construing *all* plural quantifier phrases as quantifiers over arbitrary *sets*
[Italics Carlson's] of those objects which form the range of the corresponding singular
quantifier phrases." (Carlson, 1982) is a recent interesting article in which this claim
is made once again. He is by no means the sole linguist with this belief. Carlson does
not face the question of what is to be done when the corresponding singular quantifier
phrase is "some set."

synonymous with the trivial truth "There are some sets and every set that is not a member of itself is a set," and therefore does not entail the existence of an overly large set.

Two worries of a different kind are that the construction "there are some [plural count noun] that are such that . . . they . . . " is unintelligible if the individuals in question do not form a "surveyable" set and that our understanding of this construction does not justify acceptance of full comprehension. I cannot deal with these worries here; I shall only remark that it seems likely that not much of ordinary, first-order, set theory would survive should either worry prove correct.

We have now arrived at the following view: Second-order formulas in which the individual variables are taken as ranging over all sets can be intelligibly interpreted by means of constructions available to us in a language we already understand; these constructions do not themselves need to be understood as quantifying over any sort of "big" objects which have members and which "would be" sets "but for" their size. There can thus be no objection on the score of unintelligibility or of the introduction of unwanted objects to our regarding ZF as more suitably formulated as a finitely axiomatized second-order theory than as an infinitely axiomatized first-order theory, whose axioms are the instances of a finite number of schemata, as is usual. (Of course, in the presence of the usual other first-order axioms of ZF, i.e., the axioms of extensionality, foundation, pairing, power set, union, infinity, and choice, only the one second-order axiom, Replacement:

$$\forall X (\forall x \forall y \forall z [X \langle x, y \rangle \wedge X \langle x, z \rangle \rightarrow y = z] \rightarrow \forall u \exists v \forall y [y \in v \leftrightarrow \exists x (x \in u \wedge X \langle x, y \rangle)])$$

would be needed.) The great virtue of such a second-order formulation of ZF is that it would permit us to express as single sentences and take as axioms of the theory certain general principles that we actually believe. The underlying logic of such a formulation would be any standard axiomatic system of second-order logic, e.g., the system indicated, if not given with perfect precision, in Frege's *Begriffsschrift*.[17] The logic would deliver the comprehension principles $\exists X \forall x [Xx \leftrightarrow A(x)]$ (which are needed for the derivation of the infinitely many axioms of the first-order version of ZF from the finitely many second-order axioms) either through explicit postulation of the comprehension schema, as in Joel Robbin's *Mathematical Logic*,[18] or via a rule of substitution, like the rule given in Chapter 5 of Alonzo Church's

[17](Frege, 1879).
[18](Robbin, 1969). Sec. 56 of Robbin's book contains a presentation of the version of set theory here advocated. It is noted there that this theory is "essentially the same as" Morse–Kelley set theory (MK), but the difficulties of interpretation faced either by MK or by a set theory in the ZF family for which the underlying logic is (axiomatic) second-order logic are not discussed.

Introduction[19] or the one implicit in the *Begriffsschrift*. The interpretation of this version of ZF would be given in a manner similar to that in which the interpretation of the usual formulation of ZF is given, by translation into English in the manner previously described.

Entities are not to be multiplied beyond necessity. One might doubt, for example, that there is such a thing as the set of Cheerios in the (other) bowl on the table. There are, of course, quite a lot of Cheerios in the bowl, well over two hundred of them. But is there, in addition to the Cheerios, also a set of them all? And what about the $> 10^{60}$ subsets of that set? (And don't forget the sets of sets of Cheerios in the bowl.) It is haywire to think that when you have some Cheerios, you are eating a *set*—what you're doing is: eating THE CHEERIOS. Maybe there are some reasons for thinking there is such a set—there are, after all, $> 10^{60}$ ways to divide the Cheerios into two portions—but it doesn't follow just from the fact that there are some Cheerios in the bowl that, as some who theorize about the semantics of plurals would have it, there is also a set of them all.

The lesson to be drawn from the foregoing reflections on plurals and second-order logic is that neither the use of plurals nor the employment of second-order logic commits us to the existence of extra items beyond those to which we are already committed. We need not construe second-order quantifiers as ranging over anything other than the objects over which our first-order quantifiers range, and, in the absence of other reasons for thinking so, we need not think that there are collections of (say) Cheerios, in addition to the Cheerios. Ontological commitment is carried by our *first-* order quantifiers; a second-order quantifier needn't be taken to be a kind of first-order quantifier in disguise, having items of a special kind, collections, in its range. It is not as though there were two sorts of things in the world, individuals, and collections of them, which our first- and second-order variables, respectively, range over and which our singular and plural forms, respectively, denote. There are, rather, two (at least) different ways of referring to the same things, among which there may well be many, many collections.

Leibniz once said, "Whatever is, is one."

Russell replied, "And whatever are, are many."[20]

---

[19](Church, 1956).
[20](Russell, 1982), p. 132.

5

Nominalist Platonism

Frege's definition of "x is an ancestor of y" is: x is in every class that contains y's parents and also contains the parents of any member. A philosopher whom I shall call N. once asked me, "Do you mean to say that because I believe that Napoleon was not one of my ancestors, I am committed to such philosophically dubious entities as *classes*?" Although it is certain that Frege's definition, whose logical utility, fruitfulness, and interest have been established beyond doubt, cannot be dismissed for such an utterly crazy reason, it is not at all easy to see what a good answer to N.'s question might be.

The germ of an answer may lie in the observation that there are sentences containing *plural* forms such as "are" and "them" whose logical forms look as though they ought to be representable in first-order logic, but which cannot be so represented, because of the plural forms the sentences contain. An example is "There are some horses all of which are faster than Zev and all of which are faster than the sire of any horse that is slower than all of them."[1] In contrast, the sentence "There is a horse that is faster than Zev

Reprinted with the kind permission of the editors from *The Philosophical Review* 94 (1985): 327–344. Copyright ©1985 Cornell University.

I want to thank Martin Davis, Michael Dummett, Harold Hodes, David Lewis, John McDowell, Robert Stalnaker, Linda Wetzel, and the referee for helpful comments. This paper was written while I was on a Fellowship for Independent Study and Research from the National Endowment for the Humanities.

[1] (Boolos, 1984c). To see that this sentence cannot be symbolized by a first-order formula, first substitute "number," "greater," "zero," "successor," and "slower" for "horse," "faster," "Zev," "sire," and "smaller," obtaining "There are some numbers all of which are greater than zero and all of which are also greater than the successor of any number that is smaller than all of them," and then notice that any correct symbolization of the latter sentence (using a constant 0 for "zero," a function sign s for "successor," and relation letters $>$ and $<$ for "greater than" and "less than") is a sentence true in a model M of the set of all first-order truths of arithmetic if and only if M is non-standard, that is, not isomorphic to the standard model of arithmetic. Since non-standard models exist, and since (trivially) every first-order sentence has the same truth-value in any model of the set of all first-order truths of arithmetic as in any other, no such correct symbolization

and that is faster than the sire of any horse that is slower than it," which differs from the other in containing singular forms in place of plural, can be symbolized in the notation of first order logic:

$$\exists x(x > 0 \wedge \forall y(y < x \rightarrow x > sy)).$$

If "some" is understood to mean "one or more" this sentence is *stronger* than the former: the horses of whose existence the former informs us are guaranteed to be faster than the sire of y only if y is slower than *all* of them; unlike the latter, it does not imply that there is any one horse x that is faster than the sire of y whenever y is slower than x.

Geach and Kaplan gave an earlier example of a nonfirstorderizable sentence, that is, a sentence not expressible in first-order notation, containing no numerical or quasinumerical words like "more" or "most": some critics admire only one another. This sentence is supposed to mean: there is a non-empty class of critics, each of whose members admires someone only if that person is someone else in the class. The meaning can also be put: there are some critics each of whom admires a person only if that person is one of them and none of whom admires himself. If we explain the meaning of the Geach–Kaplan sentence in this second way, we do not, it appears, quantify over classes of critics. We can also put the meaning: there are some critics who are such that (a) each of them admires a person only if he is one of them and (b) none of them admires himself.

Here then is an answer to N.'s question: in a similar vein, we may say that Napoleon is not an ancestor of N. because either no one is a parent of N. or there are some people who are such that (a) each of N.'s parents is one of them, (b) each parent of any one of them is also one of them, and (c) Napoleon is not one of them. The response that this definition commits one to such philosophically dubious entities as classes now seems wholly out of place; classes aren't mentioned anywhere in the paraphrase. By using plural forms in English quantifiers and employing the construction "one of them" we are able to define *ancestor of* in a way that preserves the essence of Frege's idea and, *at least at first blush,* avoids commitment to such "philosophically dubious entities" as classes: x is an ancestor of y if and only if (I) someone is a parent of y and (II) it is not the case that there are some people who are such that (a) each parent of y is one of them, (b) each parent of any one of them is also one of them, and (c) x is not one of them.[2] This definition may be symbolized by means of a second-order

---

can be a first-order sentence. For an account of non-standard models of arithmetic, see for example, (Boolos and Jeffrey, 1985), ch. 17.

[2]Since (II) is true if (I) is false and y alone exists, (I) is indispensable to the definition.

formula:

$$\exists w\, wPy \land \neg \exists X(\exists w\, Xw \land \forall w(wPy \to Xw) \land \forall w \forall z(wPz \land Xz \to Xw) \land \neg Xx),$$

which is equivalent to the familiar:

$$\forall X(\forall w(wPy \to Xw) \land \forall w \forall z(wPz \land Xz \to Xw) \to Xx)$$

The first question I want to discuss is whether the use of expressions like "there are some people who are such that ... they ... " or monadic second-order quantifiers like "$\exists X$" does in fact commit one to classes, despite appearances. The Geach–Kaplan sentence suggests that we consider the following case. Suppose that I assert that there are some critics, none of whom admires himself, and each of whom admires someone only if that person is one of those critics. Suppose further that I write down, with assertive intent:

$$\exists X(\exists x\, Xx \land \forall x \forall y(Xx \land x \text{ admires } y \to x \neq y \land Xy)).$$

(The first-order variables are intended to range over all critics.)

In doing either of these things have I committed myself to the existence of a class, of critics, none of whom etc.?

Let us deal first with the formula. On the usual treatment of second-order formulae, I *would* have committed myself to the existence of such a class. The formula is normally read and understood to mean, "There is a non-empty class X of critics each of whose members x admires a person y only if x is other than y and y is in the class X." If that is what the formula means, then in writing it down with assertive intent, I would, I suppose, have committed myself to the existence of a class as thoroughly as I would have committed myself had I simply said "There is a non-empty class X etc." But suppose that the formula isn't to be understood as meaning "There is a non-empty class X etc." Does my writing it down then commit me to the existence of a class? The answer, obviously, is that whether or not it does depends on what the formula is supposed to mean: if the formula means something that does not commit one to classes, then it doesn't; if something that does, then it does.

Suppose now that it is said that the formula means: there are some critics, none of whom admires himself, and each of whom admires someone only if that person is one of those critics. On this understanding, interpretation, or reading of it, does the formula commit one to classes?

Before attempting to answer this question, let us note that there is a systematic way to utilize plural forms to translate into English all formulae of second-order logic in which the second-order variables are monadic.[3]

---

[3]In many of the most important applications of second-order logic, a pairing function

Some features of this translation scheme are that it is an extension of the usual scheme for translating first-order formulae into English, and thus respects the propositional connectives, first-order quantifiers $\exists x$, and the equals-sign; that atomic formulae Xy are translated (more or less) as "it is one of them" and that second-order quantifiers $\exists X$ are translated (roughly) as "there are some objects that are such that ..." As in the translation of first-order formulae such as $\exists x \exists y \exists z (xLy \wedge yLz \wedge zLx)$, some devices like "the former," "the latter," "party of the third part," must be employed to do the cross-referencing done in a formal language by the identity and difference of variables; one convenient way to accomplish this is to introduce into English pronouns such as "it$_x$," "them$_Y$," "that$_z$" and "that$_Z$," to which variables of the formal language have been attached as subscripts. On this scheme, the translation of any instance of the comprehension principle of second-order logic is a truism. Thus the translation of the notorious $\exists X \forall x (Xx \leftrightarrow x$ is not a member of $x)$, where the first-order variables are taken to range over absolutely all sets is "(If there is a set that is not a member of itself, then) there are some sets that are such that each set that is not a member of itself is one of them and each set that is one of them is not a member of itself," as vacuous an assertion about sets as can be made, as desired. I have set out the details of the translation scheme elsewhere and will not repeat them here.[4] On this scheme, "There are some critics, none of whom etc." turns out to be an abbreviation of the translation of the formula: $\exists X (\exists x\, Xx \wedge \forall x \forall y (Xx \wedge x$ admires $y \rightarrow x \neq y \wedge Xy))$.

We are thus thrown back to answering the first question: does asserting "There are some critics, none of whom etc." commit one to the existence of a non-empty class of critics? The difficulty with this question is that the ground rules for answering it appear to have been laid down, by Professor Quine.

According to Quine, to determine whether or not "There are some critics etc." commits us to classes, we translate it into logical notation, and then see whether the variables contained in the translation must be supposed to range over classes, to have classes as their values. Now the Geach–Kaplan sentence "There are some critics etc." can be translated into standard logical notation, that is, the notation of *first*-order logic, only if one introduces special variables ranging over classes (or properties or other "dubious" entities). And then the sentence will be translated:

$$\exists a(a \text{ is a class } \wedge \exists x\, x \in a \wedge \forall x \forall y (x \in a \wedge x \text{ admires } y \rightarrow x \neq y \wedge$$
$$y \in a)).$$

will be available and monadic variables can then be made to do the work of all second-order variables.

[4]See Articles 4 and 10 in this volume.

Since this sentence cannot be true unless there is a suitable class to assign
as value to the variable a, it follows, according to Quine, that assertively
uttering the Geach–Kaplan sentence commits one to the existence of classes.

But this is a weird outcome; that was N.'s point. It shouldn't turn out
that I'm committed to classes if I state that there are some critics etc.
What I ought to be committed to is *some critics*, but not to a class of
critics. Furthermore, I would have thought, I ought not to be committed,
on any reasonable sense of the word "committed," to a class containing
all infinite classes if I say, "There are some classes such that every infinite
class is one of them," which is, as I suppose, only an awkward way of saying
"There are some classes and every infinite class is one of them" or "There
are some classes and every infinite class is a class."[5]

Our problem arises from the thought that if we wish to assess the commitment of a theory, we must first put it into first-order notation as well
as we can (sometimes this will not be possible) and then determine what
the variables must be assumed to range over. There are two suggestions
we should resist at this point: that if we are concerned with the *ontological* commitment of a theory couched in some natural language, we must
first translate it into a *first-order* formalism and that we must suppose that
second-order variables in a formula must range over, or have as values,
classes of objects over which the first-order variables of that formula range,
or have as values.

With regard to the first suggestion, we want to ask: what does translation into a first-order language have to do with "ontological commitment"?
"There are some critics etc." doesn't, it seems, commit us, in any ordinary
sense of the word "commit," to the existence of a class of critics; what it
commits us to, one would have thought, is, as we have noted, some critics
none of whom etc. We are forced by Quine's criterion to say that it commits
us to a class; but why, we should ask, should we accept the criterion? If it
is answered: because it's Quine's phrase, and he is at liberty to define it as
he pleases, then we should rejoin: if so, Quine is defining a relation which
holds between us and certain objects (in this case, classes) whose existence
the normal use of our words does not force us to admit, and hence a relation
that ought not to be called "ontological commitment." If it is said that by

---

[5]It is sometimes alleged that there are certain set-like objects, which have elements,
but which are "too big" to be sets; the term "class" is used in set theory to apply to
each such gigantic element-container, as well as to each set. A *proper class* is a class
that is not a set. Every current theory admitting the existence of proper classes denies
that there is a class that contains all infinite classes. If the existence of proper classes is
denied, then "class" and "set" become coextensive, but every current set theory denies
that there is a set containing all infinite sets. Of course, class theories typically imply
the claim there there is a class containing all infinite sets. On the distinction between
sets and classes, see (Parsons, 1983c).

admitting that there are some critics etc. we are *ipso facto* committed to a class, we must ask how this is supposed to have been shown.

We ought to recall that logicians have devoted attention to quantifiers other than the usual *for all x* and *for some x*. Among these less familiar quantifiers, which cannot be defined by means of the apparatus of first-order logic, are *for most x*, *for infinitely many x*, *for uncountably many x*, and *for at least as many x as there are objects (in the domain)*. To claim that a statement to the effect that there are infinitely many objects of a certain kind, made with the aid of the quantifier *for infinitely many x, implies the existence* (on the customary acceptation of those words) of an (infinite) class solely on the ground that the only way to utilize more familiar logical vocabulary to eliminate the unfamiliar quantifier is to employ a quantifier ranging over classes is to invite the response: what makes first-order logic the touchstone by which the ontological or existential commitments of these statements are to be assessed? The statements do not appear to commit us to classes; why believe that it is their translation into the notation of first-order logic augmented with variables ranging over classes that determines what they are actually committed to?

In the case of quantifiers like *for infinitely many x* or *for most x*, it is comparatively easy to hold one's ground in maintaining that assertions involving them need not be taken as *committing* one to the existence of classes; the variable *x* is, after all, a first-order variable. To see that second-order quantifiers are analogous in this regard to the less familiar quantifiers containing first-order variables, we must rebut the second suggestion mentioned above, that in any formula, the second-order variables have to be understood as ranging over (or having as values) classes of objects over which its first-order variables range.

This suggestion is less easily rebutted. The difficulty was well put in a recent, highly interesting article by Harold Hodes. Hodes writes, "Unless we posit such further entities [as Fregean concepts], second-order variables are without values, and quantificational expressions binding such variables can't be interpreted referentially."[6] It will become clear that I disagree with this claim. We needn't posit concepts, classes, Cantorian inconsistent totalities, etc. in order to interpret second-order quantification referentially.

The heart of the matter is this: it is only with respect to a truth-"definition" of the standard sort, a Tarski-style truth-theory for a first-order language, that the notion *value of a variable* is defined. In the case of a second-order language, such as the second-order language of set theory, there are at least two different sorts of truth-theory that can be given: on one of these, it would be quite natural to define "value" so that the second-

---

[6](Hodes, 1984), p. 130.

order variables would turn out to have classes as values; on the other, it
would not. Before we present the two theories, we need to review the usual
truth-definition for the usual first-order language of set theory. (It is because the difficulty of interpreting second-order quantification is most acute
when the underlying language is the language of set theory that we examine
truth-theories for set-theoretic languages.)

A *sequence* is a function that assigns a set to each first-order variable
of the language; we inductively define satisfaction of a formula F by a
sequence s as follows:

If F is $u \in v$, then s satisfies F iff $s(u) \in s(v)$;

if F is $u = y$, then s satisfies F iff $s(u) = s(v)$;

if F is $\neg G$, then s satisfies F iff $\neg(s$ satisfies $G)$;

if F is $(G \wedge H)$, then s satisfies F iff $(s$ satisfies $G \wedge s$ satisfies $H)$;

if F is $\exists v G$, then s satisfies F iff $\exists x \exists t(t$ is a sequence $\wedge \, t(v) =$
$x \wedge \forall u(u$ is a variable $\wedge \, u \neq v \rightarrow t(u) = s(u)) \wedge t$ satisfies $G)$.

Having given this definition, we may prove a lemma stating that if s and
t are sequences that assign the same sets to the free variables of a formula
F, then s satisfies F if and only if t satisfies F. Since a sentence contains
no free variables, it follows from the lemma that if one sequence satisfies a
sentence, all do. We may thus define truth as satisfaction by all, or by some,
sequences. Finally we may demonstrate that the Tarski biconditionals are
provable from this definition, with the aid of a small amount of set theory.

The truth-theory provides an obvious way to define the notion *value of a
variable*. We may say, simply, that x is the value of the variable v relative
to the sequence s if and only if $s(v) = x$. Notice that we wish to give a
relative definition. And we may say that x is a value of v if x is a *value* of
v relative to some sequence.

There is an obvious way to extend this development to the second-order
case. The resulting theory is the first of the two theories mentioned earlier.
We define a sequence to be a function from the set of first- and second-
order variables whose value for each first-order variable v as argument is a
set and whose value for each second-order variable V as argument is a class
(the existence of suitable sequences will of course have to be guaranteed by
principles not available to us in standard set theory). We then inductively
define satisfaction of a formula F by a sequence s as follows:

If F is $u \in v$, then s satisfies F iff $s(u) \in s(v)$;

if F is $u = v$, then s satisfies F iff $s(u) = s(v)$;

if F is Vv, then s satisfies F iff $s(v) \in s(V)$;

if F is $\neg G$, then s satisfies F iff $\neg(s$ satisfies $G)$;

if F is $(G \wedge H)$, then s satisfies F iff $(s$ satisfies $G \wedge s$ satisfies $H)$;

if F is $\exists v G$, then s satisfies F iff $\exists x[x$ is a set $\wedge \exists t(t$ is a sequence $\wedge t(v) = x \wedge \forall u(u$ is a variable $\wedge u \neq v \rightarrow t(u) = s(u)) \wedge t$ satisfies $G)]$;

if F is $\exists V G$, then s satisfies F iff $\exists x[x$ is a class $\wedge \exists t(t$ is a sequence $\wedge t(V) = x \wedge \forall u(u$ is a variable $\wedge u \neq V \rightarrow t(u) = s(u)) \wedge t$ satisfies $G)]$

As before we may define "x is the value of the (first- or second-) order variable v relative to the sequence s" as $s(v) = x$, and define "x is a value of v" as x is the value of v relative to some sequence. Thus if we have this sort of truth-definition in mind, we may say, speaking informally, that in the second-order language of set theory, classes are values of second-order variables.

But there is another sort of truth-theory that can be given for the second-order language of set theory in which no mention is made of sequences any of whose values are (proper) classes. Unlike the previous theories, this theory is formulated in a second-order language, the second-order language of set theory together with a new predicate containing two first-order variables "s" and "F" and one second-order variable "R": R and the sequence s satisfy the formula F. In this new theory a sequence is what it was in the case of the first theory, a function from the set of first-order variables whose values are all sets. The key clauses of the theory are:

If F is $u \in v$, then R and s satisfy F iff $s(u) \in s(v)$;

if F is $u = v$, then R and s satisfy F iff $s(u) = s(v)$;

if F is Vv, then R and s satisfy F iff $R\langle V, s(v)\rangle$;

if F is $\neg G$, then R and s satisfy F iff $\neg(R$ and s satisfy $G)$;

if F is $(G \wedge H)$, then R and s satisfy F iff $(R$ and s satisfy $G \wedge R$ and s satisfy $H)$;

if F is $\exists v G$, then R and s satisfy F iff $\exists x \exists t(t$ is a sequence $\wedge t(v) = x \wedge \forall u(u$ is a first-order variable $\wedge u \neq v \rightarrow t(u) = s(u)) \wedge R$ and t satisfy $G)$;

if F is $\exists V G$, then R and s satisfy F iff $\exists X \exists T(\forall x(Xx \leftrightarrow T\langle V, x\rangle) \wedge \forall U(U$ is a second-order variable $\wedge U \neq V \rightarrow \forall x(T\langle U, x\rangle \leftrightarrow R\langle U, x\rangle)) \wedge T$ and s satisfy $G)$.

("$\langle \ , \ \rangle$" is the ordered-pair function sign.)

In this theory it is reasonable both to define "x is a value of the first-order variable v with respect to the sequence s" as $s(v) = x$ and to say that sets are values of the first-order variables of the second-order language of set theory, since the sequences s mentioned in the new predicate "R and s satisfy F" are functions whose values are all sets. The present theory, however, makes no explicit mention of sequences whose values are (proper) classes. It does not proceed by introducing functions that assign to each second-order variable a unique class, possibly proper. Instead it employs a new predicate which, as one may say, is true or false relative to an assignment of a formula to the first-order variable F, a sequence to the first-order variable s, and some (or perhaps no) ordered pairs of second-order variables and sets to the second-order variable R. There is, however, no need to take the theory as assigning classes, or collections, of those sets, to the second-order variables. Of course one might attempt to argue for the claim that classes are values of the second-order variables of the original language, even according to the present truth-theory, by claiming that they are also values of the second-order variables of that theory. That claim, however, is also one that we *needn't* suppose to be true; we needn't interpret either our original second-order language of set theory or the new truth theory for this language in this manner. Friends of classes will insist that our latest theory may be so reinterpreted and will (reasonably enough) claim that so reinterpreted it is different from the second theory only in an inessential way; but foes of (proper) classes—those who believe that enough is enough, already—will reject the second theory and accordingly resist the suggestion that the third theory may be reinterpreted in the way that the friends suggest. The point of our third truth-theory is to show that the foes of classes have a satisfactory way to define truth for the second-order language of set theory.

A foe of classes may also utilize the third theory to define "x is a value of the second-order variable V with respect to R" as: $R\langle V, x \rangle$. If he does so, he may then say that sets are values of second-order variables as well as of first-. On this way of speaking, it will not in general be the case that if sets x and y are values of V with respect to R, then $x = y$. But, the foe will then emphasize, if one adopts this definition, to say that second-order variables have values is not at all the same thing as to say that their values are classes (or concepts, etc.) The foe will be at pains to reject the suggestion that in general there is any one object whose members are all and only the sets x such that $R\langle V, x \rangle$. Alternatively, instead of defining "x is a value of V with respect to R" at all, he could say that second-order variables have no values, perhaps on the ground that the truth-theory makes no mention of functions that assign objects to second-order variables.

The liar paradox prevents us from explicitly defining truth, or satisfaction, in either the first- or the second-order language itself. But in the first-order case we can expand the language by adding a single new primitive predicate for the notion of satisfaction and then axiomatically characterize satisfaction for the old language by adjoining to (a rather weak) set theory a finite number of axioms containing the new primitive. Truth of a sentence of the first-order language of set-theory may be defined in this theory, and important facts concerning satisfaction and truth can be deduced, including the Tarski biconditionals for satisfaction and truth and many "laws" of truth, for example, the statement that a conjunction is true if and only if both conjuncts are, etc. In setting up our third truth-theory we have proceeded in like manner: we have expanded the language for which we wish to define truth by adjoining to it a single new predicate, and laid down axioms containing this predicate, but have otherwise exceeded the resources of the language in no way. In particular, we have made no additional ontological assumptions not made in the original theory. And with the aid of standard axiomatic second-order logic (e.g., the system of the *Begriffsschrift*) we can prove in the third theory the usual lemmas about free variables, make the usual definition of "true sentence," and derive desired laws of truth.

One technical point deserves mention: in the first-order case, we need a guarantee that $\forall s \forall v \forall x (s$ is a sequence $\wedge\, v$ is a variable $\rightarrow\, \exists t(t$ is a sequence $\wedge\, t(v) = x \wedge \forall u(u$ is a variable $\wedge\, u \neq v \rightarrow t(u) = s(u))))$ in order to derive the consequences we desire from the axioms of the truth-theory. A small amount of set theory provides us with this guarantee. We need a like guarantee in the present case; we need to be able to show that $\forall R \forall V \forall X [V$ is a second-order variable $\rightarrow \exists T(\forall x[Xx \leftrightarrow T\langle V, x\rangle] \wedge \forall U[U$ is a second-order variable $\wedge U \neq V \rightarrow \forall x(T\langle U, x\rangle \leftrightarrow R\langle U, x\rangle)])]$. This time the requisite guarantee is forthcoming, again with the aid of a small amount of set theory, from a comprehension principle which will be a theorem of any standard axiomatic system for second-order logic:

$$\exists T \forall z[Tz \leftrightarrow \exists x \exists U(U \text{ is a second-order variable} \wedge z = \langle U, x\rangle \wedge \\ (U = V \rightarrow Xx) \wedge (U \neq V \rightarrow Rz))]$$

A somewhat disconcerting conclusion emerges: it is not in general possible to tell by inspection of its asserted formulae alone whether or not classes are to be counted among the values of the variables of a theory formalized in a second-order language of the usual sort, even in the most favored case, in which the asserted formulae of the theory include instances of the comprehension schema $\exists X \forall x(Xx \leftrightarrow A(x))$. Since one can neither presume that the formula "Xx" must have the meaning "x is a member of the class X" or that the quantifier "$\exists X$" is to be read "there is a class x such that …," one is not entitled to "read off" a commitment to classes from the asserted statements of such a theory. The reinterpretation of the quantifiers

$\exists X$ and atomic formulae Xx given by the third truth-theory alters neither the interpretation of the apparatus of first-order logic nor the truth-values to be assigned to sentences of the original language, if the friend of classes is right after all. It is therefore no mere matter of a formalism put to a deviant use that one cannot, for example, discern commitment to a universal class even when $\exists X \forall x \, Xx$ is one of the asserted statements of a theory. The possibility of so reinterpreting second-order notions shows that assessment of the ontological costs of a theory is rather less routine a matter than we may have supposed.

Having dealt at length with the truth-conditions of sentences of the second-order language of set theory, we should wish to give an account of the validity-conditions of these sentences.[7]

A sentence of the first-order predicate calculus is called (logically) *valid*, or a *logical truth*, if it is true in all models; a model M is an ordered pair of a non-empty set D and a function F from some set L of symbols to a set of relations and functions on D of appropriate degrees. D is the *domain* or *universe* of M, L is the *language* of M, and F assigns suitable denotations or references to the symbols of L. The aspect of this familiar definition of validity that here concerns us is the set-theoretic definability of the notion of validity, a consequence of the stipulation that D and L, and therefore M as well, be *sets*.

There is no universal set; there is no set of all pairs $\langle x, y \rangle$ such that x is in y; and there is no model $\langle D, F \rangle$ in which D is the universal set and F is a function from $\{\in\}$ to the set of all such pairs. At any rate, there are no such items if Zermelo–Fraenkel set theory is correct, as we shall henceforth assume.

From the nonexistence of such a model $\langle D, F \rangle$ there arises a certain difficulty: suppose that some sentence G of the language of set theory is logically valid, true in all models. What guarantee have we that G is *true*, that is, true when its variables are taken as ranging over all the sets there are and $\in$ as applying to (arbitrary) x, y if and only if x is in y? If there were such a model $\langle D, F \rangle$, there would be no problem: G would then be true in $\langle D, F \rangle$ and therefore true period. It appears that in set theory at least, the truth of a statement does not immediately follow from its validity.

Set theory itself provides a way out of the difficulty. In fact, it provides two. The first is via the reflection principle: it is a theorem (-schema) of set theory that for each sentence G of the language of set theory there is a model M, indeed one of the form

$$\langle V_\alpha, \{\langle \in, \{\langle x, y \rangle : x, y \in V_\alpha \wedge x \in y\}\rangle\}\rangle$$

---

[7]The classical discussion of validity and second-order logic is (Kreisel, 1967). (Shapiro, 1984) discusses some of the issues raised in Kreisel's paper.

such that the sentence is true in M if and only if it is true. Thus if G is false, $\neg G$ is true, hence $\neg G$ is true in some model, and therefore G is not valid. Thus it cannot happen that a sentence of the language of set theory be valid and yet false.

The second way out is via the completeness theorem, according to which G has a proof in any standard axiomatic system of first-order logic if it is valid. Since (the universal closure of) any axiom of logic is true and the rules of inference preserve truth, any valid sentence of the language of set theory is true, since (the universal closure of) any sentence occurring in a proof is true.

But it is rather strange that appeal must apparently be made to one or another non-trivial result in order to establish what ought to be obvious: viz., that a sentence is true if it is valid.

I want to point out that in addition to the usual notion of validity, there is another notion of validity stronger than the usual one, susceptible, like the notions of truth and satisfaction, only of schematic definition, and on which it is obvious, as is fitting, that a valid sentence is true. Moreover, on this notion, the fact that whatever is valid is true is not much more than an effect of the rules of inference UI (universal instantiation) and substitution, as is also appropriate.

I shall call the notion to be defined *supervalidity*. (I do think that *it* ought to be called validity and the usual notion ought to be called subvalidity. But never mind.)

The idea of supervalidity can be informally explained as follows: a sentence of the language of set theory is supervalid if it is true, no matter what sets its variables range over (as long as there is at least one set over which they range) and no matter what pairs of sets $\in$ is taken to apply to.

I suspect that it is the mistaking of validity for supervalidity and the mathematical interest of validity, together with certain doubts about the intelligibility of supervalidity, that are responsible for the prominence of the notion of validity in logical theory.

Before defining supervalidity, I would like to mention a further odd feature of the concept of validity, or logical truth, viz. that a true sentence to the effect that another sentence is valid is not itself *valid*, but rather a true statement of set theory. Of course, in view of the celebrated "limitative" theorems of logic, the thought that we should *want* true assertions of validity to be valid may strike one as greedy, but one really should not lose the sense that it is somewhat peculiar that if G is a logical truth, then the statement that G is a logical truth does not count as a logical truth, but only as a set-theoretical truth.

The formal definition of supervalidity is this: let G be a sentence of the language of set theory. Select two monadic second-order variables X, Y.

Replace all formulas $u \in v$ in G by formulas $Y\langle u, v \rangle$. Relativize all quantifiers $\forall v$ and $\exists v$ in the result to the formula Xv; that is, replace contexts $\forall v(\ldots)$ by $\forall v(Xv \to \ldots)$ and contexts $\exists v(\ldots)$ by $\exists v(Xv \wedge \ldots)$. Quantify universally with respect to Y. Take the result as the consequent of a conditional with antecedent $\exists x\, Xx$. Finally, quantify this conditional universally with respect to X. The result is the formalization of the assertion that G is supervalid.

Thus we do not define "is supervalid" by constructing a formula with one free variable that applies to the (Gödel numbers of) supervalid sentences and only to these. It is instead defined schematically, by associating with each sentence G of the first-order language of set theory, another sentence, of the second-order language, that expresses the assertion that G is supervalid, as informally explained above. In a similar way, "is true" is schematically defined by associating with each sentence G of the language the sentence G itself, and not by constructing a single formula satisfied by the (Gödel numbers of) true sentences and nothing else.

We have defined supervalidity only for sentences of the first-order language $\{\in\}$ of set theory, but it is clear how it may be done for any formula of any first- *or second*-order language at all. We shall confine attention to sentences in which the only non-logical constant is $\in$. We shall also suppose that the only non-individual variables found in second-order formulae are *monadic*.

It is apparent that for any sentence G, the sentence expressing the truth of G, that is, G itself, can be derived in axiomatic second-order logic from the sentence G' asserting the supervalidity of G, together with a suitable axiom governing the ordered pair operation $\lambda x, y \langle x, y \rangle$. (Mention of an ordered pair axiom could have been omitted had we replaced $\in$ with a *dyadic* variable Y.) One need only instantiate $\forall X$ with the abstract $\{x : x = x\}$ in G', resolve the abstracts, discharge the antecedent $\exists x\, x = x$ via logic, instantiate $\forall Y$ with $\{z : \exists x \exists y(\langle x, y \rangle = z \wedge x \in y)\}$ and use the ordered pair axiom to resolve the abstracts; the result is trivially equivalent to G. Instantiation with abstracts is legitimated in second-order logic by the comprehension schema.

It is also apparent that the result of restricting all of the quantifiers of any supervalid sentence to some one formula (possibly containing one or more parameters, for example, a parameter for a model) satisfied (with respect to any assignment of objects to those parameters) by at least one object (=set) is true (with respect to that assignment). Consequently, any supervalid sentence is valid.

Finally, it should be apparent that the axioms of axiomatic second-order logic, including the instances of the comprehension schema $\exists X \forall x [Xx \leftrightarrow A]$, X not free in A, are all supervalid and that the rules of inference of second-order logic, including the rule of substitution (of formulas for free variables), preserve supervalidity. Substitution is a rule of inference that was used by Frege, in the *Begriffsschrift*; the deductive equivalence of the comprehension schema and the rule of substitution is well known.[8] I have of course been assuming the intelligibility and legitimacy of second-order quantification over all sets or over objects of unbounded set-theoretic rank.

Any provable first-order formula is supervalid; any supervalid first-order formula is valid; and, by the completeness theorem, any valid first-order formula is provable. Thus validity, supervalidity, and provability coincide for first-order formulae. In the absence of a completeness theorem for (real, full, standard) second-order logic, we cannot make the analogous claim for second-order formulae. We know how to produce counter-examples to the claim that any given (recursively enumerable) axiom system yields as theorems all valid sentences of second-order logic. Since the counterexamples turn out to be not only valid but supervalid, the only question about inclusion among the three notions that remains is whether all valid sentences are supervalid. Otherwise put, is every (second-order) sentence of the language of set theory that is true in all models *true*?

Many set theorists find it probable or plausible that the answer is yes. They speculate that "there is no property of the universe of sets that is not reflected by some type V_α." Thus if they are right, there could be no second-order sentence that is false but nevertheless true in all models, or even true in all models of the form $\langle V_\alpha, \{\langle \in, \{\langle x, y \rangle : x, y \in V_\alpha \wedge x \in y\}\rangle\}\rangle$. Any such sentence would show, in technical parlance, that On is Π_n^1-describable for some n; and this has seemed extremely unlikely to most set theorists who have written on axioms of infinity and the structure of the set-theoretic universe. (The claim, however, would appear to be insusceptible of anything like proof from currently accepted axioms.) Thus the notion of validity has, after all, greater interest than the foregoing, belittling, line of thought might have inclined one to suppose it has: it is plausible and a reasonable "working hypothesis" that validity coincides with supervalidity, and hence that supervalidity *can* be defined by means of a single formula, and indeed by a formula of the first-order language of set theory.

In conclusion, let us mention an apparent defect of the account of supervalidity we have given: it would seem that there is no natural or obvious way to generalize the notion of supervalidity to a notion of "superconsequence" or "supersatisfiability." What we want is a way to explain what it is for some sentences (in the first instance, of the language of set theory) to be true under some one interpretation, that is, for there to be some sets and some pairs (to assign to $\in$) under which all of those sentences are true,

---

[8]Cf. Article 10 in this volume.

without introducing classes, infinitely long sentences, or an unanalyzed notion of truth or satisfaction. We have shown above how this may be done for any given sentence. There seems no satisfactory way to do it for an infinite set of sentences, however. And although the sense of loss may be mitigated by the knowledge that many important theories such as Peano Arithmetic and Zermelo–Fraenkel Set Theory, which are not finitely axiomatizable, are axiomatizable by a finite number of schemata and have a natural second-order extension that is a finite extension of axiomatic second-order logic, there is no denying that there is a loss.

6

Iteration Again

According to the iterative, or cumulative, conception of set, sets are *formed* at *stages*; indeed, every set is formed at some stage of the following "process": at stage 0 all possible collections of individuals are formed. Individuals are objects that are not sets; for the usual sorts of reasons, we shall assume that there are no individuals. Thus at stage 0 only the null set is formed. The sets formed at stage 1 are all possible collections of sets formed at stage 0, i.e., the null set and the set whose sole member is the null set. The sets formed at stage 2 are all possible collections of sets formed at stages 0 and 1. There are 4 ($= 2^2$) of these. The sets formed at stage 3 are all possible collections of sets formed at stages 0,1, and 2. There are 16 ($=2^4$) of these. The sets formed at stage 4... In general, for any natural number n, the sets formed at stage n are all possible collections of sets formed at stages earlier than n, i.e., stages $0, 1, \ldots, n-1$.

Immediately after all stages $0, 1, 2, \ldots$, there is a stage, stage ω. The sets formed at stage ω are, similarly, all possible collections of collections of sets formed at stages earlier than ω, i.e., stages $0, 1, 2, \ldots$ After stage ω comes stage $\omega + 1$: at which ... In general, for each α, the sets formed at stage α are all possible collections of sets formed at stages earlier than α.

There is no last stage: each stage is immediately followed by another. Thus there are stages $\omega + 2, \omega + 3, \ldots$ Immediately after all of these, there is a stage $\omega + \omega$, alias $\omega \cdot 2$. Then $\omega \cdot 2 + 1, \omega \cdot 2 + 2$, etc. Immediately after all of $\omega, \omega \cdot 2, \omega \cdot 3, \ldots$ comes $\omega \cdot \omega$, alias ω^2. Then $\omega^2 + 1, \ldots$ And so it goes.[1]

Notice that on this account of the iterative conception, no set is formed at exactly one stage: each set is also formed at all stages that are later than any one at which the set is formed. We do not assume that each set is first

Reprinted with the kind permission of the editor from *Philosophical Topics* 42 (1989): 5–21.

Research for this paper was carried out under grant number SES–8808755 from the National Science Foundation. I am grateful to Richard Cartwright, Michael Hallett, and David Lewis for very helpful comments.

[1] Cf. among many other accounts Article 1 in this volume.

formed at some unique stage, however, and hence do not assume that the stages are well-ordered.

Set theory, i.e., Zermelo–Fraenkel set theory (ZF) together with the axiom of choice, is sometimes said to "express," "embody," or "articulate" the iterative conception. My aim here is to clarify some of the relations between set theory, the iterative conception, another conception of set theory ("limitation of size") due to Russell and von Neumann, and a repair to the system of Frege's *Grundgesetze der Arithmetik* that embodies that other conception. Towards the end of the paper I try to cast some doubt on the idea that there is any single conception that "underlies" the whole of set theory.

We shall begin with the methodological question: what sort of justification for set theory does the iterative conception provide?

Let us call the theory whose axioms are all of the axioms of Zermelo set theory, with the exception of the axioms of extensionality and choice, Z^-. (Z is Zermelo set theory, one of whose axioms is the axiom of extensionality; the axiom of choice is not a full-fledged axiom of either Z or ZF, which is obtained by adding the axioms of replacement to Z.) We shall soon see that Z^- can be derived from a (remarkably weak-looking) formalization of the iterative conception. It does not follow that the iterative conception shows that the theorems of the subtheory Z^- of ZF are *true*, for there is no reason to think that stages (whatever *they* might be) and sets are as the conception maintains, i.e., that the conception is correct about sets and stages. Certainly, if matters are as the conception has them, then Z^- is true, for, unexceptionably, it can be *deduced* from the iterative conception. However, no independent reason has been given to believe that sets and stages are as they are according to the iterative conception.

(It is an interesting question why we are inclined to reject the skeptical hypothesis that, in the absence of some formal defect in set theory such as simple or ω-inconsistency, the iterative conception of set might be *wrong*, at least in its broad outlines.)

The iterative conception has been called "natural." "Natural" here is not a term of aesthetic appraisal (possibly linked to the Panglossian view that a more "natural" or "simpler" theory may have a greater chance of being true) but simply means that, without prior knowledge or experience of sets, we can or do readily acquire the conception, easily understand it when it is explained to us, and find it plausible or at least conceivably true. For a view to be natural in this sense, it cannot be too much at odds with our preconceptions, like the (crazy) view obtained from the iterative conception by interchanging "earlier" and "later."

Another conception of set that is natural in this sense is the *naive* conception, which can be formulated in two ways, as the thought that any predicate has a set as its extension, and as the thought that any zero or more things are the members of some one set.[2] The trouble with the naive conception is that Russell's paradox shows it to be inconsistent: the predicate "is a set that is not a member of itself" is a predicate with no set as its extension, and the sets that are not members of themselves are not the members of any set.

A different conception of set, to be examined below, is the doctrine of "limitation of size."[3] The doctrine comes in at least two versions: On a stronger version of limitation of size, objects form a set *if* and *only if* they are not in one-one correspondence with all the objects there are. On a weaker, there is no set whose members are in one-one correspondence with all objects, but objects do form a set if they are in one-one correspondence with the members of a given set. (Under certain natural conditions, this last hypothesis can be weakened to: if there are no more of them than there are members of a given set.) The difference between the two versions is that the weaker does not guarantee that objects will always form a set if they are not in one-one correspondence with all objects.

Unlike the naive and the iterative conceptions, *limitation of size* (in either version) is not a natural view, for one would come to entertain it only after one's preconceptions had been sophisticated by knowledge of the set-theoretic antinomies, including not just Russell's paradox, but those of Cantor and Burali–Forti as well.

The iterative conception is the only natural and (apparently) consistent conception of set we have, and it implies Z^-; *that* is the justification (if that is the right word) it provides for Z^-.

Dan Leary once made the observation that the metaphor of formation of sets at stages may arise from a certain *narrative* convention or principle of good exposition: in general and *ceteris paribus*, a description of objects that are arranged in some salient manner should mention those objects in an order corresponding to the arrangement. Conformably, when describing the structure of the set-theoretic universe, one would first mention the null set, then the set containing just the null set, then the sets of all those, then the sets of all *those*, and so on. One might say: there is the null set, there is its unit set, then there are the two other sets containing only those, then there are the twelve "new" (i.e., not yet mentioned) sets containing only those, ... The fact that it takes time to give such a sketch, and that certain sets will be mentioned before others, might easily enough be (mis-)taken for a quasi-temporal feature of sets themselves, and one might be tempted to say that sets coming earlier in the description actually *come earlier*, that

---

[2]Cf. (Mates, 1981), p. 43.

[3]Cf. (Hallett, 1984), esp. chs. 4 and 8, for a thorough discussion of the different versions of *limitation of size*.

sets cannot exist *until* their members do, that they *come into being* only after their members do, and that they are *formed* after all their members are.

In any case, for the purpose of explaining the conception, the metaphor is thoroughly unnecessary, for we can say instead: there are the null set and the set containing just the null set, sets of all those, sets of all *those*, sets of all *Those*, ... There are also sets of all *THOSE*. Let us now refer to these sets as "those." Then there are sets of those, sets of *those*, ... Notice that the dots " ... " of ellipsis, like "etc.," are a demonstrative; both mean: *and so forth*, i.e., in *this* manner forth.

But I am not now concerned to eliminate the metaphor, which in any event could be accomplished in short order by taking the terms "stage," "is formed at," and "is earlier than" as primitive,[4] or by replacing them with "ordinal," "has rank," and "is less than," taken as primitive. I want rather to show how little of the iterative conception is actually required for the derivation of Z^-, i.e., to show how very simple a theory of stages there is from which the axioms of Z^- follow (and how needlessly complex the axiomatization found in "The Iterative Conception of Set" is).

Let us then consider a two-sorted first-order language $\mathcal{L}$, with variables $x, y, z, \ldots$ for sets, and variables $r, s, t, \ldots$ for stages. There are three two-place predicates in $\mathcal{L}$, a stage-stage predicate $<$, read "is earlier than," a set-stage predicate F, read "is formed at," and a set-set predicate $\in$, read as usual.

Let us abbreviate "$\exists t(t < s \wedge yFt)$" as: yBs, which may be read "y is formed before s."

Then the following sentences are the axioms of our theory S. Axioms concerning "earlier than":

Tra $\quad \forall t \forall s \forall r(t < s \wedge s < r \rightarrow t < r)$
Net $\quad \forall t \forall s \exists r(t < r \wedge s < r)$
Inf $\quad \exists r(\exists t\, t < r \wedge \forall t(t < r \rightarrow \exists s(t < s \wedge s < r)))$

Axioms concerning sets and stages:

All $\qquad \forall x \exists s\, xFs$
When $\quad \forall x \forall s(xFs \leftrightarrow \forall y(y \in x \rightarrow yBs))$

The specification axioms, one for each formula $A(y)$ of $\mathcal{L}$ (not containing the variable x free):

Spec $\quad \exists s \forall y(A(y) \rightarrow yBs) \rightarrow \exists x \forall y(y \in x \leftrightarrow A(y))$.

---

[4] As is done in Article 1 of this volume.

Some comments on these axioms: *Tra*, of course, says that *earlier-than* is transitive. One of the consequences of *Net* is: $\forall s \exists r\, s < r$, i.e., every stage is earlier than some stage. *Net* follows from $\forall s \exists r s < r$, *Tra*, and the sentence *Con* of $\mathcal{L}$ expressing the connectedness of *earlier-than*, viz., $\forall s \forall t (s < t \lor s = t \lor t < s)$; *Con*, however, is *not* one of the axioms of S.

Inf states that there is a "limit" stage, a stage later than some stage but not immediately later than any stage earlier than it: the existence of stage ω and hence of such a stage as *Inf* claims to exist is a notable feature of the conception we have described. *Inf* is too weak to capture the full strength of the claims about the existence of infinite stages made in the rough description; a further axiom would be needed to guarantee the existence of a stage $\omega + \omega$, for example. It suffices, however, for the derivation of the sentence of set theory customarily called "the axiom of infinity." *Inf*, it should be noted, is used only in the derivation of the axiom of infinity.

All states what is perhaps the most distinctive feature of the iterative conception, viz., that every set is formed at some stage of the iterative process described above. *When* amplifies *All*, by telling us that a set is formed at a stage if and only if all its members are formed at earlier stages. (Thus sets are, as we have noted, continually *re*formed.)

Because $\mathcal{L}$ is a *first*-order language, *Spec*(ification) is an axiom-schema and not an axiom. It attempts to capture the thought that the sets formed at any stage are "all *possible collections*" of sets formed at stages earlier than that one. It is not entirely clear what the force of the phrase "possible collection" is supposed to be. What is the modal term "possible" doing, and in any case how does a collection differ from a set? (*When* tells us that a set is formed at a stage iff all its members are formed at earlier stages. Of course if a set is formed at a stage, it is formed at that stage. What then does *Spec add* by saying that the sets formed at each stage are all *possible collections* of sets formed at earlier stages?) The thought can be put better if we say: for any stage s and any set$s$ (notice the plural) that have all been formed before s, there is a set to which exactly those sets belong. This thought can be perfectly expressed in a second-order language: $\forall X[\exists s \forall y (Xy \to yBs) \to \exists x \forall y (y \in x \leftrightarrow Xy)]$. Elsewhere[5] I have argued that, happily, such formulations need not be regarded as quantifying over any proper classes or other set-like objects that are not actually sets. To the extent that it is not vague what the iterative conception is (i.e., not vague how far out the stages go), the full force of the conception can be expressed in a second-order language extending $\mathcal{L}$, but not in the first-order language $\mathcal{L}$ itself.

---

[5]Cf. Articles 4 and 5 in this volume.

A useful reformulation of *Spec* is: $\forall s \exists x \forall y (y \in x \leftrightarrow (A(y) \wedge yBs))$. To see that the old version implies the new, let $A'(y) \leftrightarrow A(y) \wedge yBs$, and apply the old version to $A'(y)$; the new version immediately implies the old.

It would have been cheating to take the axiom of extensionality as an axiom of S. It may be "analytic" or "analytic-whatever-it-may-mean-to-say-so" that different sets have different members, but that they do is not actually guaranteed by the iterative conception, properly so-called. Of course it would be possible to derive the axiom of extensionality by sneaking in an "!" after "$\exists x$" in *Spec*; our aim, however, is to analyze the conception we have, and not to formulate some imperfectly motivated conception that manages to imply the axioms.

It may seem, however, that something other than that it is quasi-analytic or whatever can be said about extensionality. The thought might occur to one that a set is really nothing other than its members. That is, it *is* them, is identical with them. (This idea is doubtless responsible for the perplexity that sometimes strikes beginners when they are told that an individual and its unit set are to be distinguished.) If so, then extensionality follows from the transitivity of identity: for if every member of x is a member of y and vice versa, then the members of x *are* the members of y; therefore x, i.e., the members of x, is identical with y, i.e., the members of y, and extensionality holds.

Now, there is certainly something fishy in the suggestion that a set is identical with its members—how could *it* be *them* if they are more than two?—but it may well seem that there is also something non-fishy too. Are not John, Paul, George, and Ringo a group, Dolly, Stiva, Tanya, and Grisha a family, and were not Bird, McHale, Parish, Ainge, and Johnson a starting five? Russell once wrote, "In the present chapter we shall be concerned with *the* in the plural: the inhabitants of London, the sons of rich men, and so on. In other words, we shall be concerned with *classes*."[6] It is hard to see how he could suppose that when we are concerned with the inhabitants of London, we are concerned with the class of those inhabitants unless he supposes that the inhabitants of London are, are identical with, are the same thing as, that class. It would be thoroughly unreasonable to suppose that in this passage Russell actually thought that the class is distinct from its members but constituted by them, and that whenever we referred to the members, we also referred to something different, the class.

However, one who advocates that the Beatles are, strictly speaking, identical with some one thing, a group, and the Oblonskys with a family will have some hard questions to answer, e.g., how many are that group and that family, two or eight? how can the group be in its own unit set without

---

[6](Russell, 1919), p. 181.

the four Beatles being in that unit set? Best, perhaps, not to expect this account of extensionality to succeed.

The following argument might be thought to show that extensionality is evident on the iterative conception, and that it would therefore have been fair to take $\forall x \forall y (\forall z (z \in x \leftrightarrow z \in y) \rightarrow x = y)$ as one of the axioms of S: Observe the uniqueness claims implicit in the use, in the first paragraph, of such phrases as "*the* null set" and "*the* unit set of the null set," and the other claims made there concerning the number of sets formed at early stages, e.g., that at stage 0 only the null set is formed, and that at stage 3, 16 sets are formed. These claims presuppose the truth of extensionality, which ought therefore to have been an axiom of S.

In reply it can be said: Notice that extensionality was immediately applied to calculate the number of sets formed even at stage 0, before all but a small part of the conception was given. That sets are identical if their members are the same would therefore seem to be a principle for whose evidence the iterative conception is not responsible, but rather one whose truth is perfectly obvious (for whatever reason) to us in advance of our forming the iterative conception. However, say if you wish that it is part of the iterative conception precisely because of its obviousness, but notice then how "detachable" it is from the rest of the conception: were $\forall x \forall y (\forall z (z \in x \leftrightarrow z \in y) \rightarrow x = y)$ to be taken as a further axiom, it would not be used in the derivation of any of the other axioms of Z^-, nor, unlike the axiom of infinity, would any other axiom of Z^- be needed in its derivation.

The axioms of S having been set out and discussed, the time has come to derive Zermelo set theory minus extensionality and choice from these axioms (for most of the details, see the appendix). Remarkably, all the axioms of Z^- can be derived from S, even if these are taken to include the axioms of *regularity* or *foundation*, i.e., the formulae $\exists x\, A(x) \rightarrow \exists x (A(x) \land \forall y (y \in x \rightarrow \neg A(y)))$ of the language of set theory. Among these axioms is the formula that is sometimes called *the* axiom of regularity: $\exists x\, x \in z \rightarrow \exists x (x \in z \land \forall y (y \in x \rightarrow \neg y \in z))$. The remarkable fact that these are derivable in S *even in the absence from the axioms of S of an axiom schema expressing an induction principle for stages* was first observed by Dana Scott.[7] Indeed, all formulas $\exists s\, P(s) \rightarrow \exists s (P(s) \land \forall t (t < s \rightarrow \neg P(t)))$—call the schema with these formulas as instances "induction for stages"—can be proved in S, and the axioms of regularity derived from these.

Notice that the axioms of S, even taken together, do not have the "look" of an induction principle. The derivability in S of induction for stages and the axiom-schema of regularity is therefore most surprising in view of

---

[7](Scott, 1974).

general logical experience, which tends to confirm the view that one cannot infer a principle of induction without assuming a principle of induction explicitly or implicitly. For example, if one tries to show that the (true) natural numbers satisfy mathematical induction, one typically EITHER *defines* them as objects satisfying some sort of inductive condition—as in Frege and Russell's work, where they are characterized as the members of all classes containing zero and closed under successor—in which case one needs to use induction (outside the theory) to show that the true natural numbers have all the interesting properties enjoyed by the objects defined to satisfy the condition, OR one postulates an induction principle in the theory in which one is attempting to demonstrate that the numbers satisfy induction, as when one assumes the axiom of regularity in set theory, defines the ordinals as transitive sets whose members are all transitive, defines the natural numbers as zero or successor ordinals whose members are all zero or successor ordinals, and then uses regularity to infer the well-foundedness of the ordinals, and hence that of the natural numbers, so defined. The idea that induction is always needed to derive induction may also be fostered by an acquaintance with Hume's reflections on the justification of "empirical" induction (and other skeptical philosophical writings whose tendency is that no important philosophical principle, e.g., the existence of material objects, can be proved from assumptions that appear to be weaker), perhaps also by Lewis Carroll's "Achilles and the Tortoise" or the writings of Poincaré, Quine and Wittgenstein, and certainly by the common knowledge that formal systems of arithmetic lacking induction are impossibly weak. Despite all this common knowledge and good sense, there is, as we are about to see, a derivation of an induction principle from principles that simply cannot themselves be characterized as induction principles. (Philosophers are to be predicted to claim that *Spec* is "really" a disguised induction principle.) Moral: sometimes you *can* get induction out without first putting it in.[8]

Here is the way the derivation proceeds; we follow Shoenfield's *Handbook* article:[9]

Definition *y is a* minimal member *of x if $y \in x$ and $\forall z \neg (z \in x \land z \in y)$.*

Definition *y is* grounded *if every set containing y has a minimal member.*

If every member of y is grounded, then y itself is grounded. (Logic: Suppose $y \in x$. If for some z, $z \in x$ and $z \in y$, then z is grounded, and x has a minimal member. Otherwise, $\forall z \neg (z \in x \land z \in y)$; but then y is a minimal member of x.)

---

[8]Cf. Article 24 in this volume.
[9](Shoenfield, 1978), esp. p. 327.

Definition aRs iff $\forall y(y \in a \leftrightarrow y$ is grounded $\wedge yBs)$.

Miscellaneous facts:

1. By *Spec*, for every s, there is an a such that aRs.

2. If aRs, then since all members of a are grounded, a is grounded.

3. By *When*, if aRs, then aFs.

4. Thus if $t < s$, aRs, and bRt, then b is grounded by 2, bFt by 3, bBs, and $b \in a$.

Induction for stages: $\exists s\, P(s) \rightarrow \exists s(P(s) \wedge \forall t(t < s \rightarrow \neg P(t)))$.

Proof. Suppose $P(r)$. If for all u such that $u < r$, $\neg P(u)$, then done. So suppose $u < r$ and $P(u)$. By *Spec*, for some x, $\forall a(a \in x \leftrightarrow \exists s(s < r \wedge aRs \wedge P(s) \wedge aBr))$. By 3 and the definition of "B," $\forall a(a \in x \leftrightarrow \exists s(s < r \wedge aRs \wedge P(s)))$. Since $u < r$ and $P(u)$, x is nonempty by 1. By 2, all members of x are grounded. Thus x has a minimal member a, and for some s, $s < r$, aRs, and $P(s)$. Now suppose $t < s$. By 1, for some b, bRt. By 4, $b \in a$. By *Tra*, $t < r$. If $P(t)$, then $b \in x$, a contradiction as a and x are disjoint; thus $\neg P(t)$. ■

Regularity, $\exists x\, A(x) \rightarrow \exists x(A(x) \wedge \forall y(y \in x \rightarrow \neg A(y)))$, follows directly from induction for stages: Suppose $A(x)$. By *All*, for some s, xFs. Thus $\exists s\exists x(A(x) \wedge xFs)$. By induction for stages, with $P(s) \leftrightarrow \exists x(A(x) \wedge xFs)$, $\exists s(\exists x(A(x) \wedge xFs) \wedge \forall t(t < s \rightarrow \neg\exists x(A(x) \wedge xFt)))$. Pick such s and x. Then $A(x)$ and xFs. Now suppose $y \in x$. By *When*, yBt, i.e., for some t, $t < s$ and yFt. Thus $\neg A(y)$.

The derivations of the other axioms of Z^-, pairing, union, power, the Aussonderungsschema, and infinity, are routine and relegated to the appendix. The remaining axioms of set theory are the axiom of choice and the axioms of replacement; we briefly discuss these.

The following argument might be thought to show that the axiom of choice follows from the iterative conception: Suppose that x is a set of disjoint non-empty sets. We want to show that there is a set y having exactly one member in common with each of the members of x. Let s be a stage at which x is formed. Then the members of x are formed before s and by transitivity *their* members are also formed before s. Now, it is apparent that

> (∗) there are some sets such that each of them is a member of a member of x, no two of them are members of the same member of x, and among those sets there is at least one member of each member of x.

Since those sets are all members of members of x, they are all formed before s, and thus there is a set y that contains them and no others.

The difficulty in supposing that this argument shows choice to follow is that its acceptability depends crucially upon that of $(*)$. Apparent though $(*)$ may be, a sceptic about choice would immediately be skeptical about the truth of $(*)$; one inclined to think that there need not be a set having exactly one member in common with each member of x would hardly suppose that there need be any such sets as are claimed to exist in $(*)$. $(*)$ may be perfectly obvious, but it is not *the iterative conception* that shows $(*)$, or choice, to hold. With or without the iterative conception, $(*)$ would still be apparent. And without $(*)$, all that the argument shows is that any choice set y for x that there might be will be formed no later than x itself, not that there is any such choice set. I conclude that the iterative conception provides no sort of justification at all for the axiom of choice.

In "The Iterative Conception of Set" I claimed that not even the existence of a stage corresponding to the first non-recursive ordinal is guaranteed by a formalization of the iterative conception and therefore that replacement does not follow from the iterative conception. (It is certainly not implied by S, but S formalizes only a part of the content of the conception.) The arguments found in the literature to the effect that replacement can be derived from the iterative conception without the aid of some further principle still strike me as unsatisfactory, but I shall not review them here.

One way to extend S so as to yield replacement is to exploit the idea, familiar from category theory, that *being included in* is a species of *being injectible into*. Thus suppose that, working in a second-order version of S, we change the antecedent of *Spec* from $\forall y(Xy \rightarrow yBs)$, which expresses that the sets X are included in those formed before stage s, to a formula expressing that the sets X are injectible into those so formed: $\exists R(\forall y \forall y' \forall z \forall z'(Ryz \wedge Ry'z' \rightarrow (y = y' \leftrightarrow z = z')) \wedge \forall y(\exists z\, Ryz \leftrightarrow Xy) \wedge \forall z(\exists y\, Ryz \rightarrow zBs))$. Call the resulting theory S^+. Then *Spec* is immediately recoverable: instantiate R with the identity relation on X. In a first-order version of S^+, the existential quantifier $\exists R$ is dropped and R becomes a schematic letter. The set-theoretic schema of "one-one replacement," in which the hypothesis of replacement that the relevant formula defines a function on sets is strengthened to the hypothesis that it defines a *one-one* function then immediately follows in S^+. Since (ordinary) replacement follows from one-one replacement, Aussonderung, power and extensionality, replacement is obtainable in S^+ plus extensionality.

Whether some such strengthening of *Spec* can be plausibly thought not to involve a *new* principle that is not really part of the iterative conception seems most doubtful. In any event, we turn now to a completely different conception of set, from which replacement immediately follows.

The conception is Frege's, modified to avoid the antinomies.

According to Frege, with every *concept F* there is associated a certain object $'F$, the extension of F. Furthermore, according to rule (V) of Frege's *Basic Laws of Arithmetic*, concepts are coextensive if and only if their extensions are identical: $'F = 'G \leftrightarrow \forall x(Fx \leftrightarrow Gx)$. Russell showed rule (V) inconsistent. (Frege's proof: Let F be $[x : \exists G(x = 'G \land \neg Gx)]$. Then if $\neg F\,'F, \forall G('F = 'G \to G\,'F)$, whence $F\,'F$; but then for some G, $'F = 'G$ and $\neg G\,'F$. By the left-right direction of (V), $\forall x(Fx \leftrightarrow Gx)$, and therefore $G\,'F$, contradiction.) We can conveniently simulate the Fregean framework of objects, concepts and extensions in second-order logic (as we have already begun to do). We shall suppose that *, like $'$, is an operation-sign which when attached to a concept (second-order) variable yields a term of the type of an object (first-order) variable, and lay down a suitable modification of rule (V) governing *. The modification we give incorporates the idea of limitation of size, due to Cantor, Russell, von Neumann, and Bernays that objects with too many members may behave in deviant ways, perhaps by belonging to nothing, perhaps by not existing. According to our modified version of rule (V), all such overpopulated objects will turn out to be identical.

Let F and G be concepts. We shall say that F *goes into* G if the objects falling under F are in one-one correspondence with some or all of those falling under G: i.e., if $\exists R\, R : F \to_{1-1} G$.[10]

Let V be the concept $[x : x = x]$, under which all objects fall. Every concept goes into V. We shall say that a concept F is *small* if V does not go into F. Of course V is not small. If F is small and $\forall x(Gx \to Fx)$, then G is small. Though we shall not make use of the fact, it can be proved (via a version of one proof of the Schröder–Bernstein theorem) that if V goes into F, then the objects falling under F are in one-one correspondence with all objects.

Call concepts F and G *coextensive* if the same objects fall under them: $\forall x(Fx \leftrightarrow Gx)$. Say that F is *similar* to G, in symbols: $F \sim G$, iff either both F and G are not small or F and G are coextensive; i.e. iff (F is small $\lor G$ is small $\to \forall x(Fx \leftrightarrow Gx)$). Similarity is obviously symmetric and reflexive. It is also transitive: Suppose $F \sim G$ and $G \sim H$. Then if F is small, then $\forall x(Fx \leftrightarrow Gx)$, G is small, $\forall x(Gx \leftrightarrow Hx)$, and therefore $\forall x(Fx \leftrightarrow Hx)$; if H is small, then likewise $\forall x(Fx \leftrightarrow Hx)$. Thus similarity is an equivalence relation that respects smallness.

We now associate with any concept F an object $*F$, which we shall call the *subtension* of F. We suppose that subtensions obey the following

---

[10] "$R : F \to_{1-1} G$" abbreviates: $\forall y \forall y' \forall z \forall z'(Ryz \land Ry'z' \to (y = y' \leftrightarrow z = z')) \land \forall y(Fy \leftrightarrow \exists z(Gz \land Ryz))$.

modification (new V) of Frege's rule (V):

$$\forall F \forall G(^*F = {}^*G \text{ iff } F \sim G).$$

We call the second-order theory that results when (new V) is adjoined to standard axiomatic second-order logic, FN (for Frege–von Neumann).

Let us quickly remove any doubts there might be about the consistency of FN by showing it to have a model M. The domain of M is the set of natural numbers, and * is interpreted thus: if finitely many objects fall under F, let $^*F$ be $n + 1$, where n is the number in whose binary representation there is a 1 at the 2^ks place iff k falls under F; but if infinitely many objects fall under F, let $^*F$ be zero. Then (new V) holds in M; moreover, F satisfies "is small" in M iff finitely many objects fall under F.

Let $\emptyset$ be the concept $[x : x \neq x]$, and let $0 = {}^*\emptyset$. Since there is at least one object (e.g. $^*V$ or $^*\emptyset$), $\emptyset$ is small, $\emptyset \not\sim V$, and $0 \neq {}^*V$. Thus there are at least two objects. For any object y, exactly one object falls under $[x : x = y]$; thus $[x : x = y]$ is small. Let $sy = {}^*[x : x = y]$. Then for any y, $0 \neq sy$, since $[x : x \neq x]$ is small but $\neg \forall x(x \neq x \leftrightarrow x = y)$; and if $sy = sz$, then $y = z$, since $[x : x = y]$ is small, and therefore $\forall x(x = y \leftrightarrow x = z)$. Arithmetic can therefore be carried out in FN, e.g., as in Dedekind's *Was sind und was sollen die Zahlen?* Following Frege–Russell, let N be $[x : \forall F((F0 \wedge \forall y(Fy \rightarrow Fsy)) \rightarrow Fx)]$.

We now want to develop a certain amount of set theory in FN. First define: $y \in x$ iff $\exists F(x = {}^*F \wedge Fy)$. Then $^*V \in {}^*V$. $y \in x$ may be read as usual ("y is a member of x," "x contains y," etc.).

Suppose that F is small. Then if $y \in {}^*F$, for some G, $^*F = {}^*G$ and Gy; but then $F \sim G$, $\forall y(Fy \leftrightarrow Gy)$ and Fy. Conversely, if Fy, then certainly $y \in {}^*F$. Thus if F is small, then $y \in {}^*F$ iff Fy.

If F is not small, then since F and V are both not small, $F \sim V$, and $^*F = {}^*V$; and then since $^*V \in {}^*V$, $^*V \in {}^*F$, $^*F \in {}^*V$, and $^*F \in {}^*F$.

Thus if F is $[x : x \neq {}^*V]$, then $\neg F^*V$. But since V goes into F (map V^* to 0, each x such that Nx to sx, and any other object to itself), F is not small, and $^*V \in {}^*F$. In general, if F is not small and not coextensive with V, then $\neg \forall x(Fx \leftrightarrow Vx)$, but $\forall x(x \in {}^*F \leftrightarrow x \in {}^*V)$.

Define: x is a *set* iff $\exists F(F$ is small $\wedge x = {}^*F)$; sets are thus subtensions of small concepts. 0 is a set, but $^*V$ is not. If $^*F$ is a set, then for some small G, $^*F = {}^*G$, and F is small; thus $z \in {}^*F$ iff Fz.

If x is a set, say $x = {}^*F$, F small, then F and $[z : z \in {}^*F]$ are coextensive and small, and thus $x = {}^*F = {}^*[z : z \in {}^*F] = {}^*[z : z \in x]$. Therefore if x and y are sets with the same members, then $[z : z \in x]$ and $[z : z \in y]$ are coextensive and small, $x = {}^*[z : z \in x] = {}^*[z : z \in y] = y$, and extensionality holds.

So does Aussonderung: Let z be a set, say $z = {}^*F$. Let $G = [y :$

$y \in z \wedge Xy]$. Then $\forall y(Gy \to Fy)$ and G is small. Let $x = {}^*G$. Thus $\forall y(y \in x \leftrightarrow y \in z \wedge Xy)$.

And so does the statement that for any object w and any set x, there is a set $x + w$ whose members are just w and the members of x, which is sometimes called the axiom of *adjunction*: Suppose that V goes into $[y : Fy \vee y = w]$, i.e., for some R, $R : V \to_{1-1} [y : Fy \vee y = w]$. Then V also goes into F. For after interchanging no more than two values of R, we may assume that $R(0) = w$, and then we readily see that $[y : y \neq 0]$ goes into F. But since $[xy : y = sx] : V \to_{1-1} [y : y \neq 0]$, V goes into F. Thus if $x = {}^*F$ and F is small, $[y : Fy \vee y = w]$ is also small, whence adjunction.

It follows from adjunction that for any set x, there is a set (the von Neumann successor of x) whose members are just x and the members of x. The axiom of pairing, which states that for any objects w and z, there is a set $\{w, z\}$ whose members are just w and z, is also an immediate consequence of adjunction: $\{w, z\} = (0 + w) + z$.

Notice that although ${}^*V$ is not a set, s^*V is. Thus some non-empty sets do not contain any sets at all. It follows that the axiom of unions, which states that for any set z there is a set whose members are just the members of the members of z, fails: s^*V is a counterexample. It is a surprising result, due to Lévy,[11] that a suitably modified version of unions is actually a consequence of FN. To arrive at this modification, and to derive a satisfactory theory of sets within FN, we need the notion of a *pure* object.

Abbreviate: $\exists F\, x = {}^*F$ by: Sx (x is a subtension). Thus if Sx, x is a set iff $x \neq {}^*V$.

Say that F is *closed* if $\forall y(Sy \wedge \forall z(z \in y \to Fz) \to Fy)$.

Say that x is *pure* iff $\forall F(F$ is closed $\to Fx)$.

Theorem 1 *Suppose that Sx and $\forall y(y \in x \to y$ is pure$)$. Then x is pure.*

Proof. Let F be closed. Show Fx. All $y \in x$ are pure; thus for all $y \in x$, Fy. Since Sx and F is closed, Fx. ∎

Theorem 2 *Suppose that x is pure. Then x is a set (and hence not $= {}^*V$) and all members of x are pure.*

Proof. Let G be $[x : x$ is a set $\wedge \forall z(z \in x \to z$ is a set $\wedge z$ is pure$)]$. Show G closed. Suppose Sy and $\forall z(z \in y \to Gz)$. Show Gy. Suppose y is not a set; then, since Sy, $y = {}^*V$, $y \in y$, Gy and y is a set. So y *is* a set. Suppose $z \in y$. Then Gz, so z is a set. Show z pure. Let F be closed. Show Fz. Since Gz, $\forall a(a \in z \to a$ is pure$)$. Since z is a set, Sz. By Theorem 1, z is

[11](Lévy, 1968), pp. 762–763.

pure. Thus Gy and G is closed. Since x is pure, Gx, and therefore x is a set and all members of x are pure. ∎

It follows from Theorems 1 and 2 that x is pure iff x is a set and all members of x are pure. $*V$ is not pure; neither are $s*V$, $ss*V$, etc.

If x and y are pure and for all pure sets z, $z \in x$ iff $z \in y$, then for *all* z, $z \in x$ iff $z \in y$, and by extensionality, $x = y$. That is, extensionality holds when relativized to the pure sets, as do Aussonderung and adjunction.

Since all members of pure sets are pure, an induction principle for pure sets can now be seen to hold:

$$\exists x(\text{Pure } x \wedge Gx) \rightarrow \exists x(\text{Pure } x \wedge Gx \wedge \forall y(y \in x \rightarrow \neg Gy)).$$

Proof. If $\forall x(\forall y(y \in x \rightarrow Fy) \rightarrow Fx)$, then F is certainly closed and so $\forall x(\text{Pure } x \rightarrow Fx)$. Thus if for some x, Pure x and Gx, then for some x, Pure x and (Pure x and Gx), whence by substituting: $\neg(\text{Pure } x \wedge Gx)$ for: Fx, we have that for some x, Pure x and Gx and $\forall y(y \in x \rightarrow \neg(\text{Pure } y \wedge Gy))$. Since all members of x are pure, $\forall y(y \in x \rightarrow \neg Gy)$. ∎

Regularity (even as a schema) thus holds when relativized to the pure sets. $s*V$ is a counterexample to unrelativized regularity, which states that any nonempty set x contains a member with no member in common with x.

It follows from relativized regularity that no pure set is a member of itself; otherwise some pure set is a member of itself, but no member of it is a member of itself.

Now say that x is *transitive* if all members of members of x are members of x: $\forall z \forall y(z \in y \in x \rightarrow z \in x)$. And say that x is an *ordinal* if x is pure, x is transitive, and all members of x are transitive.

Theorem 3 *Suppose x is an ordinal and $y \in x$. Then y is an ordinal.*

Proof. Since x is pure, y is pure. Since all members of x are transitive, y is transitive. If $z \in y$, then by the transitivity of x, $z \in x$, and z is transitive. Thus all members of y are transitive. ∎

Since ordinals contain only ordinals, induction for pure sets yields an induction principle for ordinals:

$$\exists x(x \text{ is an ordinal } \wedge Gx) \rightarrow \exists x(x \text{ is an ordinal } \wedge Gx \\ \wedge \forall y(y \in x \rightarrow \neg Gy)).$$

The usual double induction can now be used to show that $\in$ is connected on the ordinals; since ordinals are transitive, $\in$ is transitive on the ordinals as well. Since $\in$ is also irreflexive on the ordinals, the ordinals are strongly well-ordered by $\in$.

We now make use of the argumentation leading to the Burali–Forti paradox.

Let On be $[y : y$ is an ordinal].

Theorem 4 On *is not small.*

Proof. Suppose the contrary. Let $x = {}^*\text{On}$. Then x is a set, and therefore for all y, $y \in x$ iff y is an ordinal. All members of x are pure and Sx; by theorem 1, x is pure. If $z \in y \in x$, then y is an ordinal, z is an ordinal, and $z \in x$; thus x is transitive. And if $y \in x$, then y is an ordinal, and y is transitive; therefore all members of x are transitive. It follows that x is an ordinal, and therefore that $x \in x$, which is impossible, as x is pure. ∎

Since On is not small, for some R, $R : V \to_{1-1} \text{On}$. And since the ordinals are well-ordered by $\in$, the axiom of global choice follows immediately (von Neumann). Various versions of the usual axiom of choice ("local choice") follow from global choice and Aussonderung, as do their relativizations to pure sets.

Replacement, as well as its relativization, is immediate too. Let w be a set and F a functional relation. Suppose $R : V \to_{1-1} [z : \exists y(y \in w \wedge Fyz)]$. By (local) choice, for some S, $S : V \to_{1-1} [y : y \in w]$, which is impossible, as w is a set. Thus $[z : \exists y(y \in w \wedge Fyz)]$ is small. Let $x = {}^*[z : \exists y(y \in w \wedge Fyz)]$. Then $\forall z(z \in x \leftrightarrow \exists y(y \in w \wedge Fyz))$.

Lévy's startling proof that the axiom of unions is redundant in von Neumann's system of set theory can readily be adapted to show that the relativization of that axiom to the pure sets is a theorem of FN. Thus according to FN, for any pure set z there is a pure set whose members are just those of the members of z. (By theorem 2, all members of a pure set are pure.) For the proof, recall that FN proves the existence of the von Neumann successor; the rest of the proof is as in Lévy's article, cited above.

To sum up, let us compare the iterative conception and FN with respect to each of the axioms and axiom-schemata of set theory.

Extensionality: Evident, but, arguably, not evident on the iterative conception. An immediate consequence of FN.

Null set: Evident on the iterative conception. An immediate consequence of FN.

Pairing: Evident on the iterative conception. An immediate consequence of FN.

Regularity: Evident on the iterative conception. (What is not evident is that regularity is derivable from the weak-looking axiomatization S of the iterative conception that we gave.) Unrelativized regularity is refutable in FN $(s*V)$; it is the inductive character of the relativizing predicate "is pure" that is responsible for the derivability in FN of relativized regularity.

Choice: Evident, but not evident on the iterative conception. The derivability of global choice in FN is not surprising in view of one of the leading ideas behind FN, that there is only one "size" things can be and still not form a set, and the well-known facts that the ordinals are well-ordered and do not form a set.

Replacement: Not evident on the iterative conception. Easily derivable from choice in FN.

Aussonderung: Evident on the iterative conception. An easy logical consequence of replacement.

Union: Evident on the iterative conception. Unrelativized union is refutable in FN $(s*V)$; it is a deep and surprising result that relativized union is provable in FN.

Infinity: Evident on the iterative conception. Not even a theorem of FN+power (power is true but infinity false in the model M given above). To obtain infinity, one may supplement FN ad hoc with the smallness principle: N is small.

Power: Evident on the iterative conception. Not even a theorem of FN + infinity (as can be shown by tinkering with the set of hereditarily countable sets). To obtain power, one may similarly add to FN a principle about smallness: F is small $\rightarrow [*G : \forall x(Gx \rightarrow Fx)]$ is small.

FN thus embodies a view of sets altogether different from the iterative conception. Each view accounts for a large part of set theory but also omits much of importance. One moral to be drawn is that *it is a mistake to think that set theory,* i.e., ZF with choice, *on the whole follows from the iterative conception.* The axioms that do not follow are crucial to any reasonable development of set theory (without choice the theory of cardinality is fragmentary), and there is an alternative theory of which those axioms are consequences (but from which two important axioms of ZF do not follow). Perhaps one may conclude that there are at least two thoughts "behind" set theory.

Appendix

Pairing: $\forall z \forall w \exists x \forall y (y \in x \leftrightarrow (y = z \lor y = w))$. By *All*, for some s and t, zFs and wFt. By *Net*, for some r, $s < r$ and $t < r$. Thus zBr and wBr. But by *Spec*, $\exists x \forall y (y \in x \leftrightarrow ((y = z \lor y = w) \land yBr))$.

Union: $\forall z \exists x \forall y (y \in x \leftrightarrow \exists w(y \in w \land w \in z))$. By *All*, for some r, zFr. If $w \in z$, then by *When*, for some s, $s < r$ and wFs. If $y \in w$, then by *When* again, for some t, $t < s$ and yFt, and by *Tra*, $t < r$, whence yBr. But by *Spec*, $\exists x \forall y (y \in x \leftrightarrow (\exists w (y \in w \wedge w \in z) \wedge yBr))$.

Power: $\forall z \exists x \forall y (y \in x \leftrightarrow \forall w (w \in y \rightarrow w \in z))$. By *All*, for some s, zFs. By *When*, $\forall x (\forall w (w \in y \rightarrow w \in z) \rightarrow yFs)$. By *Net*, for some r, $s < r$. Thus if $\forall w (w \in y \rightarrow w \in z)$, yBr. But by *Spec*, $\exists x \forall y (y \in x \leftrightarrow (\forall w (w \in y \rightarrow w \in z) \wedge yBr))$.

Aussonderung: $\forall z \exists x \forall y (y \in x \leftrightarrow (y \in z \wedge A(y)))$. By *All*, for some s, zFs. By *When*, if $y \in z$, then yBs. But by *Spec*, $\exists x \forall y (y \in x \leftrightarrow ((y \in z \wedge A(y)) \wedge yBs))$.

Null set, $\exists x \forall y \neg y \in x$, follows from Aussonderung by taking $A(y) = \neg y = y$. (We take $\exists x\, x = x$, and $\exists s\, s = s$ as well, to hold by logic.)

y is *null* if $\forall z \neg z \in y$.

z is a *successor* of y if $\forall w (w \in z \leftrightarrow (w \in y \vee w = y))$.

Infinity: $\exists x (\exists y (y \in x \wedge y$ is null $) \wedge \forall y (y \in x \rightarrow \exists z (z \in x \wedge z$ is a successor of $y)))$. By Null set, a null set exists. By Pairing and Union, every set has a successor. By *When*, every null set is formed at every stage. By *When* and *Tra*, if yFt, $t < s$, z is a successor of y, then zFs. [Suppose yFt and $t < s$. By *When*, if $w \in y$, then for some u, $u < t$ and wFu. By *Tra*, $u < s$. And if $w = y$, then $t < s$ and wFt. By *When*, zFs.] By *Inf*, for some r, $\exists t\, t < r$ and $\forall t (t < r \rightarrow \exists s (t < s \wedge s < r))$. By *Spec*, $\exists x \forall y (y \in x \leftrightarrow yBr)$. Done.

7

Introductory Note to Kurt
Gödel's "Some Basic Theorems
on the Foundations of
Mathematics and their
Implications"

Historical information and overview

On 26 December 1951, at a meeting of the American Mathematical Society
at Brown University, Gödel delivered the twenty-fifth Josiah Willard Gibbs
Lecture, "Some basic theorems on the foundations of mathematics and their
implications." It is not known when he received the invitation to give this
lecture. It is probable, as Wang suggests on pp. 117–118 of (Wang, 1987),
that the lecture was the main project Gödel worked on in the fall of 1951.
In letters to Rita Dickstein (21 March 1953) and Yehoshua Bar-Hillel (7
January 1954), preserved in Gödel's *Nachlass*, he expressed his intention to
publish the lecture in the *Bulletin of the American Mathematical Society*.
These letters lend some support to the conjecture that he continued to work
on the text after 1951. The lecture was included on a list Gödel made up
bearing the title "Was ich publizieren könnte" ("What I could publish")
and also preserved in the *Nachlass*. No correspondence with the editors of
the *Bulletin* is known, however, and the only text we have is handwritten

From *Kurt Gödel, Collected Works,* Volume III, *Unpublished Essays and Lectures,*
Solomon Feferman et al., eds., Oxford: Oxford University Press, 1995, pp. 290–304,
where this essay introduced the first publication of Gödel's Gibbs Lecture. Copyright ©
1995 Oxford University Press, Inc. Used by permission of Oxford University Press, Inc.
All unspecified page references are to the page numbers in Gödel's manuscript.
   I am grateful to Cheryl Dawson, John Dawson, Solomon Feferman, Warren Goldfarb,
and Charles Parsons for much editorial and philosophical advice.

(and of a rather intricate structure; see the Textual Notes in (Feferman et al., 1995)). Since other papers of Gödel survive in typescripts—in the cases of (Gödel, 1995d) and (Gödel, 1995a) in several versions—it may also be conjectured that he did not come close to sending it off for publication.

Gödel's lecture may be divided into two parts, the first of which is an exposition of certain logical results and of philosophical views that he regards as direct consequences of those results. In this part of the lecture Gödel tries to establish that the results show mathematics to be "incompletable" or "inexhaustible," and that one of them demonstrates that *"either ... the human mind (even within the realm of pure mathematics) infinitely surpasses the powers of any finite machine, or else there exist absolutely unsolvable diophantine problems"* (13). (It will be explained below what Gödel understands by a "diophantine problem.") By an "absolutely undecidable" problem, Gödel means one that is undecidable, "not just within some particular axiomatic system, but by *any* mathematical proof the human mind can conceive" (13).

In the second, more avowedly philosophical, part of the lecture, Gödel's main concern is to adduce a number of considerations favoring the standpoint called realism or Platonism, which can be defined, in Gödel's own words, as the view that mathematical objects and "concepts form an objective reality of their own, which we cannot create or change, but only perceive and describe" (30).

Set theory and the incompletability of mathematics

The attempt to axiomatize set theory is the first of two illustrations Gödel provides of what he means by the inexhaustibility of mathematics. Gödel claims that in order to avoid the paradoxes "without bringing in something entirely extraneous[1] to actual mathematical procedure, the concept of set must be axiomatized in a stepwise manner" (3). He then proceeds to lay out the "iterative" or "cumulative" hierarchy of sets: we begin with the integers and iterate the power-set operation through the finite ordinals. This iteration is an instance of a general procedure for obtaining sets from a set A and well-ordering R: starting with A, iterate the power-set operation through all ordinals less than the order-type of R (taking unions at limit ordinals). Specializing R to a well-ordering of A (perhaps one whose ordinal is the cardinality of A) yields a new operation whose value at any set A is the set of all sets obtained from A at some stage of this procedure, a set far larger than the power-set of A. We can require that this new operation, and indeed *any* set-theoretic operation, can be so iterated, and that there

[1]It is conceivable that he may have had in mind Quine's set theories NF and ML, in which whether a formula counts as an axiom depends on whether it satisfies a somewhat artificial syntactical restriction.

should also always exist a set closed under our iterative procedure when applied to any such operation.

Axioms can be formulated to describe the sets formed at various stages of this process. But as there is no end to the sequence of operations to which this iterative procedure can be applied, there is none to the formation of axioms. " . . . nor can there ever be an end to *this* procedure of forming the axioms, because the very formulation of the axioms up to a certain stage gives rise to the next axiom" (5).

The elaboration of Gödel's views on the iterative concept of set found in (Wang, 1974) makes it clear that the axioms we thus formulate will imply all those of ZF, including the axioms of replacement. An interesting conclusion is immediate: on Gödel's view, the iterative concept of set is only *partially* embodied in the theory ZF.

Gödel seems never to have wavered from the view that ZF only partially characterizes the concept of set. In (Gödel, 1995b) he speaks of " . . . an infinity of systems, and whichever system you choose out of this infinity, there is one more comprehensive, i.e., one whose axioms are stronger" (10). And as late as (Gödel, 1964), footnote 20, he states that Mahlo's axioms, which assert the existence of Mahlo cardinals but which cannot be proved in ZF, are "implied by the general concept of set."

Gödel observes that higher-level set-theoretic axioms will entail the solution of certain Diophantine problems of level 0 left undecided by the preceding axioms; the problems, moreover, take a particularly simple form, viz., to determine the truth or falsity of sentences $\forall \mathbf{x} \exists \mathbf{y}\, P(\mathbf{x}, \mathbf{y}) = 0$, where $\mathbf{x}$ and $\mathbf{y}$ are sequences of integer variables and $P(\mathbf{x}, \mathbf{y})$ is a polynomial with integer coefficients. Let us call this class of sentences "class A." (For Gödel's proof that undecidable sentences can be taken to be in class A, see (Gödel, 1995e)).

The incompleteness theorems and incompletability

Not surprisingly, Gödel's own incompleteness theorems provide his second illustration of the incompletability of mathematics. Invoking the notion of a Turing machine, he states that the first theorem "is equivalent to the the fact that there exists no finite procedure for the systematic decision of all Diophantine problems of the type specified" (9); little further mention is then made of the first theorem, since it is the *second* theorem (10) that he thinks makes the incompletability of mathematics particularly evident.

> *For any well-defined system of axioms and rules . . . the proposition stating their consistency (or rather the equivalent number-theoretical proposition) is undemonstrable from these axioms and rules, provided these axioms and rules are consistent and*

suffice to derive a certain portion of the finitistic arithmetic of integers.

Gödel's argument that his second theorem shows the incompletability of mathematics runs as follows: No one can set up a formal system and consistently state about it that he perceives (with mathematical certitude) that its axioms and rules are correct and that he believes that they contain all of mathematics, for anyone who claims to perceive the correctness of the axioms and rules must also claim to perceive their consistency; but since the consistency of the axioms is not provable in the system, the person is claiming to perceive the truth of something that cannot be proved in the system, and is therefore obliged to abandon the claim that the system contains all of mathematics.

Gödel moves immediately to prevent a possible misunderstanding. He distinguishes the system of all true mathematical propositions from that of all demonstrable mathematical propositions, calling these mathematics in the objective and subjective senses, respectively, and claims that it is only objective mathematics that no axiom system can fully comprise. He adds that we could not, however, know of any finite rule that might happen to produce all of subjective mathematics that it is correct. The ground for both claims is the indemonstrability of the assertion of consistency. To be sure, we could successively come to recognize, of each proposition produced by subjective mathematics, that that proposition is correct; but we could not know the general proposition that they are *all* correct.

Were there to be such a rule, Gödel says, the mind would be "equivalent to a finite machine that, however, is unable to understand completely its own functioning" (12), again on the ground that the insight that the brain produces only "correct (or only consistent) results would surpass the powers of human reason" (footnote 14). Gödel supposes that if a (consistent) machine "completely understands" its own functioning, then it can recognize its own consistency.

Gödel also holds that if the human mind is "equivalent to a finite machine" (12), then there is a finite rule producing all the evident axioms of demonstrable mathematics. Since the assertion of consistency can be recast as a sentence in class A, he takes it that it follows that either the human mind surpasses the powers of a finite machine or there exist simple problems about the natural numbers not decidable by any proof the human mind can conceive. He calls his conclusion a "mathematically established fact" (13) that seems to him of great philosophical interest.

There is a gap between the proposition that no finite machine meeting certain weak conditions can print a certain formal sentence (which will depend on the machine) and the statement that if the human mind is a finite machine, there exist truths that cannot be established by any proof the human mind can conceive. It is not that no proposition about the "human mind" or human beings or brains can ever be validly inferred from a mathematical proposition. (On the contrary: since 91 is composite, no human being will ever come to know that 91 is prime.) What may be found problematic in Gödel's judgment that his conclusion is of philosophical interest is that it is certainly not obvious what it means to say that the human mind, or even the mind of some one human being, *is* a finite machine, e.g., a Turing machine. And to say that the mind (at least in its theorem-proving aspect), or *a* mind, may be represented by a Turing machine is to leave entirely open just *how* it is so represented. Nevertheless, the following statement about minds, replete with vagueness though it may be, would indeed seem to be a consequence of the second theorem: If there is a Turing machine whose output is the set of sentences that express just those propositions that can be proved by a mind that can understand all propositions expressed by a sentence in class *A*, then there is a true proposition expressed by a sentence in class *A* that cannot be proved by that mind.

Apart from the difficulties involved in deriving from the second incompleteness theorem the disjunctive claim that either the mind is not a finite machine or there exist absolutely undecidable mathematical propositions, a further problem for Gödel's view is that the supposition that the second alternative holds does not seem particularly surprising or remarkable at present. (Of course, it may well be that the existence of propositions whose truth we could never recognize is unremarkable precisely *because* we have come to understand the incompleteness theorems so well.) Why, we may wonder, should there *not* be mathematical truths that cannot be given any proof that human minds can comprehend? It may be noted that there are many persons who, influenced by the picture of the mind as a Turing machine, find the falsity of the first and the truth of the second alternative a pair of propositions they are quite willing to maintain. Others, while reserving judgment on the question whether (the mathematical abilities of) a mind can be (represented by) a Turing machine, simply find it extremely plausible that there are mathematical truths unprovable by any humanly comprehensible proof.[2]

According to Wang (Wang, 1974), pages 324–326, Gödel believed that Hilbert was right to reject the second alternative. Otherwise, by asking unanswerable questions while asserting that only reason can answer them, reason would be irrational. (This view may derive from Kant's opinion that "there are sciences the very nature of which requires that every question arising within their domain should be completely answerable in terms of

---

[2]In their introductory note to Remark 2 of (Gödel, 1990), Feferman and Solovay suggest one possible example. Cf. (Feferman et al., 1990), p. 292.

what is known, inasmuch as the answer must issue from the same sources from which the question proceeds" [A 476/B 504, translation from (Kant, 1933)]. Kant cites pure mathematics as one such science [A480/B508].[3]) Not only did Gödel reject the second alternative, he appears to have thought (at least late in his life) that there were independent reasons for accepting the first as well: Remark 3 of (Gödel, 1990) is an argument against Turing's view that "mental procedures cannot go beyond mechanical procedures" (306).[4]

Gödel's disjunctive conclusion concerning the significance of his incompleteness theorems stands in contrast with the conclusion drawn by writers such as Ernest Nagel and James R. Newman,[5] J. R. Lucas,[6] and Roger Penrose[7] to the effect that the theorems show outright that the mind is not a Turing machine, since, as they suppose, the mind can see with mathematical certainty that any Turing machine that it might be alleged to be (or be represented by) is actually consistent, and can therefore prove a proposition not provable by that machine. The classic reply to these views was given by (Putnam, 1960): merely to find from a given machine M, a statement S for which it can be proved that M, if consistent, cannot prove S is not to *prove S*—even if M *is* consistent. It is fair to say that the arguments of these writers have as yet obtained little credence.

Before we turn to the more philosophical part of Gödel's lecture, let us mention some questions that his discussion suggests. Do the impossibility of axiomatizing the concept of set and that of axiomatizing the whole of mathematics bear any interesting relation to each other? Indeed, is there a significant general phenomenon of inexhaustibility or incompleteness of which they are both examples (and if so, what is it)? Is there even a third instance of the incompletability or inexhaustibility of mathematics to be cited?

Realism, or Platonism

Gödel remarks that if either the mind is not a finite machine or there exist absolutely undecidable propositions, then the philosophical conclusions to be drawn are "very decidedly opposed to materialistic philosophy" (15). If the first alternative holds and the mind's operations cannot be reduced to those of the brain, which is made out of a finite number of neurons and

---

[3]I am grateful to Carl Posy and Sally Sedgwick for calling these passages in the *Critique of Pure Reason* to my attention.

[4]A critical assessment of Gödel's argumentation is given in unpublished work of Warren Goldfarb.

[5](Nagel and Newman, 1958).

[6](Lucas, 1961).

[7](Penrose, 1989).

their connections, then vitalism, he states, would seem to be inescapable. Gödel claims that this alternative is not known to be false and that some of the "leading men in brain and nerve physiology" (17) deny the possibility of a purely mechanistic explanation of mental processes.

The second alternative, which, he says, "seems to disprove the view that mathematics is only our own creation" (15), appears to imply some version of realism or Platonism about the objects of mathematics and gives Gödel considerably more to say.

A creator, he says, "necessarily knows all properties of his creatures, because they can't have any others except those he has given to them" (16). Gödel considers poor the objection that the constructor need not know *every* property of what he constructs, that, e.g., we cannot predict the complete behavior of machines we make (or, one might now add, of software we write). His reply to this objection is to argue that if it were correct, it would provide further support for Platonism in mathematics, because we build machines " . . . out of some given material. If the situation were similar in mathematics, then this material or basis for our constructions . . . would force some realistic viewpoint upon us even if certain other ingredients of mathematics were our own creation" (18).

Gödel's claim that a creator must know all properties of the things he creates, since they can have no others except those the creator gives them, may strike the reader as a far-fetched defense of the quite plausible claim that mathematics cannot be only (i.e., entirely) our own creation, at least not if our capacity for proving facts about the natural numbers can be adequately represented by a Turing machine. For how, one might wonder, could it have been *we* who brought about the truth of any true proposition in the absence of a proof of that proposition that we could produce? It might be said that the truth of the proposition is a consequence of stipulations we have made concerning the natural numbers. For this reply to be explanatory, however, "consequence" must mean "deductive consequence" and not (say) "higher-order semantic consequence"; but that is precisely what is *not* the case with regard to an undecidable proposition. In any case, the incompleteness theorems suggest that it is doubtful that the view that mathematics is entirely our own creation can be successfully elaborated. (Gödel does not discuss the objection to the other half of his claim, that objects might in fact acquire properties not bestowed upon them by their creator, for example, as a result of being perceived by others.)

To the objection that the meaning of a proposition about all integers can consist only in the existence of a proof of it, and therefore that neither an undecidable proposition nor its negation is true, Gödel makes a particularly interesting response. He suggests that the abhorrence mathematicians display towards inductive methods in mathematics may be "due to the very prejudice that mathematical objects somehow have no real existence. If mathematics describes an objective world ... there is no reason why inductive methods should not be applied in mathematics" (20). Thus his second alternative, that there exist absolutely undecidable propositions, favors the standpoint of empiricism in one respect.

As to what such empirical methods might look like, Gödel offers no concrete suggestion; but, in a footnote, he gives an example of a proposition where probabilities, he says, can be estimated even now: The probability that for each n there is at least one digit $\neq 0$ between the nth and the n^2-th digits of the decimal expansion of π converges toward 1 as one goes on verifying it for greater and greater n. One may, however, be uncertain whether it makes sense to ask what the probability is of that general statement, given that it has not been falsified below $n = 1,000,000$, or to ask for which n the probability would exceed .999.

Gödel then gives three arguments supporting the view he calls conceptual realism (or Platonism) and directed against the view that mathematics is our own creation.

According to the first of these, the attainment of great clarity in the foundations of mathematics has helped us little in the solution of mathematical problems; but this, says Gödel, would be impossible were mathematics our "free creation," for then mathematical ignorance could be due only to failure to understand what we have created (or to computational complexity), and would have to disappear once we attained "perfect clearness."

But, it might be replied, there is no reason to suppose that perfect *clarity* about one of our creations should yield perfect knowledge of it. What is it about creation that guarantees that once we know exactly what a creation of ours is, we must know everything about it? Gödel seems to identify progress in understanding the foundations of mathematics with the attainment of ever greater clarity about mathematics; but, one might think, mathematics might be our own creation and we might have attained perfect clarity about the fundamental properties of what we have created, but nevertheless be rather ignorant about non-fundamental properties. There is no reason to suppose that even perfect clarity with respect to all the fundamental properties of our creations must yield *complete* knowledge of those creations.

Gödel's second argument against the view that mathematics is our own creation is that mathematicians cannot create the validity of theorems at will. "If anything like creation exists at all in mathematics, then what any theorem does is exactly to restrict the freedom of creation" (22). This consideration is often thought to be a powerful argument on behalf of a realist view of mathematics of the type Gödel wishes to espouse. It is perhaps presented most forcefully as a claim to the effect that the contrary position is confused or incredible: that once it has been made clear exactly *which* objects (including operations, properties and relations) are *in question*, i.e., being talked about, which, all may concede, may well be a matter for choice or decision, the suggestion that there is still room for a decision whether or not those objects have those properties, stand in those relations, etc. cannot be believed to be true. (One might think: Once it is certain that it is 9, 4, 36, multiplication, and equality that are under consideration, how could it possibly be *up to us* whether or not the product of 9 and 4 is 36?) If the creation could not have turned out otherwise, Gödel is arguing, in what sense is there *creation* at all?

Gödel's third argument is that in order to demonstrate certain propositions about the integers, we must employ the concept of a set of integers; but the creation of integers does not "necessitate" that of sets of integers. Thus we appear to be in the "very strange situation indeed" (23) of having to make a further creation in order to determine what properties we have given to the integers, which were supposed to be our creation.

This consideration may perhaps best be taken as a "plausibility" argument: Confronted with these facts about integers, sets of integers and our knowledge of the properties of integers, how can we find even slightly plausible the suggestion that mathematics is our own creation?

Whether or not it follows from the view that mathematics is not our own creation that the objects of mathematics have an objective existence that is independent of us will of course depend on how the concepts "objective existence" and "independence" are to be understood: it may be argued that we lack an interpretation of the key terms in this putative consequence under which it is true but not trivially true.

Against conventionalism

Conceding that "free creation" is a vague term, Gödel then undertakes to give a more specific refutation of what he takes to be the most precise articulation of that suggestion, the view usually called mathematical conventionalism (though Gödel often refers to it as nominalism), according to which mathematical propositions express only certain aspects of linguistic conventions, "that is, they simply repeat parts of these conventions" (23). His discussion is intricate and, in view of the six drafts he made of a projected paper on the philosophy of Rudolf Carnap (at one time the preeminent advocate of conventionalism in mathematics), it is highly probable that Gödel was never able to formulate his objections to Carnap's view to his own complete satisfaction. Annotations to the manuscript strongly suggest that he did not intend to read this section of the lecture to his audience in Providence.

He begins by quickly disposing of what he takes to be the simplest form of conventionalism: the view that the truth of mathematical propositions is due solely to the definitions of the terms they contain. Gödel understands this to mean that there is a mechanical method for converting any mathematical truth (and no mathematical falsehood) to an explicit tautology of the form $a = a$ by systematically replacing terms by their definitions. Since any such conversion method would yield a decision procedure for arithmetical truth, this simplest version of conventionalism fails: there is no such decision procedure.

Refined versions, he claims, fare no better. He then attempts to refute the claim that "every demonstrable[8] mathematical proposition can be deduced from the rules about the truth and falsehood of sentences alone (that is, without using or knowing anything else except these rules)" (25).

Gödel's argument is that in order to derive the truth of the axioms of mathematics from rules about the truth and falsity of sentences (as, for example, the truth of $p \lor \neg p$ *is* derivable from the usual rules for truth and falsity of disjunctions and negations), one must apply mathematical and logical concepts and axioms to symbols, sets of symbols, sets of sets of symbols, etc. Thus, one who wants to explain mathematical truth as a species of tautology will find that the explanation cannot proceed without the aid of the axioms of mathematics themselves. Mathematical induction provides the central illustration of Gödel's point: any proof that all instances of mathematical induction are true will appeal, in some way or other, to a form of the principle of mathematical induction itself, or to even stronger set-theoretical principles that cannot plausibly be regarded as rules about the truth and falsity of sentences.

He writes, "while the original idea of this viewpoint was to make the truth of the mathematical axioms understandable by showing that they are tautologies, it ends up with just the opposite, i.e., the truth of the axioms must *first* be assumed and *then* it can be shown that, in a suitably chosen language, they are tautologies" (26–27).

By "tautology," it should be noted, Gödel does not mean "truth-functionally valid sentence," but rather something like "sentence whose truth can be deduced from rules stipulating the conditions under which sentences are true and false." Gödel's point is thus that the conventionalists' claim that the truth of true mathematical statements can be deduced from such rules is of no interest if true, since strong mathematical axioms, which can in no way be regarded as "syntactical," will have to be assumed in any valid deduction that shows those statements true.

_____

[8]Although "demonstrable" here might be thought to be a slip for "true," "dem." has been inserted and "true" crossed out at this point in the manuscript. (However, subsequent occurrences are not similarly changed, and the view under attack concerns mathematical truth.)

Gödel argues that any attempt to prove the tautological character of the axioms of mathematics would be a proof of their consistency, which, by his second theorem, cannot be achieved with means weaker than the axioms themselves. It may well be, he notes, that not all of the axioms are needed for the proof of consistency, but it is, he claims, a "practical certainty" that to prove consistency some "abstract concepts," such as "set" or "function of integers," together with the axioms governing these notions, will have to be employed in the proof. Since these notions cannot be considered to be syntactical, it follows, he claims, that syntax cannot rationally warrant our "precritical" beliefs concerning the consistency of classical mathematics.

Although some portions of the theory of abstract concepts can be nominalistically based, and fragments of arithmetic, concerning, e.g., numbers less than 1000, reduced to truth-functionally valid statements, a syntactical justification of mathematical induction is unavailable, "since this axiom itself has to be used in the syntactical considerations" (27). Thus the well-known reducibility of arithmetical identities like "5 + 7 = 12" to explicit tautologies is misleading, Gödel says, not only because this statement is contained in a tiny fragment of mathematics whose reducibility to tautology tells us nothing about the rest of mathematics—which includes statements that can be established only with the aid of induction—but also because either "+" is defined so as to refer only to numbers in some finite domain (in which case it does not refer to ordinary addition), or the concept of set, along with axioms about sets, will have to be used in the definitions and proofs.

Gödel then sums up the previous discussion: the essence of the nominalist–conventionalist view is that propositions which we believe express mathematical facts do not do so, and are true simply because of "an idle running of language," i.e., because the rules which determine when propositions are true or false determine that these propositions are true "no matter what the facts are" (29). To this view Gödel raises two objections, of which the first summarizes the main point of the foregoing discussion: in any putative proof that mathematics is tautologous or true solely by virtue of some such rules, one would have to use mathematics that is at least as complicated as that being asserted to be tautologous or thus true.

The second objection is that no justification can be given for regarding certain mathematical statements, such as complete induction, as "void of content," for one can easily construct systems in which certain empirical statements are taken as axioms. (For the notion of "content" which Gödel has in mind, see (Carnap, 1937), pages 42 and 120.) As it would clearly be unjustifiable to classify those empirical statements as therefore lacking in content, so, Gödel claims, it would be no more justifiable to regard those mathematical statements as actually *void of content*. Thus, according to

Gödel, no ground has been given for thinking that there are no such things as mathematical facts, a claim Gödel calls "the essence of this view" (29).

Realism and analyticity

Gödel is prepared to acknowledge a grain of truth in the nominalist position. "A mathematical proposition says nothing about the physical or psychical reality existing in space and time, because it is true already owing to the meaning of the terms occurring in it, irrespectively of the world of real things" (30). It is an error to think that the meanings of the terms are man-made or that they consist in semantical conventions. Meanings are concepts, which "form an objective reality of their own, which we cannot create or change, but only perceive and describe" (30).[9]

Philosophers of mathematics and other metaphysicians dispute whether the supposition that mathematical objects or concepts "form an objective reality of their own" is surrogate theology (if not outright craziness), is trivially correct, or is in profound need of philosophical clarification. The matter will not be resolved here. Gödel elaborates, " ... a mathematical proposition, although it does not say anything about space-time reality, still may have a very sound objective content, insofar as it says something about relations of concepts" (30–31). But the elaboration helps not at all to settle the dispute. To complicate matters further, it should be noted that the term "world," as in the phrase "world of real things," belongs to the same family as "reality" and "objective," and thus the assertion that mathematical truths say something "about the world" would seem to enjoy the same status (insane, trivial, or unclear) as the claim that they describe an "objective reality."

Gödel's realism takes a strong form: relations between the concepts are not "tautological," because among the axioms that govern the concepts entering into those relations, some must be assumed which are not tautological, but which "follow from the meaning of the primitive terms under consideration" (31). Gödel's thought is that statements, such as instances of the comprehension schema in analysis (second-order arithmetic), even those containing quantifiers ranging over all sets of integers, are valid "owing to the meaning of the term 'set'—one might even say they express the very meaning of the term 'set' " (32). Gödel distinguishes between truths he calls "analytic" (those true in virtue of the meanings of the terms expressing them or "owing to the nature of the concepts occurring therein") and "tautological" truths (those "devoid of content" or "true owing to our definitions"). It may be emphasized that Gödel does not restrict the term "analytic" to statements of the "oculists are eye-doctors" or "actresses are

---

[9]Cf. (Gödel, 1944) and Parsons' introductory note thereto in (Feferman et al., 1990).

female" variety. The analytic truths about sets, Gödel states, cannot be proved without appeal to the concept of set itself; and some analytic propositions might well be undecidable, since "our knowledge of the world of concepts may be as limited and incomplete as that of the world of things" (34). Gödel also discusses the notion of analyticity near the end of (Gödel, 1944).

Quine's influential attack (Quine, 1951) on the concept of analyticity appeared three months before Gödel delivered his lecture. Gödel's claim that the axioms of set theory are analytic—"true owing to the meanings of the terms they contain or the nature of the concepts those terms express"— is troubling for at least three sorts of reasons that do not entirely depend on Quine's claim that the phrase "true by virtue of meanings" has not been shown to isolate a significant class of truths.

In the first place, there is a difficulty about the truth of the axioms: a number of thoughtful writers believe that the axioms of set theory do not describe anything real, despite Gödel's later assertion (Gödel, 1964), page 271, that they force themselves upon us as being true. It is certainly a sensible view to hold both that Cantor's theory of transfinite numbers is a fantasy and that the standard theorems of elementary number theory and analysis are unquestionably true. In any case, the axioms of set theory lack the kind of obviousness one would have expected *axioms* characterized as "analytic" to enjoy.

Secondly, the axiom of extensionality would seem to be the only axiom of ZF that can be properly said to be true in virtue of the *meaning* of the word "set"; indeed, the axiom is often justified on the ground that the criterion of identity of sets it gives, viz., having the same members, is just part of what is meant by "set" (as opposed, say, to "property") and it is the only one that can be thus defended by an appeal to what "set" means. But since Gödel understands "true in virtue of meanings" as so much wider than "true owing to definitions" that it encompasses all axioms of set theory, Quine's questions re-arise: How is the notion of meaning that Gödel is using to be understood? When the axioms of set theory are said to be true *in virtue* of the meanings of their constituent terms, what more is said beyond that they are true? What is it for them to be true *in virtue of* the meanings of the terms they contain? A possible rejoinder to the effect that it is not the meaning of "set" but the nature of the concept of set that is of primary importance for Gödel is open to the reply that the last two questions remain unanswered under the replacement of "meanings of terms" by "natures of concepts."

Gödel's view raises worrisome questions of a third sort, suggested in part by later writing (Gödel, 1964) of his own: Could not the axioms of set theory be true, not in virtue of the concept of set or the meaning of "set," but simply because sets just happen to be as the axioms have it? Why, one might ask, must our knowledge of sets be mediated solely through our understanding of the *concept* of set; could we not know how matters stand with sets by "something like a perception" of them—to quote from the supplement to (Gödel, 1964)—that is as direct as our perception of the *concept* of set? Even lacking such a perception, might we not acquire quasi-empirical evidence, of a sort that Gödel himself has acknowledged may exist, that certain set-theoretic matters happen to stand one way rather than another? One wonders why a *conceptual* realism should be found any more plausible than an "objectual" realism.[10]

Since "our knowledge of the world of concepts may be as limited and incomplete as that of [the] world of things," Gödel holds that the paradoxes of set theory pose no more threat to his Platonism than the illusion of the stick in water poses to the view that there is an "outer world." The interesting implied suggestion is that we are taken in by something like an optical *illusion* when we accept the principles that lead to set-theoretic contradiction; perhaps we ought to wonder what we might learn about our mental faculties from a study of these principles.

Conclusion

Gödel concludes by claiming that although he has disproved the nominalist standpoint and adduced strong arguments against the more general view that mathematics is our own creation, he could not claim to have proved the realist viewpoint he favors, for to do so would require a survey of the alternatives, a proof that the survey was exhaustive, and a refutation of all the alternatives except realism. Among the alternatives to be refuted are Aristotelian realism, which he characterizes as the view that concepts are aspects or parts of things, and psychologism, which holds that mathematics is nothing but the psychological laws by which thoughts, presumably concerning calculation, etc., occur in us. About Aristotelian realism, Gödel says only that he does not think it tenable. His principal charge against psychologism, reminiscent of Frege's objections, is briefly given: if psychologism were correct, there would be no *mathematical* knowledge, but only knowledge that our mind is so constituted as to consider certain statements of mathematics true. His discussion is admittedly cursory, however, and Gödel gives psychologism, whatever its merits, much less attention than nominalism.

The suggestion with which he closes the lecture may seem utterly strange: that *after sufficient clarification* of the concepts in question, it will be possible to conduct the discussion of these matters "with mathematical rigor,"

---

[10](Parsons, 1995) contains further discussion of Gödel's use of the notion of analyticity.

at which time the result will be that the Platonistic view is the only one tenable. (Here he characterizes the position somewhat differently, as "the view that mathematics describes a non-sensual reality, which exists independently both of the acts and the dispositions of the human mind and is only perceived, and probably perceived very incompletely, by the human mind" (38).) What is surprising here is not the commitment to Platonism, but the suggestion, which recalls Leibniz's project for a universal characteristic,[11] that there could be a mathematically rigorous discussion of these matters, of which the correctness of any such view could be a "result." Gödel calls Platonism rather unpopular among mathematicians; it is probably rather more popular among them now, forty years after he gave his lecture, in some measure because of his advocacy of it, but perhaps more importantly because every other leading view seems to suffer from serious mathematical or philosophical defects. Gödel's idea that we shall one day achieve sufficient clarity about the concepts involved in *philosophical* discussion of mathematics to be able to prove, mathematically, the truth of some position in the philosophy of mathematics, however, appears significantly less credible at present than his Platonism.

---

[11] In his introductory note to (Gödel, 1944) in (Feferman et al., 1990), Parsons calls Gödel's view, given in the last paragraph of (Gödel, 1944), that Leibniz did not regard the *Characteristica universalis* as a utopian project "one of his most striking and enigmatic utterances."

8

Must We Believe in Set Theory?

According to set theory, by which I mean, as usual, Zermelo–Fraenkel set theory with the axioms of choice and foundation (ZFC), there is a cardinal λ that is equal to $\aleph_\lambda$. Call the least such cardinal κ. κ is the limit of $\{\aleph_0, \aleph_{\aleph_0}, \aleph_{\aleph_{\aleph_0}}, \ldots\}$, that is, the least ordinal greater than all $f(i)$, where, $f(0) = \aleph_0$ and $f(i+1) = \aleph_{f(i)}$ for all natural numbers i.

Is there such a cardinal? I assume that cardinals are ordinals and ordinals are von Neumann ordinals. Thus if κ exists, there are at least as many as κ sets. *Are* there so many sets?

Much very important and interesting work in set theory these days is concerned with cardinals far, far greater than κ, cardinals whose existence cannot be proved in set theory, and with the consequences of assuming that such large cardinals do exist, particularly those concerning objects at comparatively low levels of the set-theoretic universe. Since κ is the limit of an ω-sequence of cardinals smaller than κ, it is not (even) an inaccessible cardinal, the smallest common sort of cardinal whose existence cannot be proved in set theory, let alone measurable, huge, ineffable, or supercompact.

No, κ is quite small, indeed *teensy*, by the standards of those who study large cardinals.

But it's a *pretty big* number, by the lights of those with no previous exposure to set theory, so big, it seems to me, that it calls into question the truth of any theory one of whose assertions is the claim that there are at least κ objects.

Zermelo–Fraenkel set theory is an *interpreted* first-order formal theory. Its language has one non-logical constant, the two-place predicate letter $\in$, its variables range over all the (pure) sets there are, and a formula $x \in y$

This article was originally written for Gila Sher and Richard Tieszen, eds., *Between Logic and Intuition: Essays in Honor of Charles Parsons*, forthcoming from Cambridge University Press.

is true when sets a, b are assigned to the variables x, y if and only if a is a member of b. The logic of the theory is classical first-order predicate logic with identity, and the notion of a theorem of the theory is perfectly standard. Thus one of the theorems of ZFC is the formal sentence σ of the language of ZFC expressing the existence of κ; σ is understood to express the existence of κ because the language of ZFC is understood to be interpreted in the manner just described. σ is a consequence of ZFC; σ asserts that κ exists. If there are not as many as κ objects in existence, κ does not exist, σ is false, and set theory is not true.

Of course κ might exist and set theory be false for some other reason. But κ seems sufficiently large that the claim that it exists might plausibly be regarded as dubious. κ is no gnat; it's a lot to swallow.

Let me try to be as accurate, explicit, and forthright about my beliefs about the existence of κ as I can: It is not the case that I believe that κ exists and it is not the case that I firmly believe that κ does not exist. Without very many or very good reasons, and without strong views on the matter, I tend somewhat to think it probably doesn't exist, but I am really quite uncertain. I am also doubtful that anything could be provided that should be called a *reason* and that would settle the question.

I don't, I say, have what I regard as very good reasons for failing to assent to the existence of κ. But I guess I really don't believe in it, and so of course I think you shouldn't either. Imagine being confronted by a precocious trusting child T. who tells you that three days ago Teacher taught the class about infinity, day before yesterday taught the class that there were not only infinitely many things, numbers and so on, but also infinitely many infinite numbers, and so super-infinitely many things, yesterday taught the class that there were not only super-infinitely many things, but also super-infinitely many infinite numbers, and hence super-super infinitely many things, and today taught the class that you could iterate "super-" as many times as you like, and then asks, "Is it really true what Teacher said? Are there really infinitely and super-infinitely and super-super-infinitely and so on many things ...?" What would *you* say to T.? Not, I hope, "Certainly, of *course* there are, Teacher was absolutely right."

But I hope you would also say, "Teacher was completely right," on being told that Teacher had told the class that if $n > 2$, then $x^n + y^n \neq z^n$ (x, y, z, n positive integers), or on being told that Teacher had said that set theory *says that* there are as many things as that.

I said that my own reasons for thinking that κ doesn't exist weren't very good ones. Perhaps they amount only to the sense that there couldn't be *that* many things, that κ is, by ordinary lights, a (literally) unbelievably big number, and that any story according to which there are so many things around ought to be received rather skeptically.

Russell once quipped that there are fewer things in heaven and earth than are dreamt of in *our* philosophy. Dreams are rarely accurate. Why suppose this one correct?

Furthermore, to the best of my knowledge nothing in the rest of mathematics or science requires the existence of such high orders of infinity. The burden of proof should be, I think, on one who would adopt a theory so removed from experience and the requirements of the rest of science (including the rest of mathematics) as to claim that there are κ objects. κ is such an exorbitantly big number (by ordinary standards) that we would seem to need more reason than we now have to think a theory true that tells us that there are κ things in existence. And the apparent absence of reasons to believe in κ itself here seems like some sort of reason for believing in its non-existence.

But perhaps company can make up for the absence of reasons. So let me ask you what *you* think. Do you really think that there are as many sets as that? Really?

I am, of course, quite well aware that certain annoying questions can be put to one who is skeptical about the existence of κ. It is a theorem of ZFC that κ exists, and not a theorem that it is particularly hard to prove either. The proof of any theorem of ZFC appeals to only finitely many axioms of ZFC; and one who would question a theorem must be dubious about the conjunction of the axioms from which it logically follows. But perhaps it is not much more uncomfortable to refuse to accept the conjunction of those axioms of set theory (among them infinity, power set, union, and certain instances of replacement) needed to prove the existence of κ than to refuse to accept the existence of κ.

But what about cardinals less than κ? As it happens, I myself believe in the existence of $\aleph_0$, just as it is not the case that I believe in the existence of κ. Well then, one might ask me, what about $\aleph_1$? What about $\aleph_2$? What about $\aleph_\omega$? Is there an i such that you believe in the existence of $f(i)$, defined above, but fail to believe in the existence of $f(i+1)$? Or do you believe in the existence of all of them, but just not in κ? Or do you believe that there is a set of all the $f(i)$, but that that set has no union?

We all knew, however, that this sort of trouble could always be made. Like almost every other non-mathematical notion, *acceptance* or *belief* is vague and therefore the usual sorts of difficulties connected with the application of vague predicates to indiscriminable objects can be expected to confront one who maintains that we should not (or not yet) accept the existence of some large infinite cardinal but that belief in $\aleph_0$ is warranted. I chose κ only because it is easy to define but sufficiently large for there to be a serious doubt about its existence even for one who believes in the existence of $\aleph_0$.

Along with the more familiar transfinite cardinals $\aleph_\alpha$, the cardinals $\beth_\alpha$ are defined in set theory. Let us recall their definition: $\beth_0 = \aleph_0$, $\beth_{\alpha+1} = 2^{\beth_\alpha}$, (i.e. the cardinality of the set of all subsets of any set whose cardinality is $\beth_\alpha$) and $\beth_\lambda = \lim\{\beth_\beta : \beta < \lambda\}$, i.e., the least cardinal $\geq$ all $\beth_\beta$, $\beta < \lambda$. Thus $\beth_1$ is the cardinal number of the continuum, and $\aleph_\alpha \leq \beth_\alpha$ for all α. (So the continuum hypothesis states that $\aleph_1 = \beth_1$ and the generalized continuum hypothesis that $\aleph_\alpha = \beth_\alpha$ for all α.) As with the $\aleph$s, there is, similarly, a cardinal λ equal to $\beth_\lambda$. Let us call the least one ρ. (Since $\beth_\rho \leq \rho \leq \aleph_\rho \leq \beth_\rho$, $\beth_\rho = \aleph_\rho$.) I want a number that is problematically big even for one who believes in the existence of the set of real numbers. But since it is consistent that the set of reals has cardinality $\aleph_{(\kappa+1)}$, i.e., the next cardinal after κ, κ ought really to be redefined as ρ. But let us stick with κ.

How then should we respond to T. when he asks whether to believe Teacher? There is an obvious response to T.'s question "are there really so many things as that?" which seems to me to be strong evidence that *we* fail to believe that there are κ objects.

We say, "Well, according to set theory, there are."

It is a theorem of set theory that there are cardinals $\kappa = \aleph_\kappa$. According to set theory, there is such a cardinal as κ. Set theory proves, tells us, has it that there are such cardinals.

Are there dragons?

Legend has it that there are.

Legend usually has it wrong, but legend does have it that there are dragons. According to legend, there are dragons. Legend tells us that there are. If it could prove things, legend would prove that there are dragons.

But when we say that set theory proves that there is a least cardinal κ such that $\kappa = \aleph_\kappa$, we mean more than just that a certain proposition is entailed by the axioms or is part of the content or meaning of those axioms: we mean that there is a proof from the axioms of set theory of a certain sentence expressing the existence of κ, a finite sequence of formulas, the last of which expresses the existence of κ, and each of which is either an axiom or follows from earlier formulas by one of the standard logical rules of inference.

And whether or not κ exists, it *is* literally true that there is such a finite sequence of formulas.

Later on I shall have something to say about the response, "It is not literally true that there is such a finite sequence of formulas because formulas are sets or other abstract objects and abstract objects do not exist." Now, though, I want to compare the different responses to the questions "Does κ exist?" and "Does set theory prove that κ exists?"

It's fairly widely believed that one ought not to lie, one ought not to say what one believes to be false. (I take it that one lies who unwittingly speaks

the truth if he asserts something contrary to what he believes; if not, substitute "knows" for "believes.") Now it is a somewhat nice question what
it means to say that one ought not to lie, or that lying is impermissible,
unacceptable, out (for of course there are circumstances in which it is obligatory and praiseworthy to lie), but I don't want to enter into a discussion
of the meaning of "ought" or "impermissible."

I merely want to suggest that in whatever sense lying is out, so is saying
what one does not know to be true. If one merely believes a thing, p, but
does not know it, then one ought not to say p, it's out, not permissible, to
do so. If you don't know the way to Waltham, then you may not answer
"yes" to the question "Is this the way to Waltham?" any more than to the
question whether you know the way there. If you only think that this is
the way to Waltham, you may of course say, "I think this is the way" or
"I believe so"; but of course then you are saying something that you know
to be true, namely that you think, believe that this is the way. That you
believe that it is is indeed something you know.

Thus there seems to me to be—I don't know that I'm right about this,
but I think I am (but I know that I think so and that it does seem right to
me)—some kind of ban, and roughly the same kind of ban as there is on
lying, on saying what one does not know to be the case. You may not do
it, it's out, impermissible, wrong.

Perhaps the ban is not quite as strong as that on lying, but a speaker
who violates it may justifiably be accused of irresponsibility even if what
he says turns out to be true.

So I take it that when responsible speakers like ourselves say that set
theory proves that κ exists, we take ourselves to know that set theory
proves that it does. I also take it that when we refrain from saying that κ
exists, we are simply observing the ban on saying what we don't know to
be the case. A point follows which ought to be stated plainly: there are
theorems of mathematics to the effect that certain statements are theorems
of set theory that are far more certain than those very theorems.

There is a response to the question whether κ exists that one may expect
to hear nowadays. It is that one who asks the question does not realize
that he is asking an "external" question, where only an "internal" question
is appropriate, or is trying to adopt an "external" standpoint from which
to assess his own (or perhaps our) conceptual scheme, or is afflicted with
"metaphysical realism." The thought is that since set theory implies that κ
exists, to ask whether κ does exist is to attempt to call into question from
some external vantage point the one and only theory we have that treats
of such matters; to do so is fall into the metaphysical error of thinking
we might somehow acquire information concerning the way sets are that is
unmediated by any theory at all and then use this information to assess our

own current theory of sets. Perhaps, the suggestion continues, we think we have some sort of direct insight into the nature of sets, possibly analogous to perception of physical objects, with the deliverances of which we can see our own best theory of sets to be at variance.

Rubbish, for a number of reasons.

Whatever their strength or source may be, the plausibility considerations about how many things there are that conflict with various theorems of ZFC have as much right to be considered a part of "our conceptual scheme" as does ZFC. ZFC conflicts with certain intuitions about cardinality that we happen to have; those intuitions form part of a fragmentary, inchoate, rival theory. If we think that there may well be fewer things in existence than ZFC tells us there are, we are no more assessing ZFC from some external point of view than we are assessing our intuitions about cardinality from some external vantage point when we say they are contradicted by ZFC.

Furthermore, whose theory is ZFC anyway? The difficulty we are confronted with is that ZFC makes a claim we find implausible. To say we can't criticize ZFC since ZFC is *our* theory of sets is obviously to beg the question whether we ought to adopt it despite claims about cardinality that we might regard as exorbitant.

Finally, just exactly what is the matter with saying ZFC isn't correct because it tells us that there are κ objects and there aren't that many objects? (As I remarked above, this is not my view; I am agnostic on the question whether κ exists.) To be sure, one who says this may be asked how he knows there aren't. But the reply, "Get serious. Of course there aren't that many things in existence. I can't *prove* that there aren't, of course, any more than I can *prove* that there aren't any spirits shyly but eagerly waiting to make themselves apparent when the *Zeitgeist* is finally ready to acknowledge the possibility of their existence. But there aren't any such spirits and there aren't as many things around as κ. You know that perfectly well, and you also know that any theory that tells you otherwise is at best goofball."—that reply, although it does not *offer reasons* for thinking that there are fewer than κ objects in existence, would not seem to manifest any illusions that could be called metaphysical realist.

The question, in short, seems like a perfectly reasonable one to ask. The difficulty it presents is that it seems that there is very little to be said on either side of the matter, except that standard set theory says that the answer is yes, while common sense or whatever might be inclined to disagree. It is a frequent enough strategy in philosophy to try to dismiss or belittle a philosophical question which cannot be answered as indicating some philosophical confusion, but that there is any confusion at all in asking whether set theory is correct in asserting that κ exists seems to me just not to have been made out.

Part of the problem, of course, is that in the present case, we are considering whether to reject a *part* of a framework. Abandonment of the whole of set theory is not under consideration. It is not to be expected that philosophical theories about the nonsensicality of asking external questions where only internal ones are appropriate will help us in such a situation.

It may be of use here to examine a story often told in connection with set theory and commonly called the iterative conception of set theory.

According to the iterative conception,[1] every set is formed at some stage. There is a relation among stages, *earlier than*, which is transitive. For any two stages, there is stage later than both. There is a stage that is later than some stage, but not *immediately* later than any stage. A set is formed at a stage if and only if its members are all formed before that stage, i.e., formed at a stage earlier than that one. (Thus, on the present version of the iterative conception, each set is formed at every stage later than any one at which it is ever formed.) Any sets all of which are formed before some one stage are the members of some set.

It is a surprising discovery of Dana Scott that the principles about sets, stages, formation, membership, and *earlier than* just stated, are sufficient to imply (in second-order logic) all the axioms of set theory given by Zermelo, as well as the axiom of foundation.[2] (It is most surprising that they imply that *earlier than* is well-founded.) At least one further principle beyond those just recounted is needed to yield the axioms of replacement.

The iterative conception of set is sometimes supposed to "justify" Zermelo set theory (or, with the aid of additional principles about sets and stages, to justify ZFC). It is important to see exactly what sort of justification it does provide.

A less condensed version of the iterative conception will tell of a first stage at which the null set is formed, a second stage at which the null set is (re-)formed and the unit set of the null set is formed, a third stage, at which four sets, two of them new, are formed, ... an omegath stage, at which, for any sets all of which are formed at finite stages, a set is formed whose members are just those sets, an omegaplusoneth stage, ... an omegaplusomegath stage, an omegatimestwoth stage, ..., an omegasquaredth stage, etc.

The account, when presented in full, is a *picturesque* account of the universe of sets. It has another literary virtue. There is a salient partial ordering of sets, *having lower rank than*. The account respects a natural narrative convention ("neatness") by mentioning things that come earlier in some salient order earlier than those that come later in that order. But notice (a point made to me by Dan Leary) that the talk of stages and

---

[1]This version of the iterative conception is given in detail in my "Iteration again," reprinted as Article 6 in this volume.

[2]A proof is given in Article 6 in this volume.

formation is the most easily dispensable of tropes. One could present the iterative conception just as well by saying, "First there is the null set, then there is the unit set of the null set, then come two more sets, ..., then there are twelve more, which are the ones not already mentioned whose members .... Then after all *those* there are all the sets whose members are the ones just indicated but not mentioned, etc."

Now: as I remarked earlier, the axioms of replacement, which are needed to guarantee the existence of κ, cannot be derived *just* from the principles about sets and membership naturally inferable from the story just given, nor from the principles about sets, stages, membership, formation and earlier than inferable from certain common versions of the iterative conception.

And (the far more important point): Even if the iterative conception is supplemented so that replacement follows, what reason have we to think that any such story is *correct*? Certainly, *if* the story from which those principles about sets and stages can be read off is true, then set theory is true, but why should we believe that the story is in fact true? Perhaps after a while, the story turns false and there aren't those sets.

The interest of the iterative conception is that it shows that the axioms of Zermelo(–Fraenkel) set theory are not just a collection of principles chosen for their apparent consistency and ability to deliver desired theorems concerning arithmetic, analysis, and Cantorian transfinite numbers, but not otherwise distinguished from other equally powerful consistent theories. The conception is *natural* in the simple sense that people can and do easily understand, and readily regard as at least plausible, the view of sets it embodies. The naive conception (any zero or more things whatsoever form a set) is also natural in this sense, but it is of course inconsistent: the things that do not belong to themselves are some things that do not form a set. So the iterative conception is the only view of sets we have that is natural and, apparently, consistent.

In "The iterative conception of set," [3] I expressed the view that neither the axiom schema of replacement nor even the existence of a stage indexed by the first non-recursive ordinal seemed to me to be implied by the version of the iterative conception described there. Nor would the existence of κ be evident on that account of the conception. But I incline—for whatever accident of psychology—to find principles of set theory acceptable if they can be "read off" *that* presentation. Perhaps there are other theorems that seem evident that cannot be so read off. But I do not regard it as evident that there must be a stage corresponding to every ordinal that is the order-type of some well-ordering formed at some stage. My credulity has limits.

Russell's quip was that there are fewer things in heaven and earth that

---

[3] Article 1 in this volume.

are dreamt of in *our* philosophy. But might it be that there are fewer things period than are dreamt of in our philosophy? To vary the allusion, if we are sailors rebuilding our ship plank by plank on the open sea, then I know of some cargo we might want to jettison.

I emphasized earlier that I believed in the existence of *proofs*, e.g., of the sentence stating the existence of κ from the axioms of ZFC, and of $\aleph_0$. It will therefore come as no surprise that I am rather a fan of abstract objects, and confident of their existence. It behooves me, I think, to say why smaller numbers, sets, and functions don't offend my sense of reality the way κ does.

Five pages into the first of the *Dialogues between Hylas and Philonous*, Berkeley has Hylas reply to Philonous that we do not immediately perceive by sight any thing beside light and colors, and figures. Of course Hylas is wrong: we immediately perceive by sight some things other than light, colors, and figures, for example, sticks and stones, and baseballs. But we immediately perceive other things too, for example letters and parts of the sky, as Philonous correctly asserts for once. And still other sorts of things too, such as *The Globe* (a Boston newspaper, whose slogan used to be "Have you seen *The Globe* today?"). It would be a rather demented philosopher who would think, "Strictly speaking, you can't *see The Globe*. You can't even see an issue of *The Globe*. All you can really see, really immediately perceive, is a copy of some issue of some morning's *Globe*." To say this, however, reflects a misunderstanding of our word "see": more than a misunderstanding, really, it's a kind of lunacy to think that sound scientific philosophy demands that we think that we see ink-tracks but not words, i.e. word-types.

An observation due to Helen Cartwright is helpful here. It should be called the Helen Cartwright Theorem Theorem. As Richard Cartwright tells the story,[4] he once said that propositions can't be written down. "[S]he correctly replied that she had seen Gödel's Theorem, for instance, written on the board. I replied that to write Gödel's Theorem on the board is just to write on the board a sentence that formulates the theorem. It took me an inordinately long time to see that if it really takes no more than that, then the theorem is easily enough written down."

And for that matter, you *can* write a *number*, and not just a numeral on the blackboard. Please consider your view about the meaning of the word "on" before jumping to the conclusion that you can't write a number on the board.

Numbers do not twinkle. We do not engage in physical interactions with them, in which energy is transmitted, or whatever. But we twentieth-

---

[4]In the introduction to his (Cartwright, 1987), p. x.

century city dwellers deal with abstract objects *all the time.* We note with horror our *bank balances.* We listen to *radio programs*: *All Things Considered* is an abstract object. We read or write *reviews* of *books* and are depressed by *newspaper articles.* Some of us write *pieces of software.* Some of us compose *poems* or *palindromes.* We correct *mistakes.* And we draw *triangles* in the sand or on the board. Moreover bank balances, reviews, palindromes, and triangles are "given" to us "in experience," whatever it may mean to say that. To put the matter somewhat more carefully, no sense of "sensible" or "experience" has been shown to exist under which it is not correct to say that we can have sensible experience of such objects, such things as the zither melody in *Tales from the Vienna Woods*, the front page of the sports section of this morning's *Globe*, a broad grin, or a proof in set theory of the existence of κ.

It is thus no surprise that we should be able to reason mathematically about many of the things we experience, for they are already "abstract." It is very much a philosopher's view that the only objects there are are physical or material objects, or regions of space-time, or whatever it is that philosophers tell us the latest version of physical theory proclaims to be the ultimate constituents of matter. To maintain that there aren't any numbers at all because numbers are abstract and not physical objects seems like a demented way to show respect for physics, which everyone of course admires. But it is nuts to think Wiles could have spared himself all those years of toil if only he had realized that since there are no numbers at all, there are no natural numbers $x, y, z, n > 2 \ldots$

The existence of infinitely many natural numbers seems to me no more troubling than that of infinitely many computer programs or the existence of infinitely many sentences of English. Of course there is no longest (Basic) program: any program is shorter than the one that results when a suitable sentence of the form: [n] PRINT "Hello, world" is subjoined after its last line. Nor is there a longest sentence; any number of "very"s can be inserted into "This is tiresome." as nearly every speaker of English knows.

Irrealism about numbers seems no more tenable than irrealism about programs or sentences. It is an odd view, to say the least, that there are infinitely many programs but no, or only finitely many, natural numbers. What the most effective rebuttal to the view might be could well depend on how the position was articulated. At any rate, I find the existence of natural numbers as unproblematic as that of physical theories or of irrealist tracts in the philosophy of science. (When may we expect *Computer Science Without Programs* to appear?)

In one of the funniest passages in all of contemporary philosophy, David Lewis asks us to imagine what the reaction would be if we were to walk into a mathematician's office and announce that philosophy has discovered that classes do not exist.[5] Lewis concludes—he readily acknowledges that he has not provided an argument—that we have to believe in the existence of singletons, mysterious though he takes them to be, for the theory of classes pervades all of modern mathematics and we certainly don't wish to reject *that*.

Science-worship again. Can we not raise doubts about higher transfinite cardinals? Cannot a philosopher ask a set theorist whether in fact there really is such a cardinal as κ without seeming to be a crank? It does in fact take a small amount of nerve to ask a practicing set theorist whether there really is such a number as κ. I know it from having done so. The response I got from J., as I shall refer to him, was "I have no problem with that, George," even when I asked him what he would say to a child like T. I did have the impression that J. was perhaps not entirely speaking *in propria persona*, but rather was making an announcement, as from the standpoint of a set-theorist. I also had the sense that nothing I could think of to say could dislodge J. from that standpoint. (But maybe J. really does believe that there are κ and many many more sets in existence.)

What we are contemplating here, however, is nothing so radical as the rejection of singletons, but only the claim to be a body of *knowledge* on the part of a portion of set theory that treats of objects far removed from ordinary experience, the rest of physical science, the rest of mathematics, and the rest of a certain more "concrete" part of set theory.

In the supplement to the second version of his article "What is Cantor's Continuum Problem?," Gödel stated, famously, that " ... despite their remoteness from sense experience, we do have something like a perception of the objects of set theory, as is seen from the fact that the axioms force themselves upon us as being true."

I do not believe that all of the axioms of set theory *force* themselves upon us as being true, and, as far as I can tell, Gödel does not argue for this remarkable claim. The axiom of extensionality may do so: if it is false, there would have to be two sets with the same members, and that, we think, just could not be, not two *sets* with the same *members*. Perhaps also the pair set axiom forces itself upon us as being true: how could there not always be a set $\{x, y\}$ for any objects x, y? But I am by no means convinced that any of the axioms of infinity, union, or power so force themselves upon us or that all the axioms of replacement that we can comprehend do.

A pattern of argument in set theory, by which the existence of large sets is often proved, is to define a mapping of the members of ω (customarily proved by appeal to the axiom of infinity) onto certain objects, which then, by an axiom of replacement, form a set, and then to appeal to the axiom of

[5]In his (Lewis, 1991), p. 59.

union to infer the existence of a certain other set. It is in this manner that κ can be proved to exist. Even apart from worries about large sets, it seems to me that the axioms utilized in this sort of argument are significantly less evidently true than are extensionality and pair set.

That there are doubts about the power set axiom is of course well known. These doubts have to do not so much with the truth of the axiom but with the clarity or intelligibility of the notion "all subsets of." The diagnosis "unclear" or "unintelligible" has to be the wrong one, however. (It is to be noted that those who hold this view are likely to maintain the extremely suspicious view that "$\exists y \forall x (x \in y \to \forall w (w \in x \to w \in z))$" is, but "$\exists y \forall x (x \in y \leftarrow \forall w (w \in x \to w \in z))$" is not, clear, on the ground that in the latter but not the former the quantifier "$\forall x$" is unbounded. ("$\exists y \forall x \in y \forall w \in x \ w \in z$" vs. "$\exists y \forall x (\forall w \in x \ w \in z \to w \in x)$.")) No, there is nothing *unclear* about the power set axiom: not every epistemological drawback is a case of unclarity. There is, to be sure, a lot that we do not know about the set of all subsets of ω, and a lot that we know we cannot, in our present state of knowledge and understanding, find out about it. "Clear," as a number of philosophers have remarked, is an overworked word. What could possibly be *unclear* in, or *unintelligible* about, "$\forall z \exists y \forall x (x \in y \leftrightarrow \forall w (w \in x \to w \in z))$"?

But it does not seem to me unreasonable to think that perhaps it is not the case that for every set, there is a set of all its subsets. The axiom doesn't, I believe, force itself upon us as true, as extensionality and pair set do, and as, say, $0 \neq sn$, $m \neq n \to sm \neq sn$, and perhaps also mathematical induction do.

In his 1951 address to the American Mathematical Society (known as the "Gibbs lecture"),[6] Gödel describes both a process of arriving at axioms for set theory and a picture of the set theoretic universe. (It is of course a version of the iterative conception.) Let me quote a bit of it.

> ... evidently this procedure can be iterated beyond ω, in fact up to any transfinite ordinal. So it may be required as the next axiom that the iteration is possible for *any* ordinal, that is, for any order type belonging to some well-ordered set. But are we at an end now? By no means. For we have now a new operation of forming sets, namely forming a set out of some initial set A and some well-ordered set B by applying the operation "set of" to A as many times as the well-ordered set B indicates. And, setting B equal to some well-ordering of A, now we can iterate this new operation, and again iterate it into the transfinite. This will give rise to a new operation again, which we can treat in

---

[6](Gödel, 1995c). See also Article 7 in this volume.

the same way, and so on. So the next step will be to require that *any* operation producing sets out of sets can be iterated up to any ordinal number (that is, order type of a well-ordered set). But are we at an end now? No, ...

Does this view of how matters are with regard to sets really *force itself upon us as true*? Do we find that, on reflection, we are unable to deny in our heart of hearts that matters must be as Gödel has described them? Or do we suspect that, however it may have been at the beginning of the story, by the time we have come thus far the wheels are spinning and we are no longer listening to a description of anything that is the case?[7]

---

[7]In addition to Charles Parsons, two authors whose writings I particularly regret not having discussed in this paper, despite their evident pertinence to its topic and my admiration for them, are Penelope Maddy and Solomon Feferman.

II Frege Studies

Introduction

Frege's Failure and Success

Frege published three books on logic: the *Begriffsschrift* or *Ideography* (1879), the *Grundlagen der Arithmetik* or *Foundations of Arithmetic* (1884), and the *Grundgesetze der Arithmetik* or *Basic Laws of Arithmetic* (two vols., 1893 and 1903). The first introduces a system of symbolic logic. The second, after a critical survey of earlier views on natural number, presents what is claimed to be a derivation of arithmetic or the theory of the natural numbers from pure logic. The third transcribes the arguments of the second into symbolism, to show that there are no gaps in the logic, and extends the whole project from the natural to the real numbers.

Unfortunately, an assumption used by Frege in deriving arithmetic from logic in effect amounts to the assumption of the inconsistent naive conception of set, as mentioned in the Introduction to Part I. The fact that his system thus collapses into inconsistency has long led philosophers to regard Frege's lifework as a brilliant failure. More recently, however, scholars have come to recognize that there is a substantial mathematical result salvageable from the collapse of Frege's system.

Article 9 provides a non-technical overview of recent Frege studies that can supplement the foregoing remarks. The remaining articles treat of several themes related to Frege (with some articles sounding more than one theme): the source of the inconsistency in his system, the statement and proof of the theorem salvageable from the collapse of his system, the relationship of his work to that of his contemporaries, and the philosophical status of the premiss of his theorem.

The Inconsistency in Frege's System

Article 10 explains that the system in Frege's *Begriffsschrift* is, despite differences in notation, essentially just second-order logic, and so it is consistent, unlike the system of the Frege's *Basic Laws of Arithmetic*. The article also explains that the *Begriffsschrift* contains substantial mathematical results on the theory of relations (for instance, the transitivity of the ancestral, in the jargon of the Introduction to Part I) which must be considered significant contributions to the abstract or generalizing tendency in mathematics that began in the late nineteenth century.

A word needs to be said about differences in intended interpretation of second-order logic. What are the first-order entities, over which the first-order variables range; and what are the second-order entities, over which the second-order variables range? On Frege's reading, the first-order variables range over absolutely all "objects," while the second-order variables range over absolutely all "concepts," somewhat mysterious entities-that-are-not-objects. $\exists U$ and Ux are read "there is a concept U" and "the object x falls under the concept U." This contrasts both with what is the conventional reading today, and with the author's alternative reading in terms of plural quantification, as discussed in the Introduction to Part I. The article argues that this last reading avoids some objections to which others are open.

Two later articles consider the question of the source of the inconsistency in the system of the *Basic Laws of Arithmetic.* The short answer is that the *Basic Laws of Arithmetic* added to the *Begriffsschrift* a principle amounting to the assumption of the existence of sets, naively conceived, and that the system succumbed to the paradoxes the naive conception of sets involves. A fuller answer will have to say something more about the Fregean principle on the one hand, and about the set-theoretic paradoxes on the other hand.

Starting with the set-theoretic paradoxes, Russell's is always the first one mentioned, even though it was not the first one discovered, because it is the only one whose statement presupposes no knowledge of any non-trivial theorems of set theory. But there are several others, and one of them, *Cantor's paradox,* presupposes knowledge of only one non-trivial theorem, *Cantor's theorem.*

Cantor's theorem briefly stated says that if I is any set and H is its *power set* $\mathcal{P}(I)$, the set of all subsets of I, the set of all sets whose elements are all elements of I, then H has more elements than I. Cantor's theorem more fully stated says that there cannot be a function g from H to I that is one-to-one. There cannot be any such g because (i) if there were there would be a function f from I to H that is onto, and (ii) there cannot be any such f. As to (i), given such a g one could define such an f as follows: For any element c of I, let $f(c)$ be the unique element C of H such that $g(C)$ is c, if there is any such C, and let $f(c)$ be I if there is no such C. As to (ii), if f is any function from I to H, then there is a subset D of I such that $f(d)$ cannot be D for any element d of I. As to what D is, it is the set of all elements of c such that c is not an element of $f(c)$. As to why $f(d)$ cannot be D for any element d of I, if we assume $f(d)$ is D for some element d of I, a contradiction results when we ask whether d is an element of D. The answer seems to be that it is if and only if it isn't. This contradiction completes the proof.

Cantor's paradox results if we assume, with the naive conception of set, that in addition to the set of all things that are *not* self-identical, the set of which *nothing* is an element, or *null* set, there is also a set of all things that are self-identical, the set of which *everything* is an element, or *universal* set I. The set H of all subsets of I is then just the set of all sets. Since everything is an element of I, every element of H is an element of I, and so H cannot have more elements than I, contrary to the brief statement of Cantor's theorem. And by letting $g(C)$ be C itself for any element C of H, we get a function g from H to I that is one-to-one, contrary to the fuller statement of Cantor's theorem. If we go back and analyze the proof of Cantor's theorem as it applies to this case, we find that the set D it asks us to consider is Russell's set of all sets C such that C is not an element of itself, and the contradiction that completes the proof is Russell's contradiction. Historically, Russell arrived at his paradox by analyzing the application of Cantor's theorem to the universal set in just this way.   .

Turning to the principles of Frege's system, let there be given an *equivalence relation* $\approx$ on second-order entities, a relation that is reflexive and symmetric and transitive. A *contextual definition* is a principle positing that to each second-order entity U there can be assigned an entity $\dagger U$, to be called the *equivalence type* of U, in such a way that U and V will have the same equivalence type if and only if they are equivalent. A *Fregean* contextual definition is a one in which the equivalence types are assumed to be *first-order entities*. The general form of a Fregean contextual definition for equivalence types $\dagger U$ for an equivalence relation $\approx$ is as follows:

$$\forall U \forall V (\dagger U = \dagger V \leftrightarrow U \approx V)$$

The specific form of the Fregean contextual definition for equivalence types $\ddagger U$ for the equivalence relation $\equiv$ of coextensiveness is as follows:

$$\forall U \forall V (\ddagger U = \ddagger V \leftrightarrow U \equiv V)$$

Since $\equiv$ is being called coextensiveness, it is natural to call $\ddagger U$ the *extension* of U, and to call the foregoing principle the *extensions principle EP*. It is in fact a principle assumed by Frege, the most important special case of the *Axiom (V)* of his *Basic Laws of Arithmetic*.

The trouble is, briefly put, that Frege's "extensions of concepts" amount to sets, naively conceived; or more fully put, that Frege's EP amounts to the assumption that there is a one-to-one function $\ddagger$ from the second-order entities to the first-order entities, contrary to Cantor's theorem. It needs only to be verified that the argumentation of the paradoxes can be transcribed into Frege's symbolism, in order to establish that his system collapses into inconsistency. Thus it is the addition of assumption EP to the system of the *Begriffsschrift* that is responsible for the collapse. Or at least, such is the diagnosis endorsed by a majority of commentators.

Article 11 endorses this majority diagnosis. But the article points out that Frege made very few uses, and even fewer essential uses, of the inconsistent extensions principle EP in deriving arithmetic. And the article suggests a consistent weakened extensions principle EP⁻ that he could have used instead. (This is the principle already mentioned in Part I in connection with Article 6, under the label "small extensions principle;" in the article under discussion it is called the "subtensions principle.")

Article 14 answers a challenge to the majority diagnosis from the doyen of modern Frege studies, Michael Dummett. Dummett suggests that it is the introduction of second-order quantification that is responsible for the inconsistency (and seems to express some degree of sympathy with the view discussed in connection with Article 2 of Part I, according to which quantification only makes sense if the things over which one is quantifying form a set). And indeed Terence Parsons has shown that EP is consistent if the axiom of comprehension of second-order logic is assumed only for conditions involving no second-order variables (and his proof is outlined in the article). Actually, the author raised the question whether something more isn't true, namely that EP is consistent when assumed for conditions involving no *bound* second-order variables, the "simple predicative fragment," and beyond that the "ramified predicative fragment" whose exact definition need not detain us here. An extension of Parsons' proof does establish consistency here, and Richard G. Heck, Jr., has recently proved that a significant fragment of arithmetic can be developed in the resulting system. But the article nonetheless restates the case for the majority diagnosis in opposition to Dummett, while emphasizing that this local dissent from one of Dummett's suggestions does not diminish the author's global admiration of Dummett's contributions.

Article 21 is a brief note on an aspect of the proof of Cantor's theorem. Frege is not mentioned by name, but it is hoped that the connection with Frege will be clear from the above discussion of Cantor's paradox and its fatal bearing on Frege's system.

Frege's Theorem

But enough of the weaknesses in Frege's thought, which have long been known. It is time to consider the strengths that have only comparatively recently become apparent. Despite the existence of some precursors whose significance was only recognized in retrospect, the new direction in Frege studies may be said to have been launched in 1983 by Crispin Wright with his book *Frege's Conception of Numbers as Objects* (Wright, 1983). (Mention also should be made of the technical appendix thereto, co-authored with Neil Tennant.) The book drew attention to the Fregean contextual definition for equivalence types $\#U$ for the equivalence relation $\cong$ of equinumerosity:

$$\forall U \forall V (\#U = \#V \leftrightarrow U \cong V)$$

Since $\cong$ is being called equinumerosity, it is natural to call $\#U$ the *number* of U. Because Frege in enunciating the foregoing contextual definition cites Hume, it has come to be called *Hume's principle* or HP. Second-order logic together with HP has come to be called *Frege arithmetic* or FA^2. With regard to FA^2, the book makes two main contributions.

First, it shows that attempts to derive the usual contradictions of naive set theory within FA^2 break down, and conjectures that FA^2 is consistent. Second, it relates FA^2 to the single most intensively investigated system of formal arithmetic in the literature, *second-order Peano arithmetic* or PA^2, which consists of second-order logic together with the *Peano axiom* or PA. (The weaker system *first-order Peano arithmetic* or PA^1 has also been much studied.) PA asserts that there are some things we may call "natural numbers," a distinguished element among them we may call "zero," and a way in which they may be related which we may call "immediate succession," for which a certain list of a half-dozen or so postulates hold. The half-dozen or so postulates are stated in Articles 17 and 18, and it will not be needful to restate them here. Intensive investigation by mathematical logicians has shown that PA^2 is adequate to develop the classical theory of natural and real numbers, and probably adequate for most of mainstream mathematics (though certainly not for higher set theory). What the book indicates is that PA^2 can be derived from FA^2, or in other words, that within pure second-order logic, PA can be derived from HP.

This is the "substantial mathematical result salvageable from the collapse of Frege's system" alluded to earlier. It has come to be called *Frege's theorem* or FT. All the articles in this part are concerned directly or indirectly with it. The leading idea for getting the whole infinite sequence of natural numbers from HP is easy enough to indicate in gross outline: First we get zero as the number of things that are non-self-identical, then we get one as the number of things that are identical to zero, then we get two as the number of things that are identical either to zero or to one, and so on. The subtlety lies in how to say all this rigorously—especially the "and so on."

Article 12 aims to press the argument of Wright's book further, and this in two directions. The first aim of the article is to prove the conjecture that HP is consistent. (Actually, this in itself is quite easily done, and the article quotes a direct one-line proof. So the aim of the article needs to be stated in more precise terms, which requires the technical notion, to be explained in the Introduction to Part III, of *relative* consistency. Precisely and technically stated, the first aim of the article is to prove the consistency of FA^2 relative to PA^2.) The second aim of the article is to indicate a streamlined derivation of PA from HP, of PA^2 from FA^2, and more significantly, to indicate that such a derivation was essentially given already *by Frege himself.* Frege's only essential use of the inconsistent EP was to derive the consistent HP, whereafter he essentially derives everything from HP alone. A concise derivation in symbolic form is given in an appendix to Article 13. The attribution to Frege is in a sense qualified in Article 20.

Articles 17 and 18 both provide detailed analysis of Frege's theorem FT. They contain proofs to show various premisses imply various conclusions and models to show various other premisses do *not* imply various other conclusions. Very roughly what the detailed analysis establishes is that HP is needed only to get a certain intermediate principle that in turn is needed to get PA, where neither the implication from HP to the intermediate nor the implication from the intermediate to PA can be reversed. Ultimately, the motivation for such detailed studies must be supplied by the earlier articles that make the case for the importance of FT. But one result of the detailed examination may be mentioned here by way of motivation, the emergence from the examination of a previously unrecognized redundancy or dependency among the various clauses of PA. PA has been so intensively investigated by so many mathematical logicians for so many years that it is surprising that there is something new to be said about it.

Article 20, co-authored with Heck, deals with the derivation of PA from HP *as it appears in the Fregean text.* Transcription into ideography was supposed to catch any gaps in the purely logical structure of a derivation. What this article argues is that there is indeed such a gap in the proof of Frege's theorem in the *Foundations of Arithmetic,* which is filled in by the *Basic Laws of Arithmetic.* Along the way it is noted that it had apparently occurred to Frege that there might be true hypotheses expressible in the notation of his system but not provable from his axioms.

Frege and His Contemporaries

Only after Frege's true achievement, the proof of Frege's theorem, is appreciated can Frege's contributions be meaningfully compared with the those of his near-contemporaries, Richard Dedekind and Bertrand Russell. Article 13 is partly and Article 15 is largely devoted to carrying out the comparison with Dedekind, while Article 16 concerns Russell.

Article 13 notes, as to Dedekind, that the formulation of PA is really due to him, the name "Peano postulate" having been bestowed by readers who overlooked Peano's citation of Dedekind. Also due more to Dedekind than any other single contributor is the derivation of the basic laws of natural and ultimately of real numbers from PA^2. What was missing, however, in Dedekind's work, was any rigorous or even plausible derivation of PA. This Frege's theorem arguably supplies. Article 15 adds that some of the substantial mathematical results on the theory of relations found in the *Begriffsschrift* were arrived at independently by Dedekind.

Article 16 contrasts Frege's *proof,* assuming HP, of the existence of an infinity of objects, with Russell's *assumption* that there are infinitely many objects at the bottom of his hierarchy (as described in the Introduction to Part I). The need for this assumption makes it difficult to accept the claim that Russell achieved a reduction of mathematics to pure logic, though indeed the need for the assumption HP may make it difficult to accept the claim that Frege did so either. In the case of Frege, one can say his work contains a substantial mathematical result, even if the result did not amount to a reduction of mathematics to logic. Can one say the same for Russell? The article argues that one can.

Here is the result. A set x is *strongly finite* if for some natural number n there is a function from the set of natural numbers less than n to x that is onto, and *strongly infinite* if there is a function from the set of all natural numbers to x that is one-to-one. Without the choice axiom it cannot be proved that if x is not strongly finite, then it is strongly infinite. It cannot even be proved that if x is not strongly finite, then the power set $\mathcal{P}(x)$ of x is strongly infinite. *Russell's theorem,* which invites comparison with Frege's theorem, is that if x is not strongly finite, then at least the power set $\mathcal{P}(\mathcal{P}(x))$ of the power set $\mathcal{P}(x)$ of x is strongly infinite.

The Philosophical Status of Hume's Principle

Article 11 raises the question of what, philosophically speaking, follows from Frege's mathematical results? One may distinguish three theses:

1. Not all substantial mathematical results depend on spatiotemporal intuition.

2. Not all substantial mathematical results depend on something more than pure logic.

3. No substantial mathematical results depend on anything more than pure logic.

The first thesis of *anti-Kantianism* seems to be established by any significant contribution to the abstract or generalizing tendency in mathematics that began in the late nineteenth century, including Frege's contribution in the *Begriffsschrift.* Even the second thesis of *super-anti-Kantianism* or *sub-logicism* seems to be established if it is accepted that second-order logic is indeed *logic*, a claim argued in several papers in Part I. The third thesis of *logicism* is what Frege himself had hoped to establish (at least for mathematics minus geometry). But despite his proof of FT, he cannot be held to have established it unless his premiss HP can be considered in some sense a tautologous or analytic or conceptual truth.

Wright in his book argued that it should be so considered. His position was supported by his colleague Bob Hale, and opposed by Field, who as a nominalist rejects numbers and sets alike. All this led to a long-running exchange of views whose details need not concern us here. Article 19 is wholly, and Article 13 partly, devoted to the issue, as are incidental passages in other articles. Wright's position is criticized, but the criticism is from a non-nominalistic standpoint, and turns on features of HP more specific than that it involves "ontological commitment to abstract entities."

A concession may perhaps be made to Wright, to the effect that the question whether HP is a conceptual truth may be partly terminological. There may perhaps be principles so central to a concept that anyone who rejects them and nonetheless uses the usual name for the concept may be said to be using not the usual concept but another one of the same name. Such principles may perhaps be called "conceptual," and by those who accept them as truths, "conceptual truths." We may perhaps be hypothetically obliged to accept them as true *if* we accept the concept. But we are not categorically obliged to accept them as true, since we are not obliged to accept the concept. We may even on the contrary be obliged to *reject* the concept—and are obliged to reject it if its principles lead to inconsistencies.

What is special about a Fregean contextual definition is that it postulates that the equivalence types corresponding to an equivalence relation on second-order entities *are first-order entities*. It is this feature that creates the danger of inconsistency. We have at present no deep understanding of why some Fregean contextual definitions like EP lead to inconsistencies while others like HP do not, and one might be tempted to argue on that ground alone that we cannot be *obliged* to accept any particular such definition.

Rather than a deep understanding, we have at present a collection of examples and results. One result comes from Heck, and is reported in Article 19. It says that for any formula φ there is a Fregean contextual definition, a modification of EP, that is consistent if and only if φ is. It follows from a famous result of mathematical logic, Church's theorem, that there can be no "mechanical" procedure for determining whether or not a given Fregean contextual definition is consistent. One example comes from the author, and is reported in Article 13. It cites two Fregean contextual definitions, one of them being HP, that are each separately consistent, but whose conjunction is inconsistent. This like nothing else illustrates the complexity of the issue.

9

Gottlob Frege and the Foundations of Arithmetic

The German philosopher and mathematician Gottlob Frege (1848–1925) is widely regarded as one of the two greatest logicians since Aristotle. (The other is Kurt Gödel.) Frege is now credited with the creation of modern logic: among other accomplishments, he was the first person to investigate the logical foundations of mathematics and the first to construct a formal deductive system of logic.

Frege was also the logician in whose system of logic Bertrand Russell discovered the contradiction now called Russell's paradox. I shall discuss the paradox—it's the one about the set containing those sets that do not contain themselves—in some detail later on. The present story is a bittersweet one. When Frege came to appreciate the full force of Russell's paradox, he regarded it as utterly destructive of his entire logical work. But within the last five years or so, it has become clear that Frege grievously undervalued his actual achievement. Here I shall describe the main aims of the work of this great philosopher and explain a profound discovery of his, which philosophical scholarship has only recently brought to light, and of which he himself, sadly, was unaware.

Frege was not a philosopher with a broad range of interests. Unlike Plato, Aristotle, Hume, and Kant, he wrote nothing on ethics or social philosophy. But an interest in "value theory" is not required for greatness in a philosopher. Descartes also wrote nothing in this area, and Leibniz and Wittgenstein, an insignificant amount. Although Frege's output was somewhat small (three books and fewer than thirty articles) and the range of his concerns quite narrow, and in spite of the tragic end of his intellectual career, the profundity, inventiveness and rigor of Frege's work put him in the philosophical pantheon beside Descartes, Leibniz, and Kant.

The only branch of philosophy that interested Frege was logic, including formal logic and the philosophies of mathematics and language. His works

143

in the philosophy of language are currently the subject of intense and fruit-ful scrutiny by numerous philosophers and linguists, but they were to a certain extent spin-offs from the main body of his work, which concerned itself with the relation between logic and mathematics.

Frege's first work was a monograph, published in 1879 and only 88 pages long. It was entitled *Begriffsschrift,* translatable as "Conceptual Notation" or "Concept-Script." Nothing like it had been seen before, and the mono-graph was completely unappreciated when it was published, for reasons that are obvious as soon as one looks at a typical page. The odds that Frege's work was the production of a genius rather than a crackpot may have seemed long indeed to his colleagues and contemporaries.

However, *Begriffsschrift* contains the first rigorous system of formal logi-cal notation ever devised, and his notation is not all that hard to get used to. Unlike later formal languages, Frege's was a two-dimensional, treelike system of notation. (The now standard linear symbolism of logic, with its symbols, $\neg$ (not), $\rightarrow$ (if ... then ...), and $\forall x$ (for all x), derives from notations invented by Russell and Giuseppe Peano.) Instead of writing: $A \rightarrow B$, Frege symbolizes (if A then B):

$$\begin{array}{l} \rule{3cm}{0.4pt}\; B \\ \quad\rule[1ex]{0.4pt}{1ex}\rule{2cm}{0.4pt}\; A \end{array}$$

(Perhaps Frege chose this symbolization to suggest: A is the hypothesis on which B rests.) Not-A is written:

$$\rule{3cm}{0.4pt}\!\top\!\rule{1cm}{0.4pt}\; A$$

and For all x:

$$\rule{2cm}{0.4pt}\underset{x}{\cup}\rule{1cm}{0.4pt}$$

Thus no As are Bs (for all x, if x is an A, then x is not a B) would be written:

$$\rule{2cm}{0.4pt}\underset{x}{\cup}\rule{1cm}{0.4pt}\top\rule{1cm}{0.4pt}\; B \;/\; A$$

Unlike standard notation, Frege's two-dimensional symbolism depicts the logical structure of statements visually and vividly. "The convenience of the typesetter is certainly not the summum bonum," he wrote.

But although the pages of his book look like nothing so much as wallpa-per, his principal objective in *Begriffsschrift* was distinctly philosophical.

The main philosophical question that concerned Frege was the *justifi-cation* of mathematics. The issue can be understood with the aid of an example. Suppose we call x's father, x's father's father, x's father's fa-ther's father, etc. x's *forefathers.* (Assume that a person has at most one father.) Now suppose that y and z are two different forefathers of x. Then, as a moment's thought will make clear, either z is one of y's forefathers or y is one of z's forefathers:

$$x \to x\text{'s father} \; \to x\text{'s father's father} \; \to \ldots \to$$
$$\text{one of } y, z \to \ldots \to \text{ the other of } y, z$$
$$[x \to x\text{'s father} \; \to x\text{'s father's father} \to \ldots \to]$$
$$\text{one of } y, z \to \ldots \to \text{ the other of } y, z$$
$$\text{one of } y, z \to \ldots \to \text{ the other of } y, z$$

Above is a picture, from which it is apparent that if y and z are forefathers of x, then one of them is a forefather of the other. Are pictures like this one necessary for mathematics? Do we have to have seen one, either on the blackboard, on the page, or in the mind's eye, in order to be entitled to believe that of any two forefathers of x, one is a forefather of the other? (By the way, this statement, even though it mentions the highly non-mathematical notion of fatherhood, is every bit as much a mathematical statement as the assertion that if y and z are two distinct positive integers, then either y is less than z or z is less than y.)

The philosopher Immanuel Kant held that such mental "intuitions" *are* necessary, if we are to be justified in believing any but the simplest truths of mathematics, ones like "12=12." Kant's view was that it was such mental "seeings" that justify us in believing in, and make us believe, the truth of any statements that are not mere identities. Kant's opinion as to the necessity of certain psychological processes for the justification of mathematics deeply influenced the views of such twentieth-century mathematicians as David Hilbert, L. E. J. Brouwer, and Kurt Gödel.

Frege's aim in the *Begriffsschrift* was to show that logic could do the work of Kantian intuitions in cases like that of the statement about forefathers. Frege claimed that one could define "forefather" from "father" in purely logical terms in such a way that one could prove by logic alone, and without the aid of any intuitive pictures, that of any two forefathers of some one person, one is a forefather of the other. In order to show this, he had to show that the statement could be proved without the aid of any assumptions whose truth could not be shown by logic alone or could be guaranteed only by means of pictures. It was to insure that all assumptions of his demonstrations were evidently logical assumptions that Frege devised his formal language. Since, as Frege showed, such mathematical statements could be rigorously proved in the system of the *Begriffsschrift,* Kant's view of mathematics could be seen to be seriously deficient: intuition was not needed in many interesting cases where Kant and others would have thought it indispensable.

It was either during the writing of the *Begriffsschrift* or shortly afterwards that Frege became convinced that the whole of mathematics could be similarly reduced to logic. That is to say, the concepts of mathematics such as the integers, addition, multiplication, the real numbers, the notions of the calculus, etc. could all be defined in purely logical terms and the propositions of mathematics proved from these definitions by logical means alone. This view is called "logicism." Believing that the rigorous development of the calculus that had been carried out by Cauchy, Weierstraß, and others sufficed to reduce the rest of mathematics to arithmetic, Frege concentrated his efforts on showing that the arithmetic of the natural numbers (the non-negative integers $0, 1, 2, \ldots$) could be derived from logic and to that end published a second short (120 pp.) book.

The Foundations of Arithmetic appeared in 1884. The first half of the book was devoted to a caustic review of writings by various philosophers and mathematicians on the concept of number. Frege was not a kindly critic. Although a few of his predecessors (Leibniz, Hume) escaped unscathed, most of his contemporaries were savaged. Fate would have worse in store for Frege. In the second half of the book, Frege set forth his own "logicist" view of mathematics.

The fact that the *Foundations* is more or less devoid of mathematical symbols should not mislead one into thinking it is not at least in large part a mathematical work. Certainly it is a philosophical work, but, as Frege's work has made plain, there is no precise demarcation to be made between mathematics and philosophy. Frege was concerned to make it plausible that the propositions of arithmetic, once translated into logical terms, could be derived from logical axioms by logical means alone. This is a mathematical task, and Frege had now to give rigorous proofs of the claims whose demonstration he had only outlined in *The Foundations of Arithmetic*. Doing this would require developing a significant portion of mathematics in a formal system like that of *Begriffsschrift*.

After completing the *Foundations* Frege set to work. In 1893, the first volume of his *Basic Laws of Arithmetic* was published; the second appeared in 1903.

In mid-June of 1902 Frege was completing the second volume of *Basic Laws* when a letter arrived in the mail. It was from a then unknown English logician named Bertrand Russell. In a page or two, Russell pointed out to Frege that one of his Basic Laws, number (V), led to an inconsistency, a statement of the form p-and-not-p.

In logic inconsistency is deadly. Not only is no statement of the form p-and-not-p ever true, any statement q whatsoever—e.g., "The moon is made of green cheese"—follows from p-and-not-p. For if p-and-not-p, then p, whence p-or-q; and if p-and-not-p, then also not-p. But from not-p and p-or-q, q follows.

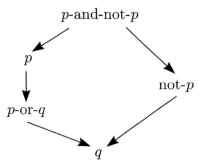

If snow is white and snow is not white, the moon is made of green cheese. Thus if an inconsistency can be derived from an axiom, so can any statement whatsoever. And that is what Russell's letter proved: that from Frege's basic laws, it was possible to derive in short order *all* propositions of mathematics, the false along with the true, and not merely those truths over whose derivations Frege had labored for more than fifteen years.

The effect of the letter on Frege was devastating. "Hardly anything more unwelcome can befall a scientific writer," he later wrote, "than for one of the foundations of his edifice to be rocked after the work is finished." He replied quickly to Russell (the dates of their letters differ by only six days), acknowledging that a defect had been discovered that would be fatal if not eradicated. After working on the problem, Frege wrote an appendix to the second volume of *Basic Laws,* in which Russell's discovery was discussed at length and the outline of a repair suggested. For a time Russell too thought that the contradiction had been resolved. "As it seems very likely that this is the true solution, the reader is strongly recommended to examine Frege's argument on the point," he wrote in his book *The Principles of Mathematics.*

They were wrong. The substitute axiom Frege proposed didn't do the work it had to do. It was logically consistent, but inconsistent with the trivially true claim that there are at least two different numbers, a statement not to be refuted by Frege's new "axiom." Soon after their books appeared, Frege and Russell came to realize that the new axiom didn't work.

A major virtue of Frege's philosophical work was its precision and clarity, which had made it vulnerable to an objection of the sort that Russell had discovered. Its discovery was more or less the end for Frege. He published a few more articles, but no more work in furtherance of his life's project, that of showing arithmetic reducible to logic. Late in life Frege speculated that perhaps arithmetic could be grounded in geometry; this later line of thought he never developed with any thoroughness.

We may explain the contradiction that Russell discovered by means of an example: Along with many, many others, one of the books listed in the catalogue of books called *Books in Print* is *Books in Print* itself. On the other hand, the catalogue of the MIT Press, a publication of the MIT Press, doesn't list itself, as it happens. So some catalogues do, and some catalogues do not, list themselves.

But no catalogue lists *all and only* those catalogues that don't list themselves: if catalogue *C* lists *all* catalogues that don't list themselves, then *C* lists itself (because if *C* doesn't list itself, then since *C* lists *all* that don't list themselves, *C* lists *C*); but then since *C* lists itself, it doesn't list *only* catalogues that don't list themselves. If it lists all non-self-listers, it doesn't list only non-self-listers. No catalogue lists all and only the non-self-listing catalogues.

No barber shaves all and only those barbers that don't shave themselves. If Figaro shaves all who don't shave themselves, then Figaro shaves himself (if he doesn't, he does), and then Figaro doesn't shave only those who don't shave themselves.

The problem that Russell discovered in Frege's *Basic Laws* is that via the right definitions, one could prove that (*p*) some set contains all and only those sets that don't contain themselves. But by logic alone, we can prove, and it could certainly be proved in Frege's *Basic Laws,* that (not-*p*) no set contains all and only those sets that don't contain themselves, just as no barber shaves all and only those barbers that don't shave themselves. Thus the inconsistency *p*-and-not-*p* could be proved in Frege's system.

It should be noted that, although logic forces us to believe that there isn't any such set, it's highly paradoxical that there isn't. For there certainly are many sets that don't contain themselves. The set of even numbers is one example. Since the set of even numbers is not an even number, it doesn't contain itself. The set of human beings is another, for that set is not a human being. Now, let me invite you to fix your attention on *the sets that don't contain themselves,* the set of evens, the set of human beings, etc. Doesn't there HAVE to be such a thing as the collection or totality of things you are thinking about, the sets that don't contain themselves? And isn't a collection or totality just the same thing as a set? How COULD there NOT be a set containing all and only the sets that don't contain themselves?

Maybe you want to say there's no *set* containing just the sets that don't contain themselves, but there is a *collection* containing just those sets. But then isn't there a collection containing all and only the collections that don't contain themselves? You can't get out of this paradox merely by substituting one word for another. (The German mathematician Ernst Zermelo also discovered this same contradiction, by the way, and wrote a letter to David Hilbert about it. Don't think we'd be living in a fool's paradise were it not for Russell.)

This version of Russell's paradox is quite well known. The version of the paradox that Frege actually treated in the appendix to his *Basic Laws* is rather more interesting, because it throws light on the discovery of Frege's I mentioned before.

What Frege discovered was that arithmetic can be derived from one single axiom that looks almost as trivial as anything in logic. Moreover, that axiom can be demonstrated to be consistent. (Or more cautiously, if it ever were found to be inconsistent, that would probably be the most surprising discovery ever made in mathematics.) Frege's discovery was obscured from us and from him by Russell's paradox, which had been thought to have invalidated the whole of Frege's formal work.

To explain Frege's discovery, I must describe some of the leading ideas of the system of Frege's *Foundations of Arithmetic*, the second, prose, book.

In the *Foundations* there are three basic notions: object, concept, and extension. Little explanation of what an object is can be given beyond saying that an object is a *thing,* but not necessarily a physical or material thing. Numbers and sets, as well as stones, electrons, cabbages and kings, are all objects. Objects, one may say, are things that can be mentioned on either side of an "identity" like *2+2=4* or *Adam = the father of Cain.*

Frege held that in addition to objects, there are entities called concepts, items very much like what the logicians of the middle ages would have called "universals" or Plato would have referred to as "forms," things that all objects of a certain kind "have in common." (Some other near-synonyms are: property, quality, characteristic, attribute.) Whenever there is a "count noun," possibly modified, like "stone," "prime number," "root of $x-5 = 0$" or "U.S. Senator," there is a corresponding concept, *being a stone, being a prime number,* etc. In Frege's terminology, objects are said to "fall under" concepts: thus Kennedy, Dole, and the ninety-eight other senators fall under the concept *being a U.S. Senator,* the number 5 is the sole object that falls under the concept *being a root of $x - 5 = 0$,* and $2, 3, 5$, and infinitely many other numbers fall under the concept *being a prime number.* There are some concepts under which *no* objects fall: *being both a dog and a cat, being identical with nothing at all, being a unicorn,* etc.

Objects and concepts are completely different sorts of entity. The final part of Frege's doctrine that is of interest to us concerns his claim that for every concept, there is a certain object called the *extension* of the concept. It will do no harm to think of the extension of a concept as the *set* of things that fall under the concept; the difference between sets and Frege's extensions is insignificant here. Every extension, to repeat, is an *object,* but of course the converse doesn't hold: not every object is an extension.

Frege's fatal axiom in the *Basic Laws,* "the set principle," was just the statement that the extension of the concept F is identical with the extension of the concept G if and only if every object falling under F falls under G and conversely, every object falling under G falls under F. More briefly: the set of Fs is the same as the set of Gs if and only if the Fs are the same as the Gs. It sounds utterly obvious, doesn't it?

But it is not consistent, and the proof Russell discovered that the set principle leads to a contradiction may be found in Box 1:

Box 1

We first define "Russellian." We say that an object x is *Russellian* if there is at least one concept F such that (a) $x =$ the extension of F and (b) x does not fall under F.

For example, if x is the extension of the concept *being a gorilla,* then x is Russellian, for since the extension x is not a gorilla (a set is not a gorilla), x does not fall under the concept *being a gorilla.*

Let $y =$ the extension of the concept *being Russellian.* Then is y Russellian or not? We shall show that from either answer, the opposite answer follows:

First, suppose that y is Russellian. Then there is at least one concept F such that $y =$ the extension of F and y does not fall under F. But by the definition of y, it is also true that $y =$ the extension of the concept *being Russellian.* So the extension of the concept $F = y =$ the extension of the concept *being Russellian.* By Frege's axiom, every object that falls under F is Russellian and every object that is Russellian falls under F. Since y does not fall under F, it follows that y is not Russellian. To sum up: if y is Russellian, y is not Russellian.

Suppose, however, that y is not Russellian. Then there is not even one concept F such that $y =$ the extension of F and y does not fall under F. But $y =$ the extension of the concept *being Russellian.* Thus y must fall under the concept *being Russellian.* (For otherwise, if y does not fall under this concept, there is at least one concept F, namely *being Russellian,* such that $y =$ the extension of F and y does not fall under F.) Since y falls under the concept *being Russellian,* y is Russellian. To sum up again: if y is not Russellian, y is Russellian.

Thus y is Russellian if and only if it is not, a contradiction.

But although Frege's principle about sets, the set of Fs = the set of Gs if and only if the Fs are the Gs, is inconsistent, a related principle about numbers, "the number principle," turns out to be consistent. The number principle runs: The number of Fs = the number of Gs if and only if the Fs and the Gs are in one-one correspondence.

The notion of one-one correspondence is familiar: The place settings at a typical dinner party are in one-one correspondence with the persons present. Since there are nine major planets and nine positions on a (National League) baseball team, the planets are in one-one correspondence with those positions: under the natural correspondence, Mercury corresponds to pitcher, ..., Saturn to short stop, ..., and Pluto to right fielder. The even integers are in one-one correspondence with the odd integers, etc., etc.

Now if one reads *The Foundations of Arithmetic* carefully, one sees that Frege uses the set principle *only* to derive the number principle, "the number of Fs = the number of Gs if and only if the Fs and the Gs are in one-one correspondence." After deriving the good number principle from the bad set principle, Frege has nothing more to do with sets. Once he has obtained the number principle, he proceeds to show how to derive arithmetic from it with the aid of nothing other than the system of logic he had set out in the *Begriffsschrift*. Moreover, the sketch of the derivation of arithmetic from the number principle that Frege gives in the *Foundations* can easily be elaborated into a completely formal derivation in the style of the *Begriffsschrift* or the *Basic Laws*. The job would be analogous to the modern-day task of translating an extremely explicit outline of a program into actual programming code.

How may we see that the number principle is consistent? It may seem obvious enough—the number of Fs equals the number of Gs if and only if the Fs and Gs are in one-one correspondence—but the number principle looks far too much like the set principle for us to take its consistency on faith. How do we know that some Super-Russell of the 22nd Century won't find some ingenious derivation of a contradiction from the number principle, the way our Russell derived a contradiction from the set principle?

In fact, on second glance the number principle may not seem altogether obvious: the even natural numbers $(0, 2, 4, ...)$ can be put in one-one correspondence with the odd natural numbers $(1, 3, 5, ...)$: $0 \leftrightarrow 1$, $2 \leftrightarrow 3$, $4 \leftrightarrow 5, ..., 2n \leftrightarrow 2n + 1 ...$ But if the number principle is true, then the number of even natural numbers ought to be the same as that of odd natural numbers. *What is this number?* Certainly not one of $0, 1, 2, ...$? There are far too many even numbers for their number to be, say, 4,369,527, or any other natural number (=non-negative whole number).

None of $0, 1, 2, ...$ is the number of even numbers there are. Thus the number principle cannot be thought to be true if the only objects we are talking about are the natural numbers themselves, for the number of even numbers is an infinite number, and not a natural number. (Natural numbers are finite numbers.)

In order to circumvent the difficulty presented by the even numbers, we shall define a new notion, *the number by Fs:* The number by $Fs = 0$ if the number of Fs is infinite, and the number by $Fs = n+1$ if the number of Fs is the non-negative integer n. Thus the number by senators is 101, the number by unicorns is 1, the number by roots of $x - 5 = 0$ is 2, and the number by even numbers is 0. Notice that if we suppose that the only objects are the natural numbers $0, 1, 2, \ldots$, then for every concept there is a natural number that is the number by that concept. (It is not true that for every concept there is a natural number that is the number of that concept: remember the evens.)

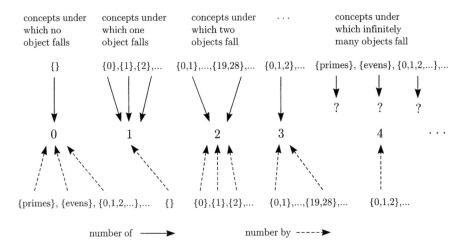

concepts under which no object falls — concepts under which one object falls — concepts under which two objects fall — · · · — concepts under which infinitely many objects fall

number of ⟶   number by ----▸

We now observe that under the assumption that the only objects are the natural numbers, the number by Fs, a notion that is defined for *all* choices of F, is the same as the number by Gs if and only if the Fs and the Gs are in one-one correspondence. To see why this is so, consult Box 2.

Box 2

Here's how to see the truth of the statement that the number by Fs = the number by Gs if and only if the Fs and Gs are in one-one correspondence. (We assume that the only objects under consideration are the natural numbers, so that the Fs are some (possibly all) of the natural numbers and the Gs some other natural numbers.)

Suppose first that the Fs and Gs are in one-one correspondence. Either there are only finitely many Fs or there are infinitely many Fs. If there are only finitely many Fs, n say, then there are also n Gs and the number by Fs = the number by Gs = $n+1$. But if there are infinitely many Fs, then there are also infinitely many Gs and the number by Fs = the number by Gs = 0. In either case, then, the number by Fs = the number by Gs. Thus if the Fs and Gs are in one-one correspondence, then the number by Fs = the number by Gs.

Conversely, suppose that the number by Fs = the number by Gs. Either that number is positive or that number is 0. If it is positive, $n+1$, say, then there are n Fs and n Gs and the Fs and Gs are in one-one correspondence. But if the number is 0, then there are infinitely many Fs and infinitely many Gs. But in this case too, the Fs and Gs are in one-one correspondence: the least F corresponds to the least G, the second least F to the second least G, the third least, etc. So in either case, the Fs and Gs are in one-one correspondence. Thus if the number by Fs = the number by Gs, then the Fs and Gs are in one-one correspondence.

Thus, although the number principle is not universally true, i.e., not true for all choices of F and G, if we take our objects to be the natural numbers and "the number of" to have its ordinary meaning, it *is* universally true when we reinterpret "number of" as "number by." Since the number principle is true of the natural numbers when thus reinterpreted, it is logically consistent; it cannot be shown false by logic. (For specialists: we have just shown how to interpret in analysis the result of adjoining the number principle to the system of *Begriffsschrift*.)

The proof we have just given of the consistency of the number principle is the same sort of proof that was first used in the nineteenth century to demonstrate that certain non-euclidean geometries are consistent. To prove these geometries consistent, we pick a domain of objects (points, great circles on a sphere, etc.), reinterpret the terms "point," "line," etc. by "point on a sphere," "great circle," etc. and show that the axioms of the non-euclidean geometry turn out true under this reinterpretation. Logicians refer to this sort of procedure as "defining a model" in Euclidean geometry for a non-euclidean geometry and it shows that if Euclidean geometry is consistent, so is the non-Euclidean geometry for which the model has been defined. The objects of our model are just the natural numbers; we have reinterpreted "number of" by "number by."

Thus the number principle, similar though it may look to the inconsistent set principle, is consistent: logic cannot show it false, for we have defined a model for it. Frege's startling discovery, of which he may or may not have been fully aware and which has been lost to view since the discovery of Russell's paradox, was that *arithmetic can be derived in a purely logical system like that of his Begriffsschrift from this consistent principle and from it alone.*

Limitations of space prevent me from explaining why this is so in full detail, but much of the flavor of Frege's derivation can be given by seeing how he proves the simple statement that the numbers 0 and 1 are not identical. Frege defines 0 as the number of the concept: *being non-self-identical.* Since everything is self-identical, no object at all falls under this concept. Frege defines 1 as the number of the concept *being identical with the number 0.* 0 and 0 alone falls under this latter concept. Since there is exactly one object falling under the latter concept and none falling under the former, the objects falling under these two concepts are NOT in one-one correspondence. Therefore by the number principle, their numbers, which are 0 and 1, are not identical.

Now that we have established that $0 \neq 1$, we can define 2 as the number of the concept *being identical with 0 or 1.* We can also define addition in the following way: if no object falls under both of the concepts F and G, m is the number of Fs and n the number of Gs, then $m+n$ is the number of $(F$–or–$G)$s. The fundamental facts about addition can be proved on the basis of this definition; multiplication can then be defined, etc., the natural numbers characterized and the rest of arithmetic carried out, all on the basis of just one simple, consistent, and trivial-looking axiom, the number principle.

What is sad is not so much that Frege's system turned out to be vulnerable to Russell's paradox as that both he and we failed to realize how valuable his actual accomplishment was. Frege proved the first great theorem of logic: arithmetic can be derived from the number principle. He might have written a book not all that different from the actual *Foundations of Arithmetic,* perhaps with the title *The Logical Analysis of Arithmetic,* in which it was shown how arithmetic can be derived by logic alone from that one axiom. Had he done so, he would have been acknowledged in his own time as having shed brilliant light on the science of arithmetic.

10

Reading the *Begriffsschrift*

The aim of the third part of the *Begriffsschrift*, Frege tells us, is:

> to give a general idea of the way in which our ideography is
> handled ... Through the present example, moreover, we see how
> pure thought, irrespective of any content given by the senses or
> even by an intuition a priori, can, solely from the content that
> results from its own constitution, bring forth judgments that at
> first sight appear to be possible only on the basis of some in-
> tuition ... The propositions about sequences developed in what
> follows far surpass in generality all those that can be derived
> from any intuition of sequences. If, therefore, one were to con-
> sider it more appropriate to use an intuitive idea of sequence as a
> basis, he should not forget that the propositions thus obtained,
> which might perhaps have the same wording as those given here,
> would still state far less than these, since they would hold only
> in the domain of precisely that intuition upon which they were
> based.[1]

He then proceeds to give a definition, Proposition 69, on which he comments, "Hence this proposition is not a judgment, and consequently *not a
synthetic judgment* either, to use the Kantian expression. I point this out
because Kant considers all judgments of mathematics to be synthetic."[2]

Reprinted by kind permission of Oxford University Press from *Mind* 94 (1985): 331–344.

I am grateful to Michael Dummett, Robin Gandy, Daniel Isaacson, David Lewis,
and Simon Blackburn for helpful comments. This paper was written while I was on a
Fellowship for Independent Study and Research from the National Endowment for the
Humanities.

[1]Gottlob Frege, *Begriffsschrift, a formula language, modeled upon that of arithmetic,
for pure thought*, §23. All references are to the Bauer–Mengelberg translation, found in
(van Heijenoort, 1967).

[2]*Begriffsschrift*, prop. 69. The remark that Kant considers all judgments of mathematics to be synthetic seems somewhat intemperate: Kant might of course agree that
69 is no *judgment*, hence no synthetic judgment.

In the preface to the *Begriffsschrift* he states, "To prevent anything intuitive [*Anschauliches*] from penetrating here unnoticed, I had to bend every effort to keep the chain of inferences free of gaps."[3] It is evident from the anti-Kantian tone of these remarks that Frege regards himself as showing the inadequacy of a certain (unspecified) Kantian view of mathematics by supplying examples of judgments that he thinks "at first sight appear to be possible only on the basis of some intuition," but which pure thought, "solely from the content that results from its own constitution," can bring forth. However an exact statement of the Kantian position under attack might run, the view is one according to which no non-trivial mathematical judgment is "possible" without "a priori intuition."

My principal aim in this paper is to examine Frege's procedure in the third part of the *Begriffsschrift* in order to see how, and how well, a Kantian view of Frege's examples might be defended and to determine to what extent Frege could claim to have shown the truth of a view that may be called *sublogicism*: the claim that there are (many) interesting examples of mathematical truths that can be reduced (in the appropriate sense) to logic. Inevitably, the uncertainties and obscurities attaching to the notions of *intuition* and *logic* will leave these matters somewhat unresolved. I will however argue that a compelling case for Frege's view can be made against a certain sort of defense of Kant.

The issue between Frege and Kant is joined over a certain technical point that arises in connection with the marginal annotations of the derivations of Part 3. If we wish to understand the issue, we cannot avoid examining the wallpaper. There is a further reason for looking at the formalism of Part 3: at least one little-known but major master-stroke is hidden there, and one of the subsidiary aims of this paper is to call attention to it, repellent though the notation in which it is cloaked may be. Another aim of the paper is simply to render Part 3 more accessible.

Before we examine Frege's achievement we must review the special notational devices which Frege introduces in Part 3. Fortunately, there are only four of them.

The first of these

$$\begin{array}{c} \delta \\ | \\ \alpha \end{array} \left(\begin{array}{c} F(\alpha) \\ \\ f(\delta, \alpha) \end{array} \right.$$

is defined in Proposition 69 to mean something that we might notate: $\forall d \forall a (Fd \land dfa \rightarrow Fa)$. (I have written "*dfa*" in place of Frege's "*f(d,a)*.") Since the relation f—Frege calls it a *procedure*—is fixed throughout Part

---

[3] *Begriffsschrift*, Preface, p. 5.

3, I shall use the abbreviation "Her(F)," suppressing "f," for this notion instead. ("Her" is for "hereditary.")

The second,

$$\underset{\beta}{\overset{\gamma}{}}f(x_\gamma, y_\beta),$$

is Frege's abbreviation for the strong ancestral of f, whose celebrated definition is presented in Proposition 76. Abbreviating "$\forall a(xfa \rightarrow Fa)$" as "In($x, F$)" (again suppressing mention of the fixed f), we may give the definition as: $\forall F(\text{Her}(F) \wedge \text{In}(x, F) \rightarrow Fy)$. We shall use: xf^*y for this notion.

The third,

$$\underset{\beta}{\overset{\gamma}{\equiv}}f(x_\gamma, z_\beta),$$

is the abbreviation for the weak ancestral, defined in Proposition 99 as $xf^*z \vee z = x$. We write this: $xf^*_{=}z$.

Finally, Frege defines

$$\underset{\varepsilon}{\overset{\delta}{I}}f(\delta, \varepsilon)$$

in Proposition 115 to mean: $\forall d\,\forall e\forall a(dfe \wedge dfa \rightarrow a = e)$. We write this: FN (for "f is a function").

We can now say what the judgments are which Frege thinks can be brought forth by pure thought solely from the content that results from its own constitution—or, as we may say, can be proved by purely logical means—but which, he thinks, appear at first sight to be possible only on the basis of some intuition. We can then take up the question whether the means used to prove them are in fact "purely logical."

If we look at the table with which the *Begriffsschrift* ends and which indicates which propositions are immediately involved in the derivations of which others, we find that there are only two propositions in the third part not used in the derivation of any others: number 98 and the last one, number 133. Since these propositions are not used to prove any others, I do not find it too far-fetched to suppose that Frege thought of these as illustrating the falsity of the Kantian view with which he is concerned.

The translation into our notation of 98 is: $xf^*y \wedge yf^*z \rightarrow xf^*z$. That of 133 is: FN $\wedge\, xf^*m \wedge xf^*y \rightarrow yf^*m \vee mf^*_{=}y$.[4] These state that the (strong) ancestral is transitive and that if the underlying relation f is a function, then the ancestral connects any two elements m and y to which

---

[4]I do not know why Frege chose to use the variable "m" here instead of (say) "w".

some one element x bears the ancestral. The analogy with the transitivity and connectedness of the less-than relation on the natural numbers, which is the ancestral of the relation *immediately precedes*, will not have escaped the reader's notice, and I dare say it did not escape Frege's.

Although Frege does not explicitly single out 98 and 133 as noteworthy in any way, it is quite reasonable to suppose that he regarded both of them as the sort of proposition that would justify the anti-Kantian viewpoint sketched above. For not only are these two the only propositions in Part 3 not used in the demonstration of others, their content can be seen as a generalization of that of familiar and fundamental mathematical principles, for the grasp of whose truth some sort of "intuition" was often supposed in Frege's time to be required. Moreover, one who attempts to convince himself of the truth of, for example 98, might well hit upon an argument that would seem to make appeal to the sort of intuition which Frege was concerned to show unnecessary. Suppose that y follows x in the f-sequence and z follows y. Then if one starts at x and proceeds along the f-sequence, one can eventually reach y. Ditto for y and z. Thus, by starting at x and proceeding along the f-sequence, one can eventually reach z, first by going to y, and thence to z. Thus z follows x in the f-sequence. Intuition, it might be suggested, discloses to us that any two paths from x to y and y to z can be combined into one single path from x to z: intuit them both and then attach in thought the beginning of the second to the end of the first. Or some such thing.

The procedure Frege employs in the derivation of 98 is of considerable interest, and we shall look at its final steps. Having arrived at

(84) $\text{Her}(F) \wedge Fx \wedge xf^*y \rightarrow Fy$

and

(96) $xf^*y \wedge yfz \rightarrow xf^*z,$

Frege generalizes upon z and y in 96 to obtain $\forall d \forall a (xf^*d \wedge dfa \rightarrow xf^*a)$. He then substitutes $\{a : xf^*a\}$ for F (as we might put it) in the definition of $\text{Her}(F)$ to obtain 97, which we can write: $\text{Her}(\{a : xf^*a\})$. He then reletters x and y as y and z in 84, substitutes $\{a : xf^*a\}$ for F in 84, and discharges $\text{Her}(\{a : xf^*a\})$ to obtain the desired 98.

Frege appears to regard the substitution of a formula for a relation letter in an already demonstrated formula as on a par with substitution of a formula for a propositional variable or relettering of a variable. Of course, in standard first-order logic, substitution of formulae for relation letters gives rise to no special worries: any formula demonstrable with the help of substitution is demonstrable without it. (Frege performs several such substitutions in Part 2, which contains none but first-order notions.) But this is emphatically not the case as regards Part 3 of the *Begriffsschrift*. The capacity to substitute formulae for relation letters gives the whole of Frege's system, which is not a system of first-order logic, significantly more power than it would otherwise have.

Although a Kantian opponent could well make an objection at this point to Frege's use of substitution, there is a more pertinent objection to be made: no one can sensibly think that *every* mathematical judgment must be based on some intuition. For certainly there are some trivial mathematical judgments which need not be so based, among them analytic judgments concerned with mathematical matters and others of a *trivial* logical nature, such as "$5+7 = 5+7$" or "if $5+7 = 12$, then $5+7 = 12$." Moreover, among such judgments are those that follow from definitions with only a *small* amount of *elementary* logical manipulation. And one of these is Frege's 98. For, let us face it, Frege's proof of 98 is unnecessarily non-elementary. One needs no rule of substitution at all to prove that if xf^*y and yf^*z, then xf^*z. For suppose xf^*y and yf^*z. We want to show xf^*z, i.e. $\forall F(\text{Her}(F) \wedge \text{In}(x, F) \rightarrow Fz)$. So suppose $\text{Her}(F)$ and $\text{In}(x, F)$. We want to show Fz. Since yf^*z, we need only show $\text{In}(y, F)$, i.e. $\forall a(yfa \rightarrow Fa)$. So suppose yfa. Since xf^*y, $\text{Her}(F)$ and $\text{In}(x, F)$, Fy. And since $\text{Her}(F)$ and yfa, Fa, QED. The trouble with 98, our Kantian might complain, is that although the *above* proof of 98 is certainly a proof by logical means alone, 98 does not *look at first sight as if* it must be based on an intuition.

Frege has not yet laid a glove on the Kantian. 98 *is* a weak example. Of course Frege's rendering of 98, "If y follows x in the f-sequence and z follows y in the f-sequence, then z follows x in the f-sequence," might have been a better choice, but the Kantian might then have been in a position to raise questions about the grounds for *reading* "xf^*y" as "y follows x in the f-sequence," plausibly arguing that this reading is itself justified only on an intuition.

No such objection can be raised against 133, $\text{FN} \wedge xf^*y \wedge xf^*m \rightarrow yf^*m \vee mf_{=}^*y$, of which an "intuitive" proof might go as follows. Suppose FN, xf^*y, and xf^*m. Since xf^*y and xf^*m, there is an f-sequence leading from x to y and an f-sequence leading from x to m. And since FN, each thing bears f to at most one thing; thus at no point along the way can either of these paths diverge from the other. Thus the paths coincide up to the point at which the shorter one gives out. Since xf^*m and xf^*y, we eventually reach both m and y; when we have done so, we will evidently have reached y before m, reached m before y, or reached m and y at the same time. In the first case, we can get from y to m along the path obtained by removing the path from x to y from the path from x to m; in the second case we can similarly get from m to y, and in the third case, $m = y$. Thus $yf^*m \vee mf_{=}^*y$. We are about to turn to Frege's derivation of 133; before

we do so, the reader might like to try his hand at giving a proper proof of
133, in the style of the proof of 98 given two paragraphs above. (One such
proof is given in the appendix.)

One significant landmark in Frege's derivation of 133 is Proposition 110:
$\forall a(yfa \rightarrow xf^*_=a) \wedge yf^*m \rightarrow xf^*_=m$. 110 is itself got from 108: $zf^*_=y \wedge yfv \rightarrow zf^*_=v$, which has a straightforward proof.[5] 108 is fairly obvious; 110 is not
at all obvious. (We cannot get yfm from yf^*m.) How does Frege get 110
from 108?

First of all, he reletters the variables in 108, replacing, z, y, and v by x,
(German) d, and (German) a, and then universally quantifies upon a and
d to get: $\forall d \forall a(xf^*_=d \wedge dfa \rightarrow xf^*_=a)$. He then takes 75: $\forall d \forall a(Fd \wedge dfa \rightarrow Fa) \rightarrow \text{Her}(F)$, which is one-half of the definition of $\text{Her}(F)$, substitutes
$\{a : xf^*_=a\}$ for F (as we would put it), and uses 108 to cut the antecedent
of the result, thereby getting 109: $\text{Her}(\{a : xf^*_=a\})$. Next he takes 78:
$\text{Her}(F) \wedge \forall a(xfa \rightarrow Fa) \wedge xf^*y \rightarrow Fy$, which is a trivial consequence of the
definition of the ancestral, respectively replaces x and y by y and m, again
substitutes $\{a : xf^*_=a\}$ for F, and drops $\text{Her}(\{a : xf^*_=a\})$ from the result by
109, to get $\forall a(yfa \rightarrow xf^*_=a) \wedge yf^*m \rightarrow xf^*_=m$, as desired.

The complexity of the definition of the substituend $\{a : xf^*_=a\}$ is noteworthy. "$xf^*_=a$" abbreviates a disjunction one of whose disjuncts is a second-
order universal quantification of a first-order formula. Were Frege merely
substituting $\{a : Ga\}$ (G a one-place relation letter) for F, i.e. reletting
F as G, we should have no qualms about his procedure. But the substitution of so complicated a formula as $xf^*_=a$ for a relation letter is a matter
considerably more problematical.

Having obtained 110, Frege straightforwardly gets 129: $\text{FN} \wedge (yf^*m \vee mf^*_=y) \wedge yfx \rightarrow (xf^*m \vee mf^*_=x)$.[6] 131: $\text{FN} \wedge \text{Her}(\{a : af^*m \vee mf^*_=a\})$
follows, again by a substitution, this time of $\{a : af^*m \vee mf^*_=a\}$ for F in
the quasi-definitional 75.

Frege then performs the same substitution to conclude the derivation.
From 131, he uses propositional logic to infer 132: $[\text{Her}(\{a : af^*m \vee mf^*_=a\}) \wedge xf^*m \wedge xf^*y \rightarrow (yf^*m \vee mf^*_=y)] \rightarrow [\text{FN} \wedge xf^*m \wedge xf^*y \rightarrow (yf^*m \vee mf^*_=y)]$. To get 133, the consequent of 132, he must obtain
the antecedent. This is how he does it. He has earlier established 81:
$Fx \wedge \text{Her}(F) \wedge xf^*y \rightarrow Fy$ (an easy consequence of the definition of the
ancestral). By propositional logic there follows 82: $(p \rightarrow Fx) \wedge \text{Her}(F) \wedge p \wedge xf^*y \rightarrow Fy$. (Frege uses "$a$" instead of "$p$.") He then substitutes hx for
p and $\{a : ha \vee ga\}$ for F in 82 ("h" and "g" are one-place relation letters,
like "F") and drops a tautologous conjunct of the antecedent to obtain 83:
$\text{Her}(\{a : ha \vee ga\}) \wedge hx \wedge xf^*y \rightarrow hy \vee gy$. The final logical move of the

---

[5]A proof is given in the appendix.
[6]A proof is given in the appendix.

Begriffsschrift is the substitution in 83 of $\{a : af^*m\}$ and $\{a : mf^*_=a\}$ for h and g, which yields the antecedent of 132.

Of course, Frege could have condensed these two substitutions for F into one, by substituting $\{a : af^*m \vee mf^*_=a\}$ for F in 81 and using propositional logic to obtain the antecedent of 132. But to prove 133, Frege has had to make two essential uses of substitution, the first being the earlier substitution of $\{a : xf^*_=a\}$ for F, the second, that of $\{a : af^*m \vee mf^*_=a\}$. It is noteworthy that the—or at any rate, one—obvious attempt to prove 133 will require the same two substitutions, in the order in which they are found in Frege's derivation.

The fact that the *Begriffsschrift* contains a subtle and ingenious double induction—for that is what Frege's pair of substitutions amounts to—used to prove a significant result in the general theory of relations is not, I think, well-known, and the distinctively mathematical talent he displayed in discovering and proving the result is certainly not adequately appreciated. Frege's accomplishment may be likened to a feat the Wright brothers did not perform: inventing the airplane *and* ending its first flight with one loop-the-loop inside another.

Our Kantian has patiently had his hand up during this discussion of Frege's method in Part 3, and it is time to give him his say.

The Kantian: "I could not agree with you more about the excellences of Proposition 133 and Frege's proof of it, but it is not a counterexample to any thesis that I hold or that a reasonable Kantian ought to hold. Indeed, if anything, it is confirming evidence for my view. I agree that 133 is precisely the sort of proposition that is possible only on the basis of an intuition. But I disagree that Frege has been able to prove it without the aid of any intuition at all. In fact, the feature of Frege's method that you have been at pains to emphasize, the substitution of formulae for relation letters, is precisely the point at which, I wish to claim, Frege appeals to intuition. I'd be prepared to concede, for the sake of avoiding an argument, that nowhere in the rest of the *Begriffsschrift* is an appeal to intuition made. But I do wish to claim that his use of the rule of substitution does involve him in just such an appeal.

"The difficulty that the rule of substitution presents can best be seen if we consider the axiom schema of comprehension: $\exists X \forall x(Xx \leftrightarrow A(x))$. It is well known that in the presence of the other standard rules of logic, the substitution rule and the comprehension schema are deductively equivalent; given either, one can derive the other. In outline, the proof of this equivalence runs as follows. From the provable $\forall x(Fx \leftrightarrow Fx)$, we obtain $\exists X \forall x(Xx \leftrightarrow Fx)$ by second-order existential generalization, whence by the substitution of $\{a : A(a)\}$ for F, we have $\exists X \forall x(Xx \leftrightarrow A(x))$. Conversely, we observe that for any formulae $P[F]$ and $A(x)$, we can prove $\forall x(Fx \leftrightarrow$

$A(x)) \to (P[F] \leftrightarrow P[\{a : A(a)\}])$; the demonstration of this is an induction on subformulae of the formula $P[F]$. Now suppose that $P[F]$ is provable. Then so is $\forall x(Fx \leftrightarrow A(x)) \to P[\{a : A(a)\}]$; and since the consequent $P[\{a : A(a)\}]$ does not contain F, $\exists X\forall x(Xx \leftrightarrow A(x)) \to P[\{a : A(a)\}]$ is also provable. Thus if we have as an axiom $\exists X\forall x(Xx \leftrightarrow A(x))$, as is guaranteed by comprehension, $P[\{a : A(a)\}]$ is provable too, QED. Thus we cannot admit substitution as a logical rule unless we are prepared to admit that all instances of the comprehension schema $\exists X\forall x(Xx \leftrightarrow A(x))$ are logical truths, and that is precisely what I wish to deny.

"For what does $\exists X\forall x(Xx \leftrightarrow A(x))$ say? If we look at the *Begriffsschrift*, we find that when Frege wishes to decipher his relation letters and second-order quantifiers, he uses the terms 'property', 'procedure', 'sequence'; he uses the terms 'result of an application of a procedure' and 'object' to tell us what sorts of things free variables like 'x' and 'y' denote. My point can be put as follows. Suppose that $A(x)$ is the formula: mf^*x. Then Frege would read the corresponding instance of the comprehension schema as 'There is a property whose instances are exactly the objects that follow m in the f-sequence'. This comprehension axiom is demonstrable in the *Begriffsschrift*. My question is: why should we believe that there is any such property? Now, *I* don't want to deny that there is such a property. I might well want to say that it's *obvious* or *evident* that there is one. And I would want to say to anyone who professed uncertainty concerning the existence of the property, 'But don't you *see* that there has to be one?' In short, it is an intuition of precisely the kind Frege thinks he has shown unnecessary that licenses the rule of substitution. Thus Frege has not dispensed with intuition; he is up to his ears in it. (I may add that the inference from $\forall x(Fx \leftrightarrow Fx)$ to $\exists X\forall x(Xx \leftrightarrow Fx)$ also strikes me as problematical, but as it is legitimated by (the second-order analogue of) the standard logical rule of existential generalization, I have agreed not to object to it.)

"Moreover there is an important difficulty connected with the interpretation of the *Begriffsschrift*.[7] Frege does not discuss the question whether properties are objects, as one might put it. It is uncertain whether Frege thinks there can, for example, be sequences of properties, whether xfy might hold when x and y are themselves properties. One would have supposed so; but then, of course, taking 'f' to mean 'Is a property that is an instance of the property' produces a Russellian problem: $\exists X\forall x(Xx \leftrightarrow \neg xfx)$ is derivable in the *Begriffsschrift*, but would be read by Frege 'There is a property whose instances are all and only those properties that are not instances of themselves', which is false, of course. Thus the system, although perhaps formally consistent, cannot be interpreted as Frege interprets it

---

[7] For an illuminating discussion of this difficulty, see (Russinoff, 1983).

in the absence of some—I think the right word is 'metaphysical'—doctrine of properties, which Frege does not supply. And what, pray, is the source of any such doctrine to be—pure logic? How then are we to interpret the *Begriffsschrift* so that its theorems all turn out to be truths that it does not require the aid of intuition to accept?

"I'm almost finished. Matters are no better and probably worse if Frege reads a second-order quantifier $\exists F$ as 'There is a set F ...' For sets clearly are 'objects'; thus the difficulty presented by Russell's paradox immediately arises if we take the range of 'F' to be all *sets*. The only escape that I can see for Frege is for him to stipulate that the *Begriffsschrift* is to be employed in formalizing a certain theory only if the theory does not speak about *all* objects. The rule of substitution would then be licensed by the *Aussonderungsschema* of set theory. But besides noting that this way out appears to be strongly at odds with his intentions in setting forth the *Begriffsschrift*, we may well wonder what justifies this appeal to the *Aussonderungsschema* if not *intuition* of some sort, for example the *picture* of the set-theoretic universe that yields the so-called 'iterative conception of set'. And now, I *am* finished."

In reply: Russell's paradox does indeed show the difficulty of taking the second-order quantifiers of the *Begriffsschrift* as ranging over all sets or all properties and reading atomic formulae like Xx as meaning "x is a member (or instance) of X." We must find another way to interpret the formalism of the *Begriffsschrift*, on which we are not committed to the existence of such entities as sets or properties, and on which the comprehension schema $\exists X \forall x (Xx \leftrightarrow A(x))$ can plausibly be claimed to be a logical law.

Interpretation of a logical formalism standardly consists in a description of the objects over which the variables of the formalism are supposed to range and a specification that states to which of those objects the various relation letters of the formalism apply. Since Frege nowhere specifies what his relation letters "f", "F", etc. apply to, it is clear, I think, that he had no one "intended" interpretation of the *Begriffsschrift* in mind: "f", for example, will have to be interpreted on each particular occasion by mentioning the pairs of objects that it is then intended to apply to. But it appears that Frege did intend the first-order variables of the *Begriffsschrift* to range over absolutely all of the "objects," or things, that there are. In any event even if Frege did envisage applications in which the first-order variables were to range over some but not all objects, it seems perfectly clear that he did allow for some applications in which they do range over absolutely all objects. And because a use of the *Begriffsschrift* in which the variables do not range over all objects that there are can, by introducing new relation letters to relativize quantifiers, be treated as one in which they do range over all objects, we shall henceforth assume that the

Begriffsschrift's first-order variables do range over all objects, whatever an object might happen to be.

But what do the second-order variables range over, if not all sets or all properties? I think that a quite satisfactory response to this question is to reject it, to say that no separate specification of items over which the second-order variables range is needed once it has been specified what the first-order variables range over.[8] Instead we must show how to give an intelligible interpretation of all the formulae of the *Begriffsschrift* that does not mention special items over which the second-order variables are supposed to range and on which Frege's rule of substitution appears as a rule of logic and the comprehension axioms appear as logical truths.

The key to such an interpretation can be found in the behavior of the logical particle "the."

If the rocks rained down, then there are some things that rained down; if each of *them* [pointing] is a K and each K is one of them, then there are some things such that each of them is a K and each K is one of them; if Stiva, Dolly, Grisha, and Tanya are unhappy with one another, then there are some people who are unhappy with one another. Existential generalization can take place on plural pronouns and definite descriptions as well as on singular, and existential generalization on plural definite descriptions is the analogue in natural language of Frege's rule of substitution. This type of inference is not adequately represented by the apparatus of standard first-order logic. However, a formalism like that of the *Begriffsschrift* can be used to schematize plural existential generalization, and our understanding of the plural forms involved in this type of inference can be appealed to in support of the claim that Frege's rule is properly regarded as a rule of logic.

By a "definite plural description" I mean either the plural form of a definite singular description, for example "the present kings of France," "the golden mountains," or a conjunction of two or more proper names, definite singular descriptions, and (shorter) definite plural descriptions, for example "Russell and Whitehead," "Russell and Whitehead and the present kings of France."

Like the familiar condition: $\exists x \forall y (Ky \leftrightarrow y = x)$ which must be satisfied by a definite singular description "The K" for its use to be legitimate, there is an analogous condition that must be satisfied by definite plural descriptions. In the simplest case, in which a definite plural description such as "the present kings of France" is the plural form of a definite singular description, the condition amounts only to there being one object or more to which the corresponding count noun in the singular description applies.

---

[8]For more on this topic, see Article 5 in this volume.

(Two or more, technically, if Moore and the Eleatic Stranger were right.) Thus like the definite singular description "The K," which has a legitimate use iff the K exists, i.e. iff there is such a thing as the K, "The Ks" has a legitimate use iff the Ks exist, i.e. iff there are such things as the Ks, iff there is at least one K.

The obvious conjecture—I do not know whether or not it is correct—is that the general condition for the legitimate use of a conjunction of proper names, definite singular descriptions, and (shorter) definite plural descriptions is simply the conjunction of the conditions for the conjoined names and descriptions. We need not worry here whether the conjecture is true; for our purposes it will suffice to consider only definite plural descriptions of the simplest sort, plural forms of definite singular descriptions.

The connection between definite plural descriptions and the comprehension principle is that the condition under which the use of "The Ks" is legitimate, viz. that there are some such things as the Ks, can also be expressed: there are some things such that each K is one of them and each one of them is a K. Thus "if there is at least one K, then there are some things such that each K is one of them and each of them is a K" expresses a logical truth. Moreover, it is a logical truth that it is quite natural to symbolize as

$$\exists x\, Kx \to \exists X(\exists x\, Xx \land \forall x(Xx \leftrightarrow Kx)),$$

which is equivalent to the instance $\exists X \forall x(Xx \leftrightarrow Kx)$ of the comprehension scheme. Thus the idea suggests itself of using the construction "there are some things such that ... them ..." to translate the second-order existential quantifier $\exists X$ so that comprehension axioms turn out to have readings of the form "if there is something ..., then there are some things such that each ... thing is one of them and each of them is something ..." Let us see how this may be done.

We begin by supposing English to be augmented by the addition of pronouns "it$_x$," "it$_y$," "it$_z$," ...; "that$_x$," "that$_y$," "that$_z$," ...; "they$_X$ / them$_X$," "they$_Y$ / them$_Y$," "they$_Z$ / them$_Z$," ...; "that$_X$," "that$_Y$," "that$_Z$," ... (For each first-order variable v of the formalism, we introduce "it" $^\frown_v$ and "that" $^\frown_v$; and for each second-order variable V, "they" $^\frown_V$, which is sometimes written "them" $^\frown_V$, and "that" $^\frown_V$.) The purpose of the subscripts is simply to disambiguate cross-reference and has nothing to do with the distinction between first- and second-order formulae or between singular and plural number. A similar augmentation would be required for translation into English of first-order formulae of the language of set theory containing multiple nested alternating quantifiers, for example formulae of the form $\forall w \exists x \forall y \exists z\, R(w, x, y, z)$. The extension of English we are contemplating is a conceptually minor one rather like lawyerese ("the former," "the latter," "the party of the seventeenth part"); our subscripts are taken for convenience to be the variables of the *Begriffsschrift* (instead of, say, numerals), but they no more *range over* any items than does "seventeen" in "the party of the seventeenth part."

We now set out a scheme of translation from the language of the *Begriffsschrift* into English augmented with these subscripted pronouns.[9] Thus we specify the conditions under which sentences of the *Begriffsschrift* are true by showing how to translate them into a language we understand.

The translation of the atomic formula Xx is $\ulcorner$it$_x$ is one of them$_X$$\urcorner$. (The corner-quotes are Quinean quasi-quotes.)

The translation of the atomic formula $x = y$ is $\ulcorner$it$_x$ is identical with it$_y$$\urcorner$. The translation of any other atomic formula, for example Fx or xfy, is determined in an analogous fashion by the intended reading of the predicate letter it contains.

Let F_* and G_* be the translations of F and G. Then the translation of $\neg F$ is $\ulcorner$Not: $F_*$$\urcorner$ and that of $(F \wedge G)$ is $\ulcorner$Both F_* and $G_*$$\urcorner$. Similarly for the other connectives of the propositional calculus.

The translation of $\exists x\, F$ is $\ulcorner$There is an object that$_x$ is such that $F_*$$\urcorner$.

To obtain the translation of $\exists X\, F$: Let H be the result of substituting an occurrence of $\neg x = x$ for each occurrence of Xx in F and let H_* be the translation of H. (H has the same number of quantifiers as F.) Then the translation of $\exists X\, F$ is $\ulcorner$Either H_* or there are some objects that$_X$ are such that $F_*$$\urcorner$.

(Since $\ulcorner$There are some objects that$_X$ are such that $F_*$$\urcorner$ properly translates not $\exists X\, F$, but $\exists X(\exists x\, Xx \wedge F)$, we need to disjoin a translation of H, which is equivalent to $\exists X(\neg \exists x\, Xx \wedge F)$, with $\ulcorner$There are some objects that$_X$ are such that $F_*$$\urcorner$ to obtain a translation of $\exists X\, F$.)

When we apply this translation scheme to the notorious $\exists X \neg \exists x \neg (Xx \leftrightarrow \neg xfx)$, with the predicate letter f given the reading: "is a member of," we obtain a long sentence that simplifies to "if some object is not a member of itself, then there are some objects (that are) such that each object is one of them iff it is not a member of itself," a trivial truth.

More generally, the translation of $\exists X \forall x (Xx \leftrightarrow A(x))$ will, as desired, be a sentence that can be simplified to one that is of the form: either there is no object such that ... it ... or there are some objects such that an arbitrary object is one of them iff ... it ... And of course, our translation scheme respects the other rules of logic in the sense that if H follows from F and G by one of these rules, and the translations F_* and G_* of (the universal closures of) F and G are true, then the translation H_* of (the universal closure of) H is also true. Our scheme, therefore, respects Frege's rule of

---

[9]This scheme was given in Article 4 in this volume.

substitution of formulae for relation letters as well.

Thus there is a way of interpreting the formulae of the *Begriffsschrift* that is faithful to the usual meanings of the logical operators and on which each comprehension axiom turns out to say something that can also be expressed by a sentence of the form "if there is something ..., then there are some things such that anything ... is one of them and any one of them is something ..." Each sentence of this form, it seems fair to say, expresses a *logical* truth if any sentence of English does. It would, of course, be folly to offer a definition of logical truth—as Jerry Fodor once said, failing to take his own advice, "Never give necessary and sufficient conditions for *anything*"—but I think one would be hard pressed to differentiate "if there is a rock, then there are some things such that any rock is one of them and any one of them is a rock" from "if there is a rock, then there is something such that if it is not a rock, then it is a rock" on the ground that the former but not the latter expresses a logical truth or on the ground that an intuition is required to see the truth of the former but not the latter.

Three final remarks about definite plural descriptions:

Valid inferences using the construction "there are some things such that ... they ..." that cannot be represented in first-order logic are not hard to come by. The interplay between this construction and definite plural descriptions is well illustrated by the inference

> Every parent of someone blue is red.
> Every parent of someone red is blue.
> Yolanda is red.
> Xavier is not red.
> It is not the case that there are some persons such that
>> Yolanda is one of them,
>> Xavier is not one of them, and
>> every parent of any one of them is also one of them.
> Therefore, Xavier is a parent of someone red.

To see that this is valid, note that it follows from the premises and denial of the conclusion that Yolanda is either red or a parent of someone red, that Xavier is not, and that every parent of anyone who is red or a parent of someone red is also red or a parent of someone red. Thus there are some people, viz. the persons who are either red or a parent of someone red, such that Yolanda is one of them, Xavier is not one of them, and every parent of any one of them is also one of them, which contradicts the last premiss.

This inference may be represented in second-order logic:

$$\forall w \forall z (Bz \wedge wPz \rightarrow Rw)$$
$$\forall w \forall z (Rz \wedge wPz \rightarrow Bw)$$
$$Ry$$
$$\neg Rx$$
$$\neg \exists X (\exists z\, Xz \wedge Xy \wedge \neg Xx \wedge \forall w \forall z (Xz \wedge wPz \rightarrow Xw))$$
Therefore, $\exists z (xPz \wedge Rz)$.

In deducing the conclusion from the premisses in the *Begriffsschrift*, one would, of course, substitute $\{a : Ra \vee \exists z(aPz \wedge Rz)\}$ for the second-order variable X, thus making a move similar to those we have seen Frege make.

It appears that not much in general can be said about "atomic" sentences that contain definite plural descriptions but do not express statements of identity. "The rocks rained down," for example, does not mean "Each of the rocks rained down." However, if the rocks rained down and the rocks under discussion are the items in pile x, then the items in pile x certainly rained down. If we have learned anything at all in philosophy, it is that it is almost certainly a waste of time to seek an analysis of "The rocks rained down" that reduces it to a first-order quantification over the rocks in question. It is highly probable that an adequate semantics for sentences like "They rained down" or "the sets possessing a rank exhaust the universe" would have to take as primitive a new sort of predication in which, for example "rained down" would be predicated not of particular objects such as this rock or that one, but rather of these rocks or those. Thus it would appear hopeless to try to say anything more about the meaning of a sentence of the form "The Ks M" other than that it means that there are some things that are such that they are the Ks and they M. The predication "they M" is probably completely intractable.

About statements of identity, though, something useful if somewhat obvious can be said: "The Ks are the Ls" is true if and only if there is at least one K, there is at least one L, and every K is an L and vice versa: $\exists x\, Kx \wedge \exists x\, Lx \wedge \forall x(Kx \leftrightarrow Lx)$. "They are the Ks" can also be naturally rendered with the aid of a free second-order variable X: $\exists x\, Xx \wedge \forall x(Xx \leftrightarrow Kx)$. And of course if some things are the Ks and are also the Ls, then the Ks are the Ls. Frege was not far wrong when he laid down Basic Law V. Of course, from time to time, there will be no set of (all) the Ks, as the sad history of Basic Law V makes plain. We cannot always pass from a predicate to an extension of the predicate, a set of things satisfying the predicate. We can, however, always pass to the things satisfying the predicate (if there is at least one), and therefore we cannot always pass from the things to a set of them.

Appendix: proofs of 108, 129, and 133

Definitions:

$$\text{Her}(F) \qquad \forall d \forall a (Fd \wedge dfa \rightarrow Fa) \qquad\qquad (69 \text{ in } \textit{Begriffsschrift})$$
$$\text{In}(x, F) \qquad \forall a (xfa \rightarrow Fa)$$
$$xf^*y \qquad \forall F(\text{Her}(F) \wedge \text{In}(x, F) \rightarrow Fy) \quad (76)$$
$$xf^*_=y \qquad xf^*y \vee x = y$$
$$\text{FN} \qquad \forall d \forall e \forall a (dfe \wedge dfa \rightarrow a = e) \qquad (115)$$

108 $zf^*_=y \wedge yfv \rightarrow zf^*_=v.$

Proof. Assume $zf^*_=y$, yfv, $\text{Her}(F)$ and $\text{In}(x, F)$. If zf^*y, then Fy, and by yfv and $\text{Her}(F)$, Fv; but if $y = z$ then $\text{In}(y, F)$ and again Fv, as yfv. Thus zf^*v, whence $zf^*_=v.$ ■

129 $\text{FN} \wedge (yf^*m \vee mf^*_=y) \wedge yfx \rightarrow (xf^*m \vee mf^*_=x).$

Proof. Assume FN, $(yf^*m \vee mf^*_=y)$, and yfx. We must show $xf^*m \vee mf^*_=x$. Suppose yf^*m. By 110 we need only show $\forall a(yfa \rightarrow xf^*_=a)$, for then $xf^*_=m$, whence xf^*m or $m = x$, and then $xf^*m \vee mf^*_=x$. So suppose yfa. Since yfx and FN, $x = a$, whence $xf^*_=a$. Now suppose $mf^*_=y$. We show mf^*x, whence $mf^*_=x$. Assume $\text{Her}(F)$ and $\text{In}(m, F)$. We are to show Fx. If mf^*y, then since $\text{Her}(F)$ and $\text{In}(m, F)$, Fy, and then, since yfx and $\text{Her}(F)$, Fx. But if $y = m$, then from yfx, mfx, whence again Fx, since $\text{In}(m, F)$. ■

The second main theorem of the *Begriffsschrift* is:

133 $\text{FN} \wedge xf^*m \wedge xf^*y \rightarrow [yf^*m \vee y = m \vee mf^*y].$

Proof after four lemmas.

Lemma 1 $bfa \rightarrow bf^*a.$ (91)

Proof. Suppose bfa. Assume $\text{Her}(F)$, $\text{In}(b, F)$; show Fa. Since bfa and $\text{In}(b, F)$, Fa. ■

Lemma 2 $cf^*d \wedge df^*a \rightarrow cf^*a.$ (98)

Proof. Suppose cf^*d and df^*a. Assume $\text{Her}(F)$ and $\text{In}(c, F)$; show Fa. Since cf^*d, $\text{Her}(F)$, and $\text{In}(c, F)$, Fd. If dfb, then since $\text{Her}(F)$, Fb; thus $\text{In}(d, F)$. Since $\text{Her}(F)$ and df^*a, Fa. ■

Lemma 3 $[c = d \vee cf^*d] \wedge dfa \rightarrow [c = a \vee cf^*a].$ (108)

Proof. Suppose $[c = d \vee cf^*d]$ and dfa. If $c = d$, then cfa, whence cf^*a by Lemma 1; if cf^*d, then since dfa, df^*a by Lemma 1, and by Lemma 2, cf^*a again. In any event, $c = a \vee cf^*a$. ∎

Lemma 4 $FN \wedge cfb \wedge cf^*m \rightarrow [b = m \vee bf^*m]$. (124)

Proof. Suppose FN and cfb. Let $F = \{z : b = z \vee bf^*z\}$. Suppose $[b = d \vee bf^*d]$ and dfa. By Lemma 3, $[b = a \vee bf^*a]$. Thus $\text{Her}(F)$. If cfa, then by FN, $b = a$, whence $b = a \vee bf^*a$; thus $\text{In}(c, F)$. Therefore if cf^*m, Fm, i.e. $b = m \vee bf^*m$. ∎

Proof of the Theorem. Suppose FN. Let $F = \{z : zf^*m \vee z = m \vee mf^*z\}$. Suppose $[df^*m \vee d = m \vee mf^*d]$ and dfa. If df^*m, then by Lemma 4, $[a = m \vee af^*m]$, whence $[af^*m \vee a = m \vee mf^*a]$; and if $d = m \vee mf^*d$, then $m = d \vee mf^*d$, and by Lemma 3, $m = a \vee mf^*a$, whence again $[af^*m \vee a = m \vee mf^*a]$. Thus $\text{Her}(F)$. Now suppose xf^*m. Assume xfa. By Lemma 4, $[a = m \vee af^*m]$, whence $[af^*m \vee a = m \vee mf^*a]$. Thus $\text{In}(x, F)$. At last, suppose xf^*y. Then Fy, i.e. $yf^*m \vee y = m \vee mf^*y$. ∎

11

Saving Frege from Contradiction

In §68 of *Die Grundlagen der Arithmetik* Frege defines the number that belongs to the concept F as the extension of the concept "equinumerous (*gleichzahlig*) with the concept F." In sections that follow he gives the needed definition of equinumerosity in terms of one-one correspondence, and in §73 attempts to demonstrate that the number belonging to F is identical with that belonging to G if and only if F is equinumerous with G. In view of Hume's well-known "standard by which we can judge of the equality and proportion of numbers,"[1] we may call the statement that the numbers belonging to F and G are equal if and only if F is equinumerous with G (or the formalization of this statement) *Hume's principle*. As we shall see, Frege's attempt to demonstrate Hume's principle, which is vital to the development of arithmetic sketched in the next ten sections of the *Grundlagen*, cannot be considered successful. We begin with a look at Frege's attempted proof before turning to our main concern, which is with two ways of repairing the damage to his work caused by the discovery of Russell's paradox.

Frege writes,

> On our definition, what has to be shown is that the extension of the concept "equinumerous with the concept F" is the same as the extension of the concept "equinumerous with the concept G," if the concept F is equinumerous with the concept G. In other words: it is to be proved that, for F equinumerous with

From *Proceedings of the Aristotelian Society*, 87 (1986/87): 137–151. Reprinted by courtesy of the Editor of the Aristotelian Society: ©1986/87.

Research for this paper was carried out under a grant from the National Science Foundation.

[1] "When two numbers are so combin'd, as that the one has always an unite answering to every unite of the other, we pronounce them equal," *Treatise*, I, III, I.

G, the following two propositions hold good universally: if the
concept H is equinumerous with the concept F, then it is also
equal to the concept G; and ... [conversely] ....

The sophisticated definition of numbers as extensions of certain concepts
of concepts and extensive use of binary relations found in the *Grundlagen*
are evidence that Frege was there committed to the existence of objects of
all finite types: "objects," the items of the lowest type 0, and, for any types
$t_1, \ldots, t_n$, relations of type $(t_1, \ldots, t_n)$ among items of types $t_1, \ldots, t_n$. An
item of type (t), for some type t, is called a concept. Concepts of type (0)
are called "first level concepts"; those of type $((0))$, "second level concepts."
The relation borne by an object x to a concept F when x falls under F
is of type $(0, (0))$; the relation η defined below is of type $((0), 0)$. It seems
clear that Frege accepted a comprehension principle governing the existence
of relations, according to which for any sequence of variables $x_1, \ldots, x_n$
of types $t_1, \ldots, t_n$ and any predicate $A(x_1, \ldots, x_n)$ (possibly containing
parameters) there is a relation of type $(t_1, \ldots, t_n)$ holding among those
items of types $t_1, \ldots, t_n$ satisfying the predicate, and only those. This
principle can be proved from the rule of substitution Frege used in the
Begriffsschrift. Thus, in view of the predicate "F is equinumerous with
G" (F a first level concept parameter, G a first level concept variable),
Frege concludes that there is a second level concept under which fall all
and only those first level concepts that are equinumerous with (the value
of the parameter) F.

It also seems clear that at the time he wrote the *Grundlagen*, Frege held
that for each concept C of *whatever* type, there is a special object $'C$, the
extension of C. Thus extensions are objects; and the number belonging
to the first level concept F is defined by Frege to be the extension of a
certain second level concept, the one under which fall all and only those
first level concepts equinumerous with F. We shall often abbreviate "(is)
equinumerous with": eq.

The announced task of §73 is to show that the number belonging to the
concept F, NF for short, $= NG$ if F eq G. Since Frege has defined NF
as $'$eq F, what must be shown is that $'$eq $F = {}'$eq G under the assumption
that F eq G. But almost all of §73 is devoted to showing that if H eq F,
then H eq G and observing that a similar proof shows that if H eq G, then
H eq F. Frege takes it that showing these two propositions is sufficient; he
writes "in other words." In a footnote he adds that a similar proof can be
given of the converse, that F eq G if $NF = NG$. And of course we know
exactly how the proof would go: "On our definition what must be shown
is that if $'$eq $F = {}'$eq G, in other words, if the following two propositions
hold good universally: if H eq F then H eq G and if H eq G then H eq F,

then F eq G. But since F eq F, by the first of these alone, F eq G."

Why did Frege suppose that one could pass so freely between "$'$eq $F = $ $'$eq G" and "for all H, H eq F iff H eq G"? It seems most implausible that any answer could be correct other than: because he thought it evident that for *any* concepts C and D of the same type (t), $'C = 'D$ if and only if for all items X of type t, CX iff DX.

Notoriously, this assumption generates Russell's paradox (in the presence of the comprehension principle, whose validity I assume). It is noteworthy that the proof Frege gave of the inconsistency of the system of his *Grundge-setze der Arithmetik* resembles Cantor's proof that there is no one-one map-ping of the power set of a set into that set rather than the version of the paradox that Russell had originally communicated to him. Of course in his second letter to Frege, well before Frege came to write the appendix to the *Grundgesetze*, where Frege's proof appears, Russell had explained to him the origins of the paradox in Cantor's work.

In the present notation, Frege's version of Russell's paradox runs: By comprehension, let R be the first level concept $[x : \exists F(x = 'F \wedge \neg Fx)]$. Consider the object $'R$, which is the extension of R. If $\neg R \, 'R$, then since for all F, $'R = 'F \rightarrow F \, 'R$, So $R \, 'R$. But then for some F, $'R = 'F$ and $\neg F \, 'R$. Thus by the principle about extensions mentioned two paragraphs back, $\forall x(Rx \leftrightarrow Fx)$. Thus $\neg R \, 'R$, contradiction.

Since Frege defines numbers as the extensions of *second* level concepts, it might be hoped that the Russell paradox does not threaten Frege's deriva-tion of arithmetic in the *Grundlagen*, for to prove the main proposition of §73, $'[H : H$ eq $F] = \, '[H : H$ eq $G]$ iff F eq G, he needs only the principle: for any second level concepts C, D, $'C = 'D$ iff for all first level H, CH iff DH. Notice the difference between this principle—call it (VI)—and the instance of (V) in which $t = 0$ that leads to Russell's paradox: for any first level concepts F, G, $'F = 'G$ iff for all objects x, Fx iff Gx. Part of the cause of the Russell paradox is that certain extensions are in the range of the quantified variable on the right side of (V). Since this is not the case with (VI), might (VI) then be consistent?

No. Define η by: $F\eta x$ iff for some second level concept D, $x = 'D$ and DF. By comprehension one level up, let $C = [F : \exists x(\neg F\eta x \wedge Fx)]$. By comprehension at the lowest level, let $X = [x : x = 'C]$. Suppose $X\eta 'C$. By the definition of η, for some D, $'C = 'D$ and DX, whence by (VI) CX. By the definition of C, for some x, $\neg X\eta x$ and Xx. By the definition of X, $x = 'C$, and therefore $\neg X\eta 'C$. Thus $\neg X\eta 'C$, whence for every D, if $'C = 'D$ then $\neg DX$. Therefore $\neg CX$. But by the definition of C, for every x such that Xx, $X\eta x$, and since $X \, 'C$, $X\eta 'C$, contradiction. As with the Russell paradox, it is the assumption that $'$ is one-one that causes the trouble.

Thus not only is (V) in full generality inconsistent, so is the apparently weaker (VI). But Frege does not need the full strength of (VI) to prove that $NF = NG$ iff F eq G. On the basis of the following proposition:

(Numbers) $\forall F \exists! x \forall H (H\eta x \leftrightarrow H \text{ eq } F))$

he can define NF as the unique object x such that for all concepts H, $H\eta x$ iff H eq F and then easily prove from this definition that $NF = NG$ iff F eq G.

Numbers expresses a proposition to whose truth Frege was committed. It is a proposition about concepts and objects couched in the language of second-order logic to which one new relation, η, has been added. ("eq" is of course definable in second-order logic in the standard way.) Thus it is involved with higher-order notions or with notions not expressible in the language of Frege's *Begriffsschrift* if at all, only in that η is a relation of concepts to objects. Notice that for any concept F the x (unique, according to Numbers) such that for all concepts H, $H\eta x$ iff H eq F will be an extension, for since F eq F, $F\eta x$, and thus for some C, $x = {}'C$ (and CF). The chief virtue of Numbers, though, is that it is formally consistent (as John Burgess,[2] Harold Hodes,[3] and the author[4] have noted).

We may see this as follows. Let the object variables in Numbers range over all natural numbers, the concept variables range over all sets of natural numbers and for all n, the n-ary relation variables range over all n-ary relations of natural numbers. (We are thus defining a "standard" model for Numbers.) Let η be true of a set S of natural numbers and a natural number n if and only if either for some natural number m, S has m members and $n = m + 1$ or S is infinite and $n = 0$. So interpreted, Numbers is true.

For let S be an arbitrary set of natural numbers. Let $n = m + 1$ if S is finite and has m members; let $n = 0$ otherwise. Then for any set U of natural numbers, $U\eta n$ holds iff either for some m, U has m members and $n = m + 1$ or U is infinite and $n = 0$, if and only if U and S have the same number of members, if and only if U eq S holds. The uniqueness of n follows from the definition of η and the fact that S eq S holds.

Much of the interest of the proof just given lies in the fact that it can be formalized in second-order arithmetic. Let $Eq(H, F)$ be the standard formula of second-order logic defining the relation "there is a one-one correspondence between U and S." The relation "S is infinite and $n = 0$ or for some m, S has m members and $n = m + 1$" can be defined by a formula $Eta(F, x)$ of second-order arithmetic in such a way that the sentence

$$\forall F \exists! x \forall H (\text{Eta}(H, x) \leftrightarrow \text{Eq}(H, F))$$

[2] (Burgess, 1984).
[3] (Hodes, 1984), p. 138.
[4] In Article 12 below.

is *provable* in second-order arithmetic. Thus we have a relative consistency proof: a proof of a contradiction in the result of adjoining the formalization

$$\forall F \exists ! x \forall H (H\eta x \leftrightarrow \text{Eq}(H, F))$$

of Numbers (with $H\eta x$ now taken as an atomic formula) to any standard axiomatic system of second-order logic could immediately ("primitive recursively") be transformed into a proof of a contradiction in second-order formal arithmetic. It is pointless to try to describe how unexpected the discovery of a contradiction in second-order arithmetic would be. Since Hume's principle is a theorem of a definitional extension of the second-order theory whose sole axiom is Numbers, it too is consistent (relative to the consistency of second-order arithmetic).

The distance between Numbers and Hume's principle is certainly not all that great: Numbers provides the justification for the introduction of the functor N, "the number belonging to"; Numbers also follows from Hume's principle when $F\eta x$ is defined as $x = NF$. One of Numbers's minor virtues is that it encapsulates the only assumption concerning the existence of extensions that Frege actually needs. For once Frege has Hume's principle in hand, he needs nothing else.

In §§74–83 of the *Grundlagen*, Frege outlines the proofs of a number of propositions concerning (what we now call) the natural numbers, including the difficult theorem that every finite number has a successor. (Formalizations of) all of these can be proved in axiomatic second-order logic from Hume's principle in more or less the manner outlined in these ten sections of the *Grundlagen*. I am uncertain whether Frege was aware that Hume's principle was all he needed; his puzzling remark at the end of the *Grundlagen* about attaching no decisive importance to the introduction of extensions of concepts may be taken as some evidence that he knew this.

It's a pity that Russell's paradox has obscured Frege's accomplishment in the *Grundlagen*. It's utterly remarkable that the whole of arithmetic can be deduced in second-order logic from this one simple principle, which might appear to be nothing more than a definition. Of course, Hume's principle isn't a definition, since "NF" and "NG" are intended to denote objects in the range of the first-order variables. (Cf. Wright's book *Frege's Conception of Numbers as Objects*.[5]) And as Frege's work shows, Hume's principle is much more powerful than we might have supposed it to be, implying, with the aid of second-order logic, the whole of second-order arithmetic (while failing to imply $\perp$).

In fact, that Hume's principle is consistent can easily come to seem like a matter of purest luck. Suppose we do for isomorphism of (binary) relations

---

[5](Wright, 1983).

what we have just done for the notion of equinumerosity of concepts: adjoin
to second-order logic an axiom

(OrdType) $^-R = {}^-S \leftrightarrow R$ iso S,

with $^-$ a function sign that takes a binary relation variable and makes a
term of the type of object variables, and R iso S some formula expressing
the order-isomorphism of the relations that are values of the variables R
and S. In other words, suppose we introduce in the obvious way what
Cantor called "order types" and Russell "relation numbers." It would, I
imagine, be the obvious guess that if Hume's principle is consistent, then
so is OrdType, which states that the order types of two relations are the
same iff the relations are order-isomorphic.

In §§85 and 86 of the *Grundlagen* Frege takes Cantor to task for having
appealed to "inner intuition" instead of providing definitions of *Number*
and *following in a series*. Frege adds that he thinks he can imagine how
these two concepts could be made precise. One would have liked to see
Frege's account of Cantor's notions; one cannot but suspect that in order
to reproduce Cantor's theory of ordinal numbers, Frege would have derived
OrdType from a (possibly tacit) appeal to (V).

Doing so would have landed him in trouble deeper than any he was in
in the *Grundlagen*, however, and not just because of the appeal to (V).
For the guess that OrdType is consistent if Hume's principle is consistent
is wrong. As Hodes has also observed, OrdType leads to a contradiction
via the reasoning of the Burali–Forti paradox. Thus although Numbers is
consistent, a principle no less definitional in appearance and rather similar
in content turns out to be inconsistent. In view of the inconsistency of (VI)
and OrdType, the consistency of Hume's principle is sheer luck.

To show that arithmetic follows from Hume's principle, or its near relation
Numbers, is to give a profound analysis of arithmetic, but it is not to base
arithmetic on a principle strikingly like Frege's Rule (V). We know from
Russell's and Cantor's paradoxes that there can be no function from (first
level) concepts to objects that assigns different objects to concepts under
which different objects fall. Identifying concepts under which the same
objects fall, we may say that there is no one-one function from concepts
into objects. But the function denoted by N is a particularly non-one-one
function. With the exception of $[x : x \neq x]$, every concept shares its number
with infinitely many other concepts. One might wonder whether one could
base arithmetic on a function assigning objects to concepts which, though
necessarily not one-one, fails to be one-one at only one of its values. We'll
see how to do this below.

In the appendix to the second volume of his *Grundgesetze,* Frege asks:

Is it always permissible to speak of the extension of a concept,

of a class? And if not, how do we recognize the exceptional
cases? Can we always infer from the extension of one concept's
coinciding with that of a second that every object which falls
under the first concept also falls under the second? These are
the questions raised by Mr. Russell's communication.

Before showing how Russell's paradox could be deduced in the system of
the *Grundgesetze,* he declares:

> Thus there is no alternative but to recognize the extensions of
> concepts, or classes, as objects in the full and proper sense of
> the word, while conceding that our interpretation hitherto of
> the words "extension of a concept" is in need of correction.

After showing that his rule (Vb) leads to Russell's paradox, Frege proves
that every function from concepts to objects assigns the same value to some
pair of concepts under which different objects fall. He observes that the
proof is "carried out without the use of propositions or notations whose
justification is in any way doubtful" and adds that

> this simply does away with extensions in the generally received
> sense of the term. We may not say that in general the expression
> "the extension of one concept coincides with that of another"
> means the same as the expression "every object falling under
> the first also falls under the second and conversely."

Frege then proposes a repair. In place of the defective rule (V), assume
(V'), which we may put: the extensions of F and G are identical iff the
same objects *other than those extensions* fall under F and G. He remarks
that "Obviously this cannot be taken as defining the extension of a concept
but merely as stating the distinctive property of this second level function."

It is well known that Frege's proposed repair fails. A particularly useful
discussion of the failure is found in Resnik's book *Frege and the Philosophy
of Mathematics*.[6] I want to consider an alternative repair to the *Grundgesetze* suggested by the second question asked in its appendix: How do we
recognize the exceptional cases?

Frege does not in fact offer an answer to his question. Although he
does discuss certain exceptional cases in the appendix, they are not the
ones referred to in his question, which are, presumably, the concepts that
lack an extension in the customary sense of the term. The exceptions
Frege discusses are not concepts but certain objects, namely, extensions of
concepts.

---

[6](Resnik, 1980).

But there is a simple answer that Frege might have given, one that uses only such notions as were available in 1908. Identification of the exceptional concepts will suggest a replacement for rule (V) which Frege might well have found perfectly acceptable, and which seems no more ad hoc than Frege's own rule (V′). The defect Russell revealed could have been repaired rather early, and by a patch that is really quite simple and closely related to ideas found in Cantor's work, some of which, at least, was familiar to Frege.

I shall not discuss the question whether the repair vindicates logicism. I doubt that anything can do that. I merely wish to claim that the repair I shall give should have been no less acceptable to Frege than the one he actually offered.

We'll begin the description of the repair with a bit of stipulation. Let's detach the existence of extensions from the term "coextensive" and say that a concept F is *coextensive with* a concept G if and only if all objects that fall under F fall under G and vice versa. Five more definitions follow, of "subconcept," "goes into," "V," "small" and "similar."

Let us call a concept F a *subconcept of* a concept G if every object that falls under F falls under G. Let us say that a concept F *goes into* G if F is equinumerous with a subconcept of G. If F is a subconcept of G, then F goes into G; if F goes into G and G goes into H, then F goes into H. It can be shown that if F and G go into each other, then they are equinumerous.

Let V be the concept, $[x : x = x]$, *identical with itself*. And let us say that a concept F is *small* if V does not go into F. V is not small. If F goes into G and G is small, then F is small; thus any subconcept of a small concept is small and any concept equinumerous or coextensive with a small concept is small. Let us say that F is *similar to* G iff (F is small $\lor$ G is small $\rightarrow$ F is coextensive with G).

We want now to see that *is similar to* is an equivalence relation. Reflexivity and symmetry are obvious. As for transitivity, suppose that F is similar to G and G to H. If F is small, then F is coextensive with G (for F is similar to G), thus G is small, and then G is coextensive with H; thus F is coextensive with H. And in like manner, but going the other way, if H is small, F is coextensive with H. Thus F is similar to H.

We now suppose that associated with each concept F, there is an object $*F$, which I will call the *subtension* of F, and that as extensions were supposed to be in one-one correspondence with equivalence classes of the equivalence relation *coextensive with*, so subtensions are in one-one correspondence with equivalence classes of the equivalence relation *similar to*; thus the principle (New V) holds: $*F = *G$ iff F is similar to G.

In view of the "Julius Caesar problem" it may be uncertain whether (New V) can be taken as *defining* subtensions, but like (V) and unlike (V′) it does not merely state the distinctive property of a certain second level function.

(V) and our replacement (New V) explain in a non-circular way, as (V′) did not, when objects given as extensions or subtensions of concepts are identical; the statements of the identity conditions do not contain expressions explicitly referring to those very extensions or subtensions.

Moreover, (New V) enables us to define the "exceptional" concepts quite easily, as those that are not small. For it follows from (New V) that for every concept F, if F is small, then for every concept G, $*F = *G$ iff F is coextensive with G. Furthermore if F is not small then there is a concept G not coextensive with F but such that $*F = *G$; of course any such G will itself fail to be small. (Since F is not small, F is equinumerous with V; but as we shall see, V is equinumerous with V $- 0$ (defined below). Thus F is equinumerous with one of its proper subconcepts G; since G is not small, $*F = *G$.)

We must now make it plausible that arithmetic can be developed in second-order logic from (New V) alone. There are many ways to do this; perhaps the easiest is to develop "finite set theory" from (New V) taking the development of arithmetic from finite set theory for granted.

Following Frege, let us say that $x \in$ (is a member of) y, if for some F, $y = *F$ and Fx. And let us call an object y a *set* if $y = *G$ for some *small* concept G. Thus if y is a set and for some concept H, $y = *H$, then $x \in y$ iff Hx.

Again, *à la* Frege, let $0 = *[x : x \neq x]$. Since 0 is an object, $[x : x \neq x]$ is small and 0 is therefore a set. For all x, not: $x \in 0$. $*V$, however, is not a set. Therefore there are at least two objects. Thus for any object y, the concept $[x : x = y]$ is small; let $\{y\} = *[x : x = y]$. For any object y, $\{y\}$ is a set. (So $\{*V\}$ is a set even though $*V$ is not.)

For any concept F and any object y, let $F + y$ be the concept $[x : Fx \lor x = y]$ and $F - y$ the concept $[x : Fx \land x \neq y]$. We now want to see that if F is small, so is $F + y$, for any object y.

We first observe since $0 \neq \{z\}$, V goes into V $- 0$ via the map which sends each object x into $\{x\}$. Suppose that $F + y$ is not small. Then V goes into $F + y$ via the map φ which, switching one or two values of φ if necessary, we may assume sends 0 to y. Then V $- 0$ goes into F via (a restriction of) φ. Since V goes into V $- 0$, V goes into F and F is not small. It follows that if F is small, so is $F + y$.

For any objects z, w, let $z + w = *[x : x \in z \lor x = w]$. Then if z is a set, so is $z + w$; $x \in z + w$ iff $x \in z \lor x = w$.

As in *Grundlagen* §83, we may define $HF = [x : \forall F(F0 \land \forall z \forall w(Fz \land Fw \rightarrow Fz + w) \rightarrow Fx)]$. An induction principle for HF follows directly: to show that all HF objects fall under a certain concept F, it suffices to show that 0 does, and that $z + w$ does whenever z and w do. Thus all HF objects together with all of their members are HF sets.

The axioms of (second-order) General Set Theory are:

Extensionality: $\forall x \forall y(\forall z(z \in x \leftrightarrow z \in y) \rightarrow x = y)$,
Adjunction: $\forall w \forall z \exists y \forall x(x \in y \leftrightarrow x \in z \lor x = w)$, and
Separation: $\forall F \forall z \exists y \forall x(x \in y \leftrightarrow x \in z \land Fx)$.

These axioms all hold when relativized to HF. For extensionality, note that two HF sets coincide if the same HF sets belong to both; separation is easily proved by induction on z. Second-order arithmetic can now be deduced in the usual way from General Set Theory. It is of some interest to note that the relativizations of the remaining axioms of Zermelo–Fraenkel set theory plus Choice minus Infinity can also be deduced from (New V).

Note also that the derivation of General Set Theory from (New V) is quite elementary, not much more difficult than it would have been from (V). We have had to check that certain subtensions were sets, but these checks were easily made. And although equinumerosity figures in the definition of the key notion of smallness, the Schröder–Bernstein theorem, or the technique of its proof, is nowhere used.

The hereditarily finite sets are the members of the smallest set A containing all finite subsets of A. Of course the null set $\emptyset$ is hereditarily finite, as are $\{\emptyset\}$, $\{\{\emptyset\}\}$, $\{\emptyset, \{\emptyset\}\}$, etc. An alternative characterization of the hereditarily finite sets is that they are the members of the smallest set containing $\emptyset$ and containing $z \cup \{w\}$ whenever it contains z and w. Our construction shows that the hereditarily finite sets can be seen as "constructed from" the relation *is similar to* as the finite cardinals arise from the relation *is equinumerous with*, and as extensions were supposed to arise from *is co-extensive with*. Truth-values arise in a similar manner from *is materially equivalent to*, via the axiom: $Vp = Vq \leftrightarrow (p \leftrightarrow q)$.

When the natural numbers or the hereditarily finite sets are thus "constructed" from equinumerosity or similarity, other objects are constructed too. We have already met the non-set $*V$. On the construction of the *Grundlagen*, along with the usual natural numbers some funny numbers arise, among them the number NV of things there are, the number $N[x : \exists F x = NF]$ of numbers there are, and the number of finite numbers there are. Frege acknowledged the last of these, dubbing it ∞_1, but he must admit all of them if he wants to define 0 as the number belonging to the concept *not identical with itself*. (It is consistent with Numbers that all three are distinct; it is also consistent that they are all identical.)

It is often said that Zermelo–Fraenkel set theory is motivated by a doctrine of "limitation of size": a collection is a set if it is "small" or "not too big," a collection being "too big" if it is equinumerous with the collection of *all* sets. The notion of smallness is sometimes taken to motivate the axioms of set theory: it is thought that if certain sets are small, then certain other sets formed from them by various operations will also be small. (Michael Hallett has effectively criticized the thought that the power set operation produces small sets from small sets.[7]) In most treatments of set theory, the idea of smallness is left at the motivational level. Our construction explicitly incorporates it into our axiom (New V) governing subtensions. Another respect in which our construction differs from that of ZF or its class-theoretic relatives is the combination of a "universal" object with the absence of a complement operation: for every x, $x \in *V$; but if for every $x \neq 0$, $x \in y$, then $0 \in y$ also.

It follows from (New V) that if F is small, then $*F = *G$ if and only if F and G are coextensive; if neither F nor G is small, then $*F = *G$ (for F and G then satisfy the definition of "similar"). Our construction, as we have noted, concentrates the non-one-oneness of the function $*$ in a single value, the object $*V$. A theorem of set theory throws some light on the question how non-one-one any function like $*$ from concepts to objects must be. It follows from the Zermelo–König inequality (which can be proved in ZF plus the axiom of choice) that for any infinite set x and function f from the power set of x into x, there is a member a of x such that there are at least as many subsets y of x such that $fy = a$ as there are subsets of x altogether. Thus (higher-order set theory implies that) any attempt to assign concepts (classes) to objects must assign to some one object as many concepts as there are concepts altogether. There is then a clear sense in which the failure of $*$ to be a one-one function is no worse than necessary and the replacement of extensions $'F$ by subtensions $*F$ is a minimal departure from the project of the *Grundgesetze*. The theorem also shows that project not to have been a near miss.

Although I have given an informal sketch of the derivation of General Set Theory from (New V), it is to be emphasized that this derivation can be carried out formally in axiomatic second-order logic in which the sole axiom (other than the standard axioms of second-order logic) is (New V). (Of course the rules of formation will guarantee that for each concept variable F, $*F$ is a term of the type of object variables.)

There remains a matter not yet attended to: the consistency of (New V). It should now be no surprise that (New V) is consistent (if second-order arithmetic is). Indeed, it is quite simple to provide a standard second-order model for (New V).

As in the proof of the consistency of Numbers, let the object variables range over all natural numbers. Since the model is standard and its domain is countably infinite, a subset X of the domain satisfies "is small" if and only if X is finite. We must now supply a suitable function τ from sets of

[7](Hallett, 1984).

natural numbers to natural numbers with which to interpret $*$.

Let D be some one-one map of all finite sets of natural numbers into the natural numbers. (The best known such D is given by: $D(X) =$ the number whose binary numeral, for every number x, contains a 1 at the 2^x's place iff $x \in X$.) Then if X and Y are finite sets of natural numbers, $1 + D(X) = 1 + D(Y)$ iff $X = Y$.

For any set X of natural numbers, let $\tau(X) = 0$ if X is infinite and $= 1 + D(X)$ if X is finite. Then $\tau(X) = \tau(Y)$ iff either X and Y are both infinite or $X = Y$, iff (X or Y is finite $\rightarrow X = Y$). Thus (New V) does indeed have a standard model: it is true over the natural numbers when $*$ is interpreted by τ. Utilizing the particular function D defined above, we can convert the foregoing argument into a proof of the consistency of (New V) (relative to that of second-order arithmetic) in the usual way.

How then does (New V) prevent Russell's paradox? Let's try to re-derive it: By comprehension, let R be the first level concept $[x : \exists F(x = *F \wedge \neg Fx)]$. If $\neg R * R$, then since for all F, $*R = *F \rightarrow F * R$, $R * R$. So $R * R$. So for some F, $*R = *F$ and $\neg F * R$. But we cannot show that $\forall x(Rx \leftrightarrow Fx)$ unless we can show that F or R is small, and this there is no way of doing if second-order arithmetic is consistent. The unsurprising conclusion is that R is not small. It is more interesting to note that since every number fails to fall under at least one concept of which it is the number, the Russellian number $N[x : \exists F(x = NF \wedge \neg Fx)]$ is (provably) identical with $N[x : \exists F\, x = NF]$, the number of numbers.

A piece of mathematics carried out in an inconsistent theory need not be vitiated by the inconsistency of the theory: it may be possible to develop the mathematics in a suitable proper subtheory. The development of arithmetic outlined in the *Grundlagen* can be carried out in the consistent theory obtained by adding Numbers to the system of *Begriffsschrift* as well as in the inconsistent system of the *Grundgesetze*. Consistent systems similar to, but stronger than, (New V) plus second-order logic can readily be given, e.g., by replacing "small" by "countable." It would be of some interest to find out how much of the mathematics done in the *Grundgesetze* can be reproduced in such systems.

12

The Consistency of Frege's *Foundations of Arithmetic*

Is Frege's *Foundations of Arithmetic* inconsistent? The question may seem to be badly posed. The *Foundations*, which appeared in 1884, contains no formal system like those found in Frege's *Begriffsschrift* (1879) and *Basic Laws of Arithmetic* (Vol. 1, 1893, Vol. 2, 1903). As is well known, Russell showed the inconsistency of the system of the *Basic Laws* by deriving therein what we now call Russell's paradox. The system of the *Begriffsschrift*, on the other hand, can plausibly be reconstructed as an axiomatic presentation of second-order logic, which is therefore happily subject to the usual consistency proof, consisting in the observation that the universal closures of the axioms and anything derivable from them by the rules of inference are true in any one-element model.[1] Since the *Foundations* contains no formal system at all, our question may be thought to need rewording before an answer to it can be given.

One might nevertheless think that, however reworded and badly posed or not, it must be answered yes. The *Basic Laws*, that is, the system thereof, *is* inconsistent and is widely held to be a formal elaboration of the mathematical program outlined in the earlier *Foundations*, which contains a more thorough development of its program than one is accustomed to find in programmatic works. Thus the inconsistency which Russell found

From *On Being and Saying: Essays in Honor of Richard Cartwright*, Judith Jarvis Thomson, ed., Cambridge: The MIT Press, 1987, pp. 3–20. Reprinted by kind permission of The MIT Press. Copyright ©1987 Massachusetts Institute of Technology.

The papers by Paul Benacerraf, Harold Hodes, and Charles Parsons cited below have been major influences on this one. I would like to thank Paul Benacerraf, Sylvain Bromberger, John Burgess, W. D. Hart, James Higginbotham, Harold Hodes, Paul Horwich, Hilary Putnam, Elisha Sacks, Thomas Scanlon, and Judith Jarvis Thomson for helpful comments. Research for this paper was carried out under grant SES–8607415 from the National Science Foundation.

[1](Russinoff, 1983).

in the later book must have been latent in the earlier one.

Moreover, the characteristic signs of inconsistency can be found in the use Frege makes in the *Foundations* of the central notions of "object," "concept," and "extension." Objects fall under concepts, but some extensions—numbers, in particular and crucially—contain concepts, and these extensions themselves are objects, according to Frege. Thus, although a division into two types of entity, concepts and objects, can be found in the *Foundations*, it is plain that Frege uses not one but two instantiation relations, "falling under" (relating some objects to some concepts) and "being in" (relating some concepts to some objects), and that both relations sometimes obtain reciprocally: The number 1 is an object that falls under "identical with 1," a concept that is in the number 1. Even more ominously (because of the single negation sign), the number 2 does not fall under "identical with 0 or 1," which is in 2. Thus the division of the *Foundations*'s entities into two types would appear to offer little protection against Russell's paradox.

It is not only Russell's paradox that threatens. Recall that Frege defines 0 as the number belonging to the concept "not identical with itself."[2] If there is such a number, would there not also have to be a number belonging to the concept "identical with itself," a *greatest* number? Cantor's paradox also threatens.

It is therefore quite plausible to suppose that it is merely through its lack of formality that the *Foundations* escapes outright inconsistency and that, when suitably formalized, the principles employed by Frege in the *Foundations* must be inconsistent.

This plausible and, I suspect, quite common supposition is mistaken, as we shall see. Although Frege freely assumes the existence of needed concepts at every turn, he by no means avails himself of extensions with equal freedom. With one or two insignificant but possibly revealing exceptions, which I discuss later, the *only* extensions whose existence Frege claims in the central sections of the *Foundations* are the extensions of higher level concepts of the form "equinumerous with concept *F*." (I use the term "equinumerous" as the translation of Frege's *gleichzahlig*.) It turns out that the claim that such extensions exist can be consistently integrated with existence claims for a wide variety of first level concepts in a way that makes possible the execution of the mathematical program described in Sections 68–83 of the *Foundations*. Indeed I shall now present a formal theory, *FA* ("Frege Arithmetic"), that captures the whole content of these

---

[2]Plurals find happy employment here, as elsewhere in the discussion of concepts: for example, instead of "the number belonging to the concept 'horse'," one can say "the number of horses." 0 is thus *defined* by Frege to be the number of things that are not self-identical. And Frege was right!

central sections and for which a simple consistency proof can be given, one that shows *why* FA is consistent.

FA is a theory whose underlying logic is standard axiomatic second-order logic written in the usual Peano–Russell logical notation. FA could have been presented as an extension of the system of Frege's *Begriffsschrift.* Indeed, there is some evidence that Frege thought of himself as translating *Begriffsschrift* notation into the vernacular when writing the *Foundations.* Not only does the later work abound with allusions and references to the earlier, along with repetitions of claims and arguments for its significance, when Frege defines the ancestral in Section 79, he uses the variables x, y, d, and F in exactly the same logical roles they had played in the *Begriffsschrift.*

FA is a system with three sorts of variable: first-order (or object) variables a, b, c, d, m, n, x, y, z, $\ldots$; unary second-order (or concept) variables F, G, H, $\ldots$; and binary second-order (or relation) variables φ, ψ,$\ldots$ The sole nonlogical symbol of the language of FA is η, a two-place predicate letter attaching to a concept variable and an object variable. (η is intended to be reminiscent of $\in$ and may be read "is in the extension." Frege's doctrine that extensions are objects receives expression in the fact that the second argument place of η is to be filled by an object variable.) Thus the atomic formulas of FA are of the forms Fx (F a concept variable), $x\varphi y$, and $F\eta x$. Formulas of FA are constructed from the atomic formulas by means of propositional connectives and quantifiers in the usual manner.

Identity can be taken to have its standard second-order definition: $x = y$ if and only if $\forall F(Fx \leftrightarrow Fy)$. Frege endorses Leibniz's definition ("$\ldots$ *potest substitui $\ldots$ salva veritate*") in Section 65 of the *Foundations* but does not actually do what he might easily have done, viz. state that Leibniz's definition of the identity of x and y can be put: y falls under every concept under which x falls (and vice versa).

The logical axioms and rules of FA are the usual ones for such a second-order system. Among the axioms we may specially mention (i) the universal closures of all formulas of the form

$$\exists F \forall x (Fx \leftrightarrow A(x)),$$

where $A(x)$ is a formula of the language of FA not containing F free; and (ii) the universal closures of all formulas of the form

$$\exists \varphi \forall x \forall y (x \varphi y \leftrightarrow B(x, y)),$$

where $B(x, y)$ is a formula of the language not containing φ free. Throughout Sections 68–83 of the *Foundations* Frege assumes, and needs to assume, the existence of various particular concepts and relations. The axioms (i) and (ii) are called comprehension axioms; these will do the work in FA of Frege's concept and relation existence assumptions.

The sole (nonlogical) axiom of the system FA is the single sentence

Numbers: $\forall F \exists! x \forall G(G\eta x \leftrightarrow F \text{ eq } G)$,

where F eq G is the obvious formula of the language of FA expressing the equinumerosity of the values of F and G, viz.

$$\exists \varphi [\forall y(Fy \rightarrow \exists! z(y\varphi z \wedge Gz)) \wedge \forall z(Gz \rightarrow \exists! y(y\varphi z \wedge Fy))].$$

Here the sign η is used for the relation that holds between a concept G and the extension of a (higher level) concept under which G falls; before we used the term "is in" for this relation and "contains" for its converse. In Section 68 Frege first asserts that F is equinumerous with G if and only if the extension of "equinumerous with F" is the same as that of "equinumerous with G" and then defines the number belonging to the concept F as the extension of the concept "equinumerous with the concept F." Since Frege, like Russell, holds that existence and uniqueness are implicit in the use of the definite article, he supposes that for any concept F, there is a unique extension of the concept "equinumerous with F." Thus the sentence Numbers expresses this supposition in the language of FA; it is the sole nonlogical assumption[3] utilized by Frege in the course of the mathematical work done in Sections 68–83.

How confident may we be that FA is consistent? Recent observations by Harold Hodes and John Burgess bear directly on this question. To explain them, it will be helpful to consider a certain formal sentence, which we shall call Hume's principle:

$$\forall F \forall G(NF = NG \leftrightarrow F \text{ eq } G).$$

Hume's principle is so called because it can be thought of as explicating a remark that Hume makes in the *Treatise* (I, III, I, par. 5), which Frege quotes in the *Foundations*:

> We are possest of a precise standard by which we can judge of the equality and proportion of numbers ... When two numbers are so combin'd, as that the one has always an unite answering to every unite of the other, we pronounce them equal ...

The symbol N in Hume's principle is a function sign which when attached to a concept variable makes a term of the same type as object variables; thus $NF = NG$ and $x = NF$ are well-formed. Taking $N \ldots$ as abbreviating "the number of $\ldots$s," we may read Hume's principle: The number of Fs is the

---

[3]It is nonlogical by my lights, though not, of course, by Frege's.

number of Gs if and only if the Fs can be put into one-one correspondence with the Gs. (As Hume said, more or less).

In his article "Logicism and the Ontological Commitments of Arithmetic,"[4] Hodes observes that a certain formula, which he calls "(D)" is satisfiable. He writes:

(D) $\begin{matrix} \forall X \exists x \\ \\ \forall Y \exists y \end{matrix}$ $(x = y \leftrightarrow X \text{ eq } Y)$

is satisfiable. In fact if we accept standard set theory, it's true.

(I have replaced Hodes's "$(Q_E z)(Xz, Yz)$" by "X eq Y." The label "(D)" . is missing from the text of his article.) Branching quantifiers, which are notoriously hard to interpret, may always be eliminated in favor of ordinary function quantifiers. Eliminating them from (D) yields the formula $\exists N \exists M \forall X \forall Y (NX = MY \leftrightarrow X \text{ eq } Y)$. Now (D) is satisfiable if and only if Hume's principle is satisfiable. For if (the function quantifier equivalent of) (D) holds in a domain U, then for some functions N, M, $\forall X \forall Y (NX = MY \leftrightarrow X \text{ eq } Y)$ holds in U; since $\forall Y (Y \text{ eq } Y)$ holds in U, so does $\forall Y (NY = MY)$, and therefore so does Hume's principle $\forall X \forall Y (NX = NY \leftrightarrow X \text{ eq } Y)$. Conversely, Hume's principle implies (D). Thus a bit of deciphering enables us to see that Hodes's claim is tantamount to the assertion that Hume's principle is satisfiable.

Hodes gives no proof that (D), or Hume's principle, is satisfiable. But Burgess, in a review of Crispin Wright's book *Frege's Conception of Numbers as Objects*,[5] shows that it is. He writes:

> Wright shows why the derivation of Russell's paradox cannot be carried out in $N^=$ [Wright's system, obtained by adjoining a version of Hume's principle to second-order logic], and ought to have pointed out that the system is pro[v]ably consistent. (It has a model whose domain of objects consists of just the cardinals zero, one, two, ... and aleph-zero.)[6]

It will not be amiss to elaborate this remark. To produce a model $\mathcal{M}$ for Hume's principle that also verifies all principles of axiomatic second-order logic, take the domain U of $\mathcal{M}$ to be the set $\{0, 1, 2, \ldots, \aleph_0\}$. To ensure that $\mathcal{M}$ is a model of axiomatic second-order logic, take the domain of the concept variables to be the set of all subsets of U, and similarly take the domain of the binary (or, more generally, n-ary) relation variables to be the

---

[4](Hodes, 1984), p. 138.

[5](Wright, 1983).

[6](Burgess, 1984). The text of the review has "probably consistent," which is an obvious misprint.

set of all binary (or n-ary) relations of U, that is, the set of sets of ordered pairs (or n-tuples) of members of U.

To complete the definition of $\mathcal{M}$, we must define the function f by which the function sign N is to be interpreted in $\mathcal{M}$. The *cardinality* of a set is the number of members it contains. U has the following important property: *The cardinality of every subset of U is a member of U.* (Notice that the set of natural numbers *lacks* this property.) Thus we may define f as the function whose value for every subset V of U is the cardinality of V. We must now see that Hume's principle is true in $\mathcal{M}$.

Observe that an assignment s of appropriate items to variables satisfies $NF = NG$ in $\mathcal{M}$ if and only if the cardinality of $s(F)$ equals the cardinality of $s(G)$ and satisfies F eq G in $\mathcal{M}$ if any only if $s(F)$ can be put into one-one correspondence with $s(G)$. Since the cardinality of $s(F)$ is the same as that of $s(G)$ if and only if $s(F)$ can be put into one-one correspondence with $s(G)$, every assignment satisfies $(NF = NG \leftrightarrow F$ eq $G)$ in $\mathcal{M}$, and $\mathcal{M}$ is a model for Hume's principle.

A similar argument shows the satisfiability of Numbers: Let the domain of $\mathcal{M}$ again be U, and let $\mathcal{M}$ specify that η is to apply to a subset V of U and a member u of U if and only if the cardinality of V is u. Then Numbers is true in $\mathcal{M}$. (On receiving the letter from Russell, Frege should have immediately checked into Hilbert's Hotel.)

(It may be of interest to recall the usual proof that the comprehension axioms (i) are true in standard models (like $\mathcal{M}$) for second-order logic: Let $A(x)$ be a formula not containing free F, and let s be an assignment. Let C be the set of objects of which $A(x)$ is true, and let s' be just like s except that $s'(F) = C$. Since $A(x)$ does not contain free F, s' satisfies $\forall x(Fx \leftrightarrow A(x))$ and s satisfies $\exists F \forall x(Fx \leftrightarrow A(x))$. Similarly for the comprehension axioms (ii).)

There is a cluster of worries or objections that might be thought to arise at this point: Does not the appeal to the natural numbers in the consistency proof vitiate Frege's program? How can one invoke the existence of the numbers in order to justify FA? There is a quick answer to this objection: You mean we *shouldn't* give a consistency proof? More fully: We are simply trying to use what we know in order to allay all suspicion that a contradiction is formally derivable in FA, about whose consistency anyone knowing the history of logic might well be quite uncertain. We are not attempting to show that FA is true.

But there is perhaps a more serious worry. At a crucial step of the proof of the consistency (with second-order logic) of the formal sentence called Hume's principle, we made an appeal to an informal principle connecting cardinality and one-one correspondence which can be symbolized as—Hume's principle. (We made this appeal when we said that the cardinality of $s(F)$ is the same as that of $s(G)$ if and only if $s(F)$ can be put into one-one correspondence with $s(G)$.) Should this argument then count as a *proof* of the consistency of Hume's principle? What assurance can any argument give us that a certain sentence is consistent, if the argument appeals to a principle one of whose formalizations is the very sentence we are trying to prove consistent?

The worry is by no means idle. We have attempted to prove the consistency of Hume's principle by arguing that a certain structure $\mathcal{M}$ is a model for Hume's principle; in proving that $\mathcal{M}$ is a model for Hume's principle we have appealed to an informal version of Hume's principle. A similar service, however, can be performed for the notoriously inconsistent naive comprehension principle $\exists y \forall x (x \in y \leftrightarrow \ldots x \ldots)$ of set theory: By informally invoking the naive comprehension principle, we can argue that all of its instances are true under the interpretation I under which the variables range over all sets that there are and $\in$ applies to a, b if and only if b is a set and a is a member of b. Let $\ldots x \ldots$ be an arbitrary formula not containing free y. (By the naive comprehension principle) let b be the set of just those sets satisfying $\ldots x \ldots$ under I. Then for every a, a and b satisfy $x \in y$ under I if and only if a satisfies $\ldots x \ldots$ under I. Therefore b satisfies $\forall x (x \in y \leftrightarrow \ldots x \ldots)$ under I, and $\exists y \forall x (x \in y \leftrightarrow \ldots x \ldots)$ is true under I. Thus I is a model of all instances of the naive comprehension principle. (Doubtless Frege convinced himself of the truth of the fatal Rule (V) of *Basic Laws* by running through some such argument.) Of course we can now see that, *pace* the principle, there is not always a set of just those sets satisfying $\ldots x \ldots$ But how certain can we be that the proof of the consistency of Hume's principle and FA does not contain some similar gross (or subtle) mistake, as does the "proof" just given of the consistency of the naive comprehension principle?

Let us first notice that the argument can be taken to show not merely that FA is consistent, but that *it is provable in standard set theory* that FA is consistent. (Standard set theory is of course ZF, Zermelo–Fraenkel set theory.) The argument can be "carried out" or "replicated" *in* ZF. Thus, if FA is inconsistent, ZF is in error. (Presumably the word "provably" in Burgess's observation refers to an informal, model-theoretic proof, which could be formalized in ZF, or to a formal ZF proof.) Thus anyone who is convinced that nothing false is provable in ZF must regard this argument as a proof that FA is consistent. Moreover, if ZF makes a false claim to the effect that FA, or any other formal theory, is consistent, then ZF is not merely in error but is itself inconsistent, for ZF will then certainly also make the correct claim that there exists a derivation of $\bot$ in FA. (Indeed systems much weaker than ZF, for example, Robinson's arithmetic Q, will then make that correct claim.)

Something even stronger may be said. We shall show that any derivation of an inconsistency in FA can immediately be turned into a derivation of an inconsistency in a well-known theory called "second-order arithmetic" or "analysis," about whose consistency there has never been the slightest doubt. In the language of analysis there are two sorts of variables, one sort ranging over (natural) numbers, the other over sets of and relations on numbers. The axioms of analysis are the usual axioms of arithmetic, a sentence expressing the principle of mathematical induction ("Every set containing 0 and the successor of every number contains every natural number"), and, for each formula of the language, a comprehension axiom expressing the existence of the set or relation defined by the formula.[7] If ZF is consistent, so is analysis; but ZF is stronger than analysis, and the consistency of analysis can be proved in ZF. It is (barely) conceivable that ZF is inconsistent; but unlike ZF, analysis did not arise as a direct response to the set-theoretic antinomies, and the discovery of the inconsistency of analysis would be the most surprising mathematical result ever obtained, precipitating a crisis in the foundations of mathematics compared with which previous "crises" would seem utterly insignificant.

Let us sketch the construction by which proofs of $\bot$ in FA can be turned into proofs of $\bot$ in analysis. The trick is to "code" $\aleph_0$ by 0 and each natural number z by $z+1$ so that the argument given may be replicated in analysis. It is easy to construct a formula $A(z, F)$ of the language of analysis that expresses the relation "exactly z natural numbers belong to the set F": Simply write down the obvious symbolization of "there exists a one-one correspondence between the natural numbers less than z and the members of F." Let $\mathrm{Eta}(F, x)$ be the formula

$$[\neg \exists z\, A(z, F) \wedge x = 0] \vee [\exists x(A(z, F) \wedge x = z + 1)].$$

Then, since $\exists! x\, \mathrm{Eta}(F, x)$ and

$$[\exists x(\mathrm{Eta}(F, x) \wedge \mathrm{Eta}(G, x)) \leftrightarrow F \text{ eq } G]$$

are theorems of analysis, so is the result

$$\forall F \exists! x \forall G(\mathrm{Eta}(G, x) \leftrightarrow F \text{ eq } G)$$

of substituting $\mathrm{Eta}(G, x)$ for $G\eta x$ in Numbers, as the following argument, which can be formalized in analysis, shows: Let F be any set of numbers. Let x be such that $\mathrm{Eta}(F, x)$ holds. Let G be any set. Then $\mathrm{Eta}(G, x)$ holds if and only if F eq G does. And since F eq F holds, x is unique. Of course each of the comprehension axioms of FA is provable in analysis under these substitutions, since they turn into comprehension axioms of analysis. Thus a proof of $\bot$ in FA immediately yields a proof of $\bot$ in analysis.

---

[7]A standard reference concerning analysis is sec. 8.5 of (Shoenfield, 1967).

It is therefore as certain as anything in mathematics that, if analysis is consistent, so is FA. Later we shall see that the converse holds. (A sketch of a major part of the proof of the converse was given by Frege, in the *Foundations*. Of course.) The connection between FA and Russell's paradox is discussed later. Since the possibility that analysis might be inconsistent at present strikes us as utterly inconceivable, we may relax in the certainty that neither Russell's nor any other contradiction is derivable in FA.

We now want to show that the definitions and theorems of Sections 68–83 of the *Foundations* can be stated and proved in FA, *in the manner indicated by Frege.* I am not sure that it is possible to appreciate the magnitude and character of Frege's accomplishment without going through at least some of the hard details of the derivation of arithmetic from Numbers, in particular those of the proof that every natural number has a successor, but readers who wish take it on faith that the derivation can be carried out in FA along a path *very* close to Frege's may skim over some of the next seventeen paragraphs. Do not forget that it is Frege himself who has made formalization of his work routine.

In the course of replicating in FA Frege's treatment of arithmetic, we shall of course make definitional extensions of FA. For example, as Frege defined the number belonging to the concept F as the extension of the concept "equinumerous to F," so we introduce a function symbol N, taking a concept variable and making a term of the type of object variables, and then define $NF = x$ to mean $\forall G(G\eta x \leftrightarrow F$ eq $G)$; the introduction of the symbol N together with this definition is of course licensed by Numbers. It will also prove convenient to introduce terms $[x : A(x)]$ for concepts: $[x : A(x)]t$ is to mean $A(t)$; $F = [x : A(x)]$ is to mean $\forall x(Fx \leftrightarrow A(x))$; $[x : A(x)]\eta y$ is to mean $\exists F(F = [x : A(x)] \wedge F\eta y)$; $[x : A(x)] = [x : B(x)]$ is to mean $\forall x(A(x) \leftrightarrow B(x))$, etc. The introduction of such terms is of course licensed by the comprehension axioms (i).

Sections 70–73 provide the familiar definition of equinumerosity. In 73, Frege proves Hume's principle. Note that the comprehension axioms (ii) provide the facts concerning equinumerosity needed for this theorem to be provable. Once Hume's principle is proved, *Frege makes no further use of extensions.*[8,9]

---

[8] See sec. VI of (Parsons, 1983a).

[9] In his estimable *Frege's Conception of Numbers as Objects*, Wright sketches a derivation of the Peano axioms in a system of higher-order logic to which a version of Hume's principle is adjoined as an axiom. Wright discusses the question of whether such a system would be consistent, attempts to reproduce various well-known paradoxes in such a system, is unsuccessful, and concludes on page 156 that "there are grounds, if not for optimism, at least for a cautious confidence that a system of the requisite sort is capable of consistent formulation." Wright's instincts are correct, as Hodes and Burgess have

In 72 Frege defines "number": "n is a number" is to mean "there exists a concept such that n is the number which belongs to it." In parallel, we make the definition in FA: $Zx \leftrightarrow \exists F(NF = x)$. In 74 Frege defines 0 as the number belonging to the concept "not identical with itself"; we define in FA: $0 = N[x : x \neq x]$. The content of 75 is given in the easy theorem of FA:

$$\forall F \forall G([\forall x \neg Fx \rightarrow ((\forall x \neg Gx \leftrightarrow F \text{ eq } G) \wedge NF = 0)]$$
$$\wedge [NF = 0 \rightarrow \forall x \neg Fx]).$$

In 76 Frege defines "the relation in which every two adjacent members of the series of natural numbers stand to each other."[10] Correspondingly, we define nSm (read "n succeeds m"):

$$\exists F \exists x \exists G(Fx \wedge NF = n \wedge \forall y(Gy \leftrightarrow Fy \wedge y \neq x) \wedge NG = m).$$

$\neg 0Sa$ immediately follows in FA from this definition: Zero succeeds nothing. In 77 Frege defines the number 1. We make the corresponding definition: $1 = N[x : x = 0]$. $1S0$ is easily derived in FA.

The theorems corresponding to those of 78 are proved without difficulty:

(1) $aS0 \rightarrow a = 1$,

(2) $NF = 1 \rightarrow \exists x\, Fx$,

(3) $NF = 1 \rightarrow (Fx \wedge Fy \rightarrow x = y)$,

(4) $\exists x\, Fx \wedge \forall x \forall y(Fx \wedge Fy \rightarrow x = y) \rightarrow NF = 1$,

(5) $\forall a \forall b \forall c \forall d(aSc \wedge bSd \rightarrow (a = b \leftrightarrow c = d))$,

(6) $\forall n(Zn \wedge n \neq 0 \rightarrow \exists m(Zm \wedge nSm))$.

Although Frege and we have now defined "number," defined 0 and 1, proved that they are different numbers, proved that "succeeds" is one-one, and proved that every non-zero number is a successor, "finite number," that is, "natural number," has not yet been defined; nor has it been shown that every natural number has a successor.

seen. It may be of interest to note that FA supplies the answer to a question raised by Wright on page 156 of his book. It is a theorem of FA that the number of numbers that fall under none of the concepts of which they are the numbers is *one*. (Zero is the only such number.)

[10]Note that, although Frege here introduces the expression "folgt in der natürlichen Zahlenreihe unmittelbar auf" for the *succeeds* relation, he will define "finite" number only at the end of Section 83.

In 79 Frege defines the ancestral of φ, "y follows x in the φ-series," as in the *Begriffsschrift*. Thus in FA we define $x\varphi^*y$:

$$\forall F(\forall a(x\varphi a \to Fa) \wedge \forall d\forall a(Fd \wedge d\varphi a \to Fa) \to Fy).$$

80 is a commentary on 79. At the beginning of 81 Frege introduces the terminology "y is a member of the φ-series beginning with x" and "x is a member of the φ-series ending with y" to mean: either y follows x in the φ-series or y is identical with x. Frege uses the phrase "in the series of natural numbers" instead of "in the φ-series" when φ is the converse of the succeeds relation. In FA we define mPn to mean nSm, $m < n$ to mean mP^*n, and $m \leq n$ to mean $m < n \vee m = n$. Frege defines "n is a finite number" only at the end of Section 83. In FA we define Fin n to mean $0 \leq n$.

In 82 and 83 Frege outlines a proof that every finite number has a successor. He adds that, in proving that a successor of n always exists (if n is finite), it will have been proved that "there is no last member of this series." (He obviously means the sequence of finite numbers.) This will certainly have been shown if it is also shown that no finite number follows itself in the series of natural numbers; in 83 Frege indicates that this proposition is necessary and how to prove it.

Frege's ingenious idea is that we can prove that every finite number has a successor by proving that if n is finite, the number of numbers less than or equal to n—in Frege's terminology "the number which belongs to the concept 'member of the series of natural numbers ending with n' "—succeeds n. Frege's outline can be expanded into a proof in FA of: Fin $n \to N[x : x \leq n]Sn$. Since $ZN[x : x \leq n]$ is provable in FA, so is (Fin $n \to \exists x(Zx \wedge xSn)$).

In 82 Frege claims that certain propositions are provable; the translations of these into FA are $aSd \wedge N[x : x \leq d]Sd \to N[x : x \leq a]Sa$ and $N[x : x \leq 0]S0$. Frege adds that the statement that for finite n the number of numbers less than or equal to n succeeds n then follows from these by applying the definition of "follows in the series of natural numbers."

$N[x : x \leq 0]S0$ is easily derived in FA: $xP^*y \to \exists a\, aPy$ follows from the definition of the ancestral; consider $[z : \exists a\, aPz]$. Since $\neg 0Sa$ and $1S0$ are theorems, so are $\neg aP0$, $\neg aP^*0$, $x \leq 0 \leftrightarrow x = 0$, and $N[x : x \leq 0] = N[x : x = 0]$, from which, together with the definition of 1, $N[x : x \leq 0]S0$ follows.

But the derivation of $aSd \wedge N[x : x \leq d]Sd \to N[x : x \leq a]Sa$ is not so easy. Frege says that, to prove it, we must prove that $a = N[x : x \leq a \wedge x \neq a]$; for which we must prove that $x \leq a \wedge x \neq a$ if and only if $x \leq d$, for which in turn we need Fin $a \to \neg a < a$. This last proposition is again to be proved, says Frege, by appeal to the definition of the ancestral; it is the fact that we need the statement that no finite number follows itself, he writes, that obliges us to attach to $N[x : x \leq n]Sn$ the antecedent Fin n.

An interpretive difficulty now arises: It is uncertain whether or not Frege is assuming the finiteness of a and d in Section 82. Although he does not say so, it would appear that he must be assuming that d, at least, is finite, for he wants to show $(aSd \wedge N[x : x \leq d]Sd \rightarrow N[x : x \leq a]Sa)$ by showing $aSd \rightarrow (x \leq a \wedge x \neq a \leftrightarrow x \leq d)$. Without assuming the finiteness of a and d, he can certainly show $aSd \rightarrow \forall x(x < a \leftrightarrow x \leq d)$. However, $\neg a < a$, or something like it, is needed to pass from $x < a$ to $(x \leq a \wedge x \neq a)$, and Frege would therefore appear to need Fin a. But since Fin 0 is trivially provable and $\forall d \forall e(dPa \wedge$ Fin $d \rightarrow$ Fin $a)$ easily follows from Propositions 91 and 98, $(xPy \rightarrow xP^*y)$ and $(xP^*y \wedge yP^*z \rightarrow xP^*z)$, of the *Begriffsschrift*, Frege's argument can be made to work in FA, provided that we take him as assuming that d (and therefore a) is finite. Let us see how.

From Propositions 91 and 98, $dPa \rightarrow (xP^*d \vee x = d \rightarrow xP^*a)$ easily follows. We also want to prove

(∗) $dPa \rightarrow (xP^*a \rightarrow xP^*d \vee x = d)$,

for which it suffices to take $F = [z : \exists d \, dPz \wedge \forall d(dPz \rightarrow xP^*d \vee x = d))$, and show $(xP^*a \rightarrow Fa)$ by showing, as usual, $(xPb \rightarrow Fb)$ and $(Fa \wedge aPb \rightarrow Fb)$.

$(xPb \rightarrow Fb)$: Suppose xPb. Then the first half of Fb is trivial; and if dPb, then by 78(5) of the *Foundations*, $x = d$, whence $xP^*d \vee x = d$. As for $(Fa \wedge aPb \rightarrow Fb)$, suppose Fa and aPb. The first half of Fb is again trivial; now suppose dPb. By 78(5), $d = a$. Since Fa, for some c, cPa, and then $xP^*c \vee x = c$. Since cPa and $d = a$, cPd. But then by 91 and 98, xP^*d, whence $xP^*d \vee x = d$. Thus $(xP^*a \rightarrow Fa)$, whence $dPa \rightarrow (xP^*a \rightarrow xP^*d \vee x = d)$ and $dPa \rightarrow (xP^*a \leftrightarrow xP^*d \vee x = d)$ follow.

We must now prove

(∗∗)   Fin $a \rightarrow \neg aP^*a$.

Since $\neg 0P^*0$, it suffices to show $0P^*a \rightarrow \neg aP^*a$. We readily prove $(0Pb \rightarrow \neg bP^*b)$ and $(\neg aP^*a \wedge aPb \rightarrow \neg bP^*b)$: If $0Pb$ and bP^*b, then by (∗), $bP^*0 \vee b = 0$, whence by 91 and 98, $0P^*0$, impossible; if $\neg aP^*a$, aPb, and bP^*b, then by (∗), $bP^*a \vee b = a$, whence by 91 and 98, aP^*a, contradiction.

Combining (∗) and (∗∗) yields

$$dPa \wedge \text{Fin } a \rightarrow ((xP^*a \vee x = a) \wedge x \neq a \leftrightarrow xP^*d \vee x = d).$$

Abbreviating, we have

$$dPa \wedge \text{Fin } a \rightarrow (x \leq a \wedge x \neq a \leftrightarrow x \leq d]).$$

and then by Hume's principle

$$dPa \wedge \text{Fin } a \rightarrow N[x : x \leq a \wedge x \neq a] = N[x : x \leq d].$$

Thus, if Fin d, $N[x : x \leq d]Sd$, and dPa, then Fin a and aSd; since $a \leq a$,

$$N[x : x \leq a]SN[x : x \leq a \wedge x \neq a] = N[x : x \leq d];$$

since aSd, by 78(5), $N[x : x \leq d] = a$, and therefore $N[x : x \leq a]Sa$. Since Fin 0 and $N[x : x \leq 0]S0$, we conclude

Fin $n \rightarrow (\text{Fin } n \wedge N[x : x \leq n]Sn)$,

whence Fin $n \rightarrow N[x : x \leq n]Sn$.

O.K., stop skimming now. One noteworthy aspect of Frege's derivation of what are in effect the Peano postulates is that so much can be derived from what appears to be so little. Whether or not Numbers is a purely logical principle is a question that we shall consider at length in what follows. I now want to consider the status of the other principles employed by Frege, which, having argued the matter elsewhere, I shall assume are properly regarded as logical. Frege shows these principles capable of yielding conditionals whose antecedent is the apparently trivial and in any event trivially consistent Numbers and whose consequents are propositions like $\forall m(\text{Fin } m \rightarrow \exists n(Zn \wedge nSm))$. The consequents would "not in any wise appear to have been thought in" Numbers; thus these conditionals at least look synthetic, and Frege himself would appear to have shown the principles and rules of logic that generate such weighty conditionals to be synthetic. But if the principles of Frege's logic count as synthetic, then a reduction of arithmetic to logic gives us no reason to think arithmetic analytic. There is a criticism of Kant to which Frege is nevertheless entitled: Kant had no conception of this sort of analysis and no idea that content could be thus created by deduction.

The hard deductions found in the *Begriffsschrift* and the *Foundations* would make evident, if it were not already so, the utter vagueness of the notions of *containment* and of *analyticity*. Even though *containment* appears to be closed under obvious consequence, it is certainly not closed under consequence; there is often no saying just when conclusions stop being contained in their premises.

In particular, the argument Frege uses to prove the existence of successors—show by induction on finite numbers n that the number belonging to the concept $[x : x \leq n]$ succeeds n—is a fine example of the way in which content is created. "Through the present example" wrote Frege in Section 23 of the *Begriffsschrift*

> we see how pure thought ... can, solely from the content that
> results from its own constitution, bring forth judgments that at
> first sight appear to be possible only on the basis of some intu-
> ition. This can be compared with condensation, through which
> it is possible to transform the air that to a child's consciousness
> appears as nothing into a visible fluid that forms drops.

That successors appear to have been condensed by Frege out of less than
thin air may well have heightened some of its readers' suspicions that the
principles employed in the *Foundations* are inconsistent.

On the other hand, Frege's construction of the natural numbers fore-
shadows von Neumann's well-known construction of them, the consistency
of which was never in doubt. Frege defines 0 as the number of things that
are non-self-identical; von Neumann defines 0 as the set of things that are
non-self-identical. Frege shows that n is succeeded by the number of numbers less than or equal to n; von Neumann defines the successor of n as the
set of numbers less than or equal to n. Peano arithmetic based on the von
Neumann definition of the natural numbers can be carried out (interpreted)
in a surprisingly weak theory of sets sometimes called General Set Theory,
the axioms of which are:

Extensionality:	$\forall x \forall y (\forall z (z \in x \leftrightarrow z \in y) \rightarrow x = y),$
Adjunction:	$\forall w \forall z \exists y \forall x (x \in y \leftrightarrow x \in z \lor x = w),$
Separation axioms:	$\forall z \exists y \forall x (x \in y \leftrightarrow x \in z \land A(x)).$

There is a familiar model for general set theory in the natural numbers:
$x \in y$ if and only if starting at zero and counting from right to left, one
finds a 1 at the xth place of the binary numeral for y. It is obvious that
extensionality, adjunction, and separation hold in this model. Thus it has
been clear all along that something *rather* like what Frege was doing in the
Foundations could consistently be done.

The results of the *Foundations* that the series of finite numbers has no
last member and that the "less than" relation on the finite numbers is
irreflexive complement those of the *Begriffsschrift*, whose main theorems,
when applied to the finite numbers, are that "less than" is transitive (98)
and connected (133). Much more of mathematics can be developed in FA
than Frege carried out in his three logic books. (It would be interesting to
know how much of the *Basic Laws* can be salvaged in FA.) Since addition
and multiplication can be defined in any of several familiar ways and their
basic properties proved from the definitions, the whole of analysis can be
proved (more precisely, interpreted) in FA. (The equiconsistency of analysis
and FA can be proved in Primitive Recursive Arithmetic.) Thus it is a vast
amount of mathematics that can be carried out in FA.

Instead of discussing this rather familiar material, I want instead to take a look at certain strange features of FA, one of which was alluded to earlier. Frege defined 0 as the number belonging to the concept "not identical with itself." What is the number belonging to the concept "identical with itself"? What is the number belonging to the concept "finite number"? Frege introduces the symbol ∞_1 to denote the latter number, shows that ∞_1 succeeds itself, and concludes that it is not finite. But, although Frege does not consider the former number and hence does not deal with the question of whether the two are identical, it is clear that he must admit the existence of such a number. The statement that there is a number that is the number of all the things there are (among them itself) is antithetical to Zermelo–Fraenkelian doctrine, but as a view of infinity it is not altogether uncommonsensical. The thought that there is only one infinite number, *infinity*, which is the number of all the things there are (and at the same time the number of *all* the finite numbers), is not much more unreasonable than the view that there is no such thing as infinity or infinite numbers. In any event the view is certainly easier to believe than the claim that there are so many infinite numbers that there is no set or number, finite or infinite, of them all.

But can we decide the question of whether these numbers are the same? Not in FA. $N[x : x = x] = N[x : \text{Fin } x]$ is true in some models of FA, for example, the one given, and false in others, as we can readily see. Let U' be the set of all ordinals $\leq \aleph_1$, and let η be true of V, u ($V \subseteq U'$, $u \in U'$) if and only if the (finite or infinite) cardinality of V is u. Numbers is then true in this structure, Fin x is satisfied by the natural numbers, $N[x : \text{Fin } x]$ denotes $\aleph_0$, but $N[x : x = x]$ denotes $\aleph_1$. $N[x : x = x] = N[x : \text{Fin } x]$ is thus an undecidable sentence of FA. Of course, so is $\exists x \neg Zx$, but $N[x : x = x] = N[x : \text{Fin } x]$ is an undecidable sentence about numbers. From Frege's somewhat sketchy remarks on Cantor, one can conjecture that Frege would have probably regarded $N[x : x = x] = N[x : \text{Fin } x]$ as false.

I now turn to the way Russell's paradox bears on the philosophical aims of the *Foundations*. My view is a more or less common one: As a result of the discovery of Russell's paradox our idea of logical truth has changed drastically, and we now see arithmetic's commitment to the existence of infinitely many objects as a greater difficulty for logicism than Russell's paradox itself.

But is not Frege committed to views that generate Russell's paradox? Does he not suppose that every predicate determines a concept and every concept has a unique extension? In Section 83 he says:

> And for this, again, it is necessary to prove that this concept has an extension identical with that of the concept "member of

the series of natural numbers ending with d."

In Section 68 he mentions the extension of the concept "line parallel to line a." And the number belonging to the concept F is defined as the extension of the concept "equinumerous with the concept F." How, in view of his avowed opinions on the existence of extensions, can he be thought to escape Russell's paradox?

The first quotation can be dealt with quickly, as a turn of phrase. Had Frege written " ... to prove that an object falls under this concept if and only if it is a member of the series of ... ," it would have made no difference to the argument. The extension of the concept "line parallel to the line a" is used merely to enable the reader to understand the point of the definition of number. (These are the insignificant but possibly revealing exceptions mentioned to the claim that the only extensions to whose existence Frege explicitly commits himself in 68–83 are those of concepts of the form "equinumerous to the concept F.") Thus, if there is a serious objection to Frege's introduction of extensions of concepts, it must concern the definition of numbers as extensions of concepts of the form "equinumerous with the concept F."

And of course there is one. According to Frege, for every concept F there is a unique object x, an "extension," such that for every concept G, G bears a certain relation, "being in," designated by η, to x if and only if the objects that fall under F are correlated one-one with those that fall under G; that is, Numbers holds. And although the language of FA, in which Numbers is expressed, is not one in which the most familiar version of Russell's contradiction $\exists x \forall y (y\eta x \leftrightarrow \neg y\eta y)$ is a well-formed sentence, it is not true that Frege is now safe from all versions of Russell's paradox.

For consider Rule (V) of Frege's *Basic Laws*:

$$\forall F \exists! x \forall G (G\eta x \leftrightarrow \forall y (Fy \leftrightarrow Gy)),$$

which yields an inconsistency in the familiar way.

Suppose Rule (V) true. By comprehension, let $F = [y : \exists G(G\eta y \wedge \neg Gy)]$. Then for some x,

(†) $\forall G(G\eta x \leftrightarrow \forall y(Fy \leftrightarrow Gy))$.

Since $\forall y(Fy \leftrightarrow Fy)$, by (†) $F\eta x$. If $\neg Fx$, then $\forall G(G\eta x \to Gx)$, whence Fx; but if Fx, then for some G, $G\eta x$ and $\neg Gx$, whence by (†), $\neg Fx$, contradiction.

Or consider the simpler

SuperRussell: $\exists x \forall G(G\eta x \leftrightarrow \exists y(Gy \wedge \neg G\eta y))$.

Suppose SuperRussell true. Let x be such that for every G, $G\eta x$ if and only if $\exists y(Gy \land \neg G\eta y)$. By comprehension, let $F = [y : y = x]$. Then, $F\eta x$ iff $\exists y(Fy \land \neg F\eta y)$, iff $\exists y(y = x \land \neg F\eta y)$, iff $\neg F\eta x$, contradiction.

SuperRussell and Rule (V) are sentences of the language of FA about the existence of extensions every bit as much as Numbers is. Just as Numbers asserts the existence (and uniqueness) of an extension containing just those concepts that are equinumerous with any given concept, so SuperRussell asserts the existence of an extension containing just those concepts that fail to be in some object falling under them and Rule (V) asserts the existence (and uniqueness) of an extension containing just those concepts under which fall the same objects as fall under any given concept. Frege must deny that SuperRussell and Rule (V) are principles of logic—if he maintains that the comprehension axioms are principles of logic. Principles of logic cannot imply falsity. But then Frege cannot maintain both that every predicate of concepts determines a higher level concept and that every higher level concept determines an extension and would thus appear to be deprived of any way at all to distinguish Numbers from SuperRussell and Rule (V) as a principle of logic.

Too bad. The principles Frege *employs* in the *Foundations* are consistent. Arithmetic can be developed on their basis in the elegant manner sketched there. And although Frege couldn't and we can't supply a reason for regarding Numbers (but nothing bad) as a logical truth, Frege was better off than he has been thought to be. After all, the major part of what he was trying to do—develop arithmetic on the basis of consistent, fundamental, and simple principles concerning objects, concepts, and extensions—can be done, in the way he indicated. The threat to the *Foundations* posed by Russell's paradox is to the philosophical significance of the mathematics therein and not at all to the mathematics itself.

It is unsurprising that we cannot regard Numbers as a purely logical principle. Consistent though it is, FA implies the existence of infinitely many objects, in a strong sense: Not only does FA imply $\exists x \exists y(x \neq y)$, $\exists x \exists y \exists z(x \neq y \land x \neq z \land y \neq z)$, etc., it implies $\exists F(\text{DedInf}F)$, where DedInf$F$ is a formula expressing that F is Dedekind infinite, for example, $\exists x \exists G(\neg Gx \land \forall y(Fy \leftrightarrow Gy \lor y = x) \land F \text{ eq } G)$. In logic we ban the empty domain as a concession to technical convenience but draw the line there: We firmly believe that the existence of even two objects, let alone infinitely many, cannot be guaranteed by logic alone. After all, logical truth is just *truth no matter what things we may be talking about and no matter what our (nonlogical) words mean.* Since there might be fewer than two items that we happen to be talking about, we cannot take even $\exists x \exists y(x \neq y)$ to be valid.

How then, we might now think, *could* logicism ever have been thought to be a mildly plausible philosophy of mathematics? Is it not obviously demonstrably inadequate? How, for example, could the theorem

$$\forall x(\neg x < x) \wedge \forall x \forall y \forall z(x < y \wedge y < z \rightarrow x < z) \wedge \forall x \exists y(x < y),$$

of (one standard formulation of) arithmetic, a statement that holds in no
finite domain but which expresses a basic fact about the standard ordering
of the natural numbers, be even a "disguised" truth of logic?[11] The axiom
of infinity was soon enough recognized by Russell as both indispensable to
his program and as damaging to the claims that could be made on behalf of
the program; and it is hard to imagine anyone now taking up even a small
cudgel for $\exists x \exists y(x \neq y)$.

I have been arguing for these claims: (1) Numbers is no logical truth;
and therefore (2) Frege did not demonstrate the truth of logicism in the
Foundations of Arithmetic. (3) Logic is synthetic if mathematics is, because
(4) there are many interesting, logically true conditionals with antecedent
Numbers whose mathematical content is not appreciably less than that of
their consequents. To these I want to add: (5) Since we have no understanding of the role of logic or mathematics in cognition, the failure of logicism
is at present quite without significance for our understanding of mentality.
Had Frege succeeded in eliminating the nonlogical residue from his *Foundations*, the question would remain what the information that arithmetic
is logic tells us about the cognitive status of arithmetic. But Frege's work
is not to be disparaged as a (failed) attempt to inform us about the role
of mathematics in thought. It is a powerful mathematical[12] analysis of the
notion of natural number, by means of which we can see how a vast body
of mathematics can be deduced from one simple and obviously consistent
principle, an analysis no less philosophical for its rigor, profundity, and
surprise.

A fantasy: After the *Begriffsschrift* Frege writes, not the *Foundations
of Arithmetic*, but another book with the same title whose main claim is
that, since arithmetic is deducible by logic alone from the triviality "the
number of Fs is the same as the number of Gs if and only if the Fs can
be correlated one-one with the Gs," arithmetic is analytic, not synthetic,
as Kant supposed. Frege then argues for the analyticity of $NF = NG \leftrightarrow
F$ eq G on the ground that both halves of the biconditional have the same
content, express the same thought. He considers an attempted defense of
Kant: Since the existence of an object can be inferred from $NF = NF$,
$NF = NF$ must be regarded as synthetic, and therefore so must $NF =
NG \leftrightarrow F$ eq G. Frege replies that $7 + 5 = 7 + 5$ is analytic.

If Frege had abandoned one of his major goals—the quest for an understanding of numbers not as objects but as "logical" objects—taken as a

---

[11]See (Benacerraf, 1960).
[12]See (Benacerraf, 1995).

starting point the self-evident and consistent $\forall F \forall G (NF = NG \leftrightarrow F \text{ eq } G)$, and worked out the consequences of this one axiom in the *Begriffsschrift*, he would have been wholly justified in claiming to have discovered a foundation for arithmetic. To do so would have been to trade a vain philosophical hope for a thoroughgoing mathematical success. Not a bad deal. He could also have plausibly claimed to demonstrate the analyticity of arithmetic. (Of course his own work completely undermines the interest of such a claim.)

Perhaps the saddest effect of Russell's paradox was to obscure from Frege and us the value of Frege's most important work. Frege stands to us as Kant stood to Frege's contemporaries. The *Basic Laws of Arithmetic* was his *magnum opus*. Are you sure there's nothing of interest in those parts of the *Basic Laws* that aren't in prose?

13

The Standard of Equality of Numbers

One of the strangest pieces of argumentation in the history of logic is found in Richard Dedekind's *Was sind und was sollen die Zahlen?*, where, in the proof of that monograph's Theorem 66, Dedekind attempts to demonstrate the existence of infinite systems. Dedekind defines a system S as *infinite* if, as we would now put it, there are a one-one function φ from S to S and an element of S not in the range of φ. Since it is now known that set theory without the axiom of choice does not imply that a set that is infinite in the usual sense is infinite in Dedekind's sense (although it does imply the converse), it is now common to prefix "Dedekind" when speaking of infinity in this stronger sense. The sets with which we shall be concerned are Dedekind infinite if they are infinite at all, however, and I shall therefore omit "Dedekind" before "infinite."

Theorem 66 of *Was sind* reads, "There are infinite systems"; the proof of it Dedekind offered runs:

> *Proof.** The world of my thoughts, i.e., the totality S of all things that can be objects of my thought, is infinite. For if s denotes an element of S, then the thought s', that s can be an object of my thought, is itself an element of S. If s' is regarded as the image $\varphi(s)$ of the element s, then the mapping φ on S determined thereby has the property that its image S' is a part of S; and indeed S' is a proper part of S, because there are elements in S (e.g., my own ego [*mein eigenes Ich*]), which are different from every such thought s' and are therefore not

From *Meaning and Method: Essays in Honor of Hilary Putnam*, George Boolos, ed., Cambridge: Cambridge University Press, 1990, pp. 261–277. Copyright ©1990 Cambridge University Press. Reprinted with the permission of Cambridge University Press.

I am grateful to Ellery Eells, Dan Leary, Thomas Ricketts, and Gabriel Segal for helpful comments. Research for this essay was carried out under grant no. SES–8808755 from the National Science Foundation.

contained in S'. Finally, it is clear that if a, b are different elements of S, then their images a', b' are also different, so that the mapping φ is distinct (similar). Consequently, S is infinite, q.e.d.

---

* A similar observation is found in §13 of (Bolzano, 1851).

It is tempting to think that Dedekind isn't in as deep a hole as his mentioning so wildly nonmathematical an item as his own ego might suggest and to suppose that he has merely chosen a bad example. Wouldn't the sentence *Berlin ist in Deutschland* and the operation of prefixing *Niemand glaubt daß* have been just as good as Dedekind's own ego and the operation that takes any object s in the world of Dedekind's thoughts to that funny thought about s? Instead of the things that can be objects of his thought (whatever these might be) couldn't he have cited (say) the set of German sentences, i.e., sentence-types, as an example of an infinite set? And had he cited that set, wouldn't he have given an obviously correct proof of Theorem 66 by giving an obviously correct example of an infinite set?

It is significant that nowhere in the remainder of *Was sind und was sollen die Zahlen?* does Dedekind appeal to Theorem 66 in the proof of any other Theorem. Why, one might wonder, did not Dedekind simply omit the theorem and its proof, the incongruity of whose argumentation and subject matter Dedekind himself could not have failed to find glaring?

Recall that the aim of *Was sind*, according to Dedekind, was to lay the foundations of that part of logic that deals with the theory of numbers—thus the theory of numbers is a part of logic—and that his answer to the title question of his monograph was that numbers are "free creations of the human mind." Some of what that saying means emerges in Section 73, where he writes:

> If in observing a simply infinite system N, ordered by a mapping φ, the special character of the elements is completely disregarded, only their distinguishability is held fixed, and account is taken of only those relations to one another in which they are placed by the mapping φ that orders them, then these elements are called *natural numbers* or *ordinal numbers* or also simply *numbers*, and the basis element 1 is called the basis-number of the number series N. With regard to this freeing of the elements from all other content (abstraction), one can justifiably call the numbers a free creation of the human mind. The relations or laws which are derived just from the conditions α, β, γ, δ[1] in 71

---

[1] α states that the successor of a number is a number; β, that 1 is a number and that mathematical induction holds of the numbers; γ, that 1 is not the successor of a number; and δ, that successor is one-one.

[these are Dedekind's versions of what have come to be known
as the "Peano Postulates"] and therefore are always the same
in all ordered simply infinite systems, however the names acci-
dentally given to the individual elements may be pronounced,
form the first object of the *Science of Numbers* or *Arithmetic*.

Thus, arithmetic is about certain objects, the numbers, abstracted from
simply (we now say "countably") infinite systems, systems satisfying the
"Peano" conditions α, β, γ, δ under some appropriate choice of base element
and successor operation. Since they have been abstracted from systems
satisfying α, β, γ, δ, the numbers too satisfy these conditions. Logic,
Dedekind would appear to be claiming, suffices for the derivation of all, or
at any rate all familiar, arithmetical facts from the mere assumption that
the numbers, together with 1 and successor, are objects that satisfy the
"Peano" conditions. (Dedekind proves that the existence of simply infinite
systems follows from that of infinite systems.)

The trouble with trying to prove Theorem 66 by mentioning the set of
sentences of German is that Dedekind would probably have regarded a
sentence (or any other abstract object) as as much a free creation out of
ink-tracks or other physical objects as a number is a free creation out of
objects. Dedekind did not cite the most obvious infinite system, the system
of the natural numbers themselves, in the proof of Theorem 66. It would
thus appear that he thought that a satisfactory proof of it must mention
some infinite set of non-abstract items, out of which the natural numbers
could have been freely created, and that he was therefore not at liberty to
cite a set of sentences or abstract objects of any other sort as an example
of an infinite set.

Dedekind's proof, however, is fallacious if thoughts are taken to be actual
physical occurrences. Ignoring worries about opacity, we may grant that if
u is a thought that s can be an object of my thought, and likewise for v and
t, then $u = v$ if and only if $s = t$. It does not follow, and it is indefensible
to assume, that for every object s, or at least for every object s that is an
object of my thought, *there is* such a thing as the thought that s can be
an object of my thought. There just aren't all those thoughts around. (As
Frege, commenting on Theorem 66, put it "Now presumably we shall not
hurt Dedekind's feelings if we assume that he has not thought infinitely
many thoughts.")[2] Dedekind makes this unwarranted assumption in the
proof by using the definite article and speaking of "*the* thought that s can
be an object of my thought." Of course, without some guarantee that all
those thoughts exist, the proof fails: Dedekind hasn't defined a function on

---

[2] In (Frege, 1897), p. 136. I am grateful to Arnold Koslow for telling me of this
quotation.

the (whole) world of his thoughts. The present king of France strikes again.

Thus Dedekind is in trouble that we do not appear to be in. We, but not he, can use the set of sentences of German as an example of an infinite set. The difficulty for us in doing so will shortly emerge.

Dedekind's notion of free creation[3] raises too many problems for us to find it satisfactory: One somewhat less obvious difficulty it poses is a "third man" difficulty, that of saying why we don't get different systems when we abstract *twice* from a system satisfying conditions α, β, γ, δ. Or *do* we get different, but isomorphic, systems? Or can we abstract only *once*? Best not to take him too seriously here.

The view can be made more appealing and more plausible if we forget about abstraction and free creation, and take Dedekind to be saying that statements about the natural numbers can be regarded as logically true statements about all systems satisfying conditions α, β, γ, δ. Charles Parsons, in an illuminating study, "The Structuralist View of Mathematical Objects,"[4] has called this the eliminative reading of *Was sind*. Perhaps we might take Dedekind to be claiming that an arithmetical statement, expressed by a sentence S in the notation, say, of second-order logic, in which all number quantifiers are relativized to the predicate letter N, 1 denotes one, and s denotes successor, has the logical form: $\forall N, s, 1(\alpha, \beta, \gamma, \delta(N, s, 1) \rightarrow S)$. (Addition and multiplication, etc., can be handled by familiar techniques due to Dedekind.) Thus the monadic predicate letter N, the monadic function sign s, and the constant 1 turn into a second-order monadic predicate variable, a second-order monadic function variable, and a (peculiarly shaped) first-order variable, which are then universally quantified upon. To complete the interpretation of the resulting sentence, we might want to add that the first-order variables range over all the things there are.

For any such arithmetical sentence S, let $D(S)$ be the second-order sentence $\forall N, s, 1(\alpha, \beta, \gamma, \delta(N, s, 1) \rightarrow S)$. $D(S)$, it will be observed, contains no non-logical constants at all. We now want to inquire into the relation between S and $D(S)$.

Suppose that S is true, i.e., true when interpreted over the natural numbers, together with successor and one, and that N', s', and $1'$ satisfy $\alpha, \beta, \gamma, \delta(N, s, 1)$. Then since the natural numbers together with successor and one also satisfy $\alpha, \beta, \gamma, \delta(N, s, 1)$, by a valid second-order argument given by Dedekind, N', s', and $1'$ are isomorphic to the natural numbers, successor, and one and therefore satisfy S. Thus $D(S)$ is a logical truth.

---

[3] For a more detailed account of this notion, see (Parsons, 1990).

[4] (Parsons, 1990).

Conversely, suppose that $D(S)$ is a logical truth. Let

(∗) $N,'$ $s,'$ and $1'$ satisfy $\alpha, \beta, \gamma, \delta(N, s, 1)$.

They therefore also satisfy S. Since the natural numbers, successor, and one also satisfy $\alpha, \beta, \gamma, \delta(N, s, 1)$, they are isomorphic to N', s', and $1'$, and therefore also satisfy S; that is, S is true. Thus, it would appear, we have shown that S is true if and only if $D(S)$ is a logical truth. Does not this argument show that arithmetical truths are logical truths disguised only by the omission of an antecedent condition and a few symbols of logic?

Of course—on the assumption (∗), true, by our lights, that *there are N'*, s', and $1'$ together satisfying $\alpha, \beta, \gamma, \delta(N, s, 1)$: We used this assumption when we argued that if $D(S)$ is a logical truth, then S is true.[5] We have had to make a true assumption, but one that we have as yet found no reason to regard as logically true, in order to show that we can effectively associate with each sentence of arithmetic, a sentence in the vocabulary of logic in such a way that with each truth and no falsehood of arithmetic there is associated a logical truth. To succeed this far in reducing arithmetic to logic we have had to make an assumption not yet certified as logically true: There are infinite systems.

Is that a difficulty? It might seem not. We make a non-logical assumption to reduce arithmetic to logic. We then throw away the ladder. But ladder or no, we have reduced arithmetic to logic, haven't we?

Parsons has pointed out a difficulty in supposing that we have.[6] He notes that if there are no infinite systems, then $D(S)$ is true, for every arithmetical sentence S, so " ... both A and $\neg A$ have true canonical forms, which amounts to the inconsistency of arithmetic."

Parsons's observation leads us to the heart of the matter. Logicism is not adequately characterized as the view that arithmetic is reducible to logic if all that is meant thereby is that there is an effective mapping of statements of arithmetic to statements of logic that assigns logical truths to all and only the truths of arithmetic. Nor is it vindicated merely by exhibiting such a mapping E. For E to vindicate logicism, it must show that arithmetic is "reducible to" logic, "really" logic, logic "in disguise," "a part of" logic. Then at least, for any arithmetical sentence S, $E(S)$ must give the content of S, must state, in the language of logic, how matters must be if and only if S is true.[7] But then E must do for falsity what it does for truth and also

---

[5] We didn't need it for the other half, since we are entitled to *assume* that there is a system satisfying $\alpha, \beta, \gamma, \delta(N, s, 1)$. *Was sind* proves that this assumption could have been replaced by the assumption that there is an infinite set.

[6] (Parsons, 1990).

[7] Is it possible to say what logicism is without using intensional notions like *part* or *content*?

assign logical falsehoods to the falsehoods of arithmetic; otherwise there will be certain truths S of arithmetic such that $E(S)$ is compatible with $E(\neg S)$, and no mapping that thus violates negation can be regarded as giving the content of statements of arithmetic in logical terms and hence as reducing arithmetic to logic.

For arithmetical sentences, like all others, come in triples: For any two arithmetical sentences there is a third, their conjunction, that, however matters may be, holds when matters are that way if and only if both sentences hold when matters are that way. A mapping E under which $E(S \wedge S')$ is not equivalent in this sense to the conjunction of $E(S)$ and $E(S')$ cannot be thought to give the content of all three of S, S' and $(S \wedge S')$ and cannot therefore count as a reduction of arithmetic to logic.

Similarly for negation: If for some arithmetical sentence S, $E(\neg S)$ is not equivalent to the negation of $E(S)$, then E does not give the content of both S and $\neg S$, and therefore does not show arithmetic reducible to logic. In advance of any possible reduction to logic certain arithmetical statements immediately (logically) imply certain others, and certain arithmetical statements are immediately incompatible with certain others. A reduction of arithmetic to logic, although it may reveal previously unrecognized implications or incompatibilities among the statements of arithmetic, cannot disclose that these immediate implications and incompatibilities actually fail to obtain.

For a mapping E to vindicate logicism, then, E must at the very least respect the truth-functional operators on closed formulae. Since $E(S)$ will always be logically true if S is true, E will respect negation if and only if $E(S)$ is always logically false for false S. It is clear that Dedekind's mapping D respects conjunction.

But D does not respect negation. $\neg \forall N, s, 1(\alpha, \beta, \gamma, \delta(N, s, 1) \rightarrow S)$ is not logically equivalent to $\forall N, s, 1(\alpha, \beta, \gamma, \delta(N, s, 1) \rightarrow \neg S)$. The latter follows logically from the former, as Dedekind showed. But the former follows from the latter in general only under the assumption that infinite systems exist. Indeed, if S is, say, $1 = 1$, then the former is equivalent to $\perp$, the latter to "there are no infinite systems," and the conditional with antecedent the latter and consequent the former is then equivalent to "there are infinite systems."

If S is the statement "there are infinite systems," a truth, but presumably not a logical truth, then $D(S)$ is a logical truth, as desired; but $D(\neg S)$ is equivalent to $\neg S$, and therefore not a logical falsehood.

A third example: The mapping D assigns to "$\neg 17 \times 14 = 228$" a sentence that is (absent logically guaranteed infinite systems) consistent with what it assigns to "$17 \times 14 = 228$." D cannot therefore count as reducing arithmetic to logic in any reasonable sense of the phrase.

All would be well, of course, if, as a matter of logic, there were infinite systems, if Theorem 66 had been established as securely as, and in the manner of, Theorems 65 and 67. But no purely logical ground has been given for thinking Theorem 66 true. That, and not the non-mathematical character of the objects it mentions, is the real problem with Dedekind's proof of Theorem 66.

We ought to mention that there can be no effective mapping of sentences of arithmetic to sentences of logic under which truths of arithmetic are mapped to logical truths and falsehoods to logical falsehoods: Otherwise arithmetic would be decidable, since the truth-value of any statement could be ascertained by calculating the truth-value of its image under the mapping in any one-element model. Nor is there a mapping of sentences of arithmetic to sentences of *first*-order logic under which the truths of arithmetic and only those are mapped to logical truths. Otherwise it would be possible to decide effectively whether an arithmetical sentence S is true or false: Effectively enumerate all first-order logical truths; then the image of S under the mapping appears in the enumeration if and only if that of $\neg S$ does not, and S is true if and only if its image appears.

Infinity is cheap. As Dedekind showed, a domain S will be infinite if there are a one-one function φ from S to S and an object in S not in the range of φ. Indeed, it's often easier than one may suspect to show a domain infinite. For example, in conjunction with the trivial truth "$\exists x \exists y\, x \neq y$," the ordered pair axiom, commonly thought to be innocuous, is an axiom of infinity. Any domain in which both hold is infinite, for if $a \neq b$, then the function that assigns to each object x in the domain the ordered pair $\langle a, x \rangle$ will be one-one and omit $\langle b, b \rangle$ from its range.

It is well-known that very weak systems of set theory guarantee that there are infinitely many objects: the conjunction of the null set and unit set axioms supply an object and a one-one function meeting Dedekind's criterion. It is thus not difficult to provide a theory committed to there being infinitely many objects. The difficulty (insuperable, I will urge) is to find a logically true theory with this commitment.

We have seen that in order to be able to claim that the function that assigns $\forall N, s, 1(\alpha, \beta, \gamma, \delta(N, s, 1) \to S)$ to any sentence S of arithmetic shows that "arithmetic is a part of logic," Dedekind needs a proof from logical truths that there are infinitely many objects. No satisfactory way has yet presented itself.

I want to consider the suggestion that a principle I call Hume's principle can be used to help Dedekind out. As we shall see, Dedekind would probably not like the suggestion. And in the end, I shall argue, we can't accept it, either.

"We are possest of a precise standard," wrote Hume, "by which we can judge of the equality and proportion of numbers; and according as they correspond or not to that standard, we determine their relations, without any possibility of error. When two numbers are so combin'd as that the one has always an unite answering to every unite of the other, we pronounce them equal; and 'tis for want of such a standard of equality in extension, that geometry can scarce be esteem'd a perfect and infallible science."[8] Reflecting on Frege's idea that statements about numbers are assertions about what he called concepts, we may formalize Hume's dictum as a second-order formula. Let "$\#F$" abbreviate "the number of objects falling under the concept F" and "$F \approx G$" express the existence of a one-one correspondence between the objects falling under F and those falling under G.[9] Then Hume's principle may be written: $\forall F \forall G (\#F = \#G \leftrightarrow F \approx G)$.

Frege attempts to prove Hume's principle in §73 of his *Foundations of Arithmetic*. The difficulty with the proof he gives there is that it appeals to the theory of concepts and objects, whose inconsistency Russell pointed out in his first letter to Frege. After having derived Hume's principle from this inconsistent theory, Frege derives the axioms of arithmetic from Hume's principle.

More exactly, in the *Foundations of Arithmetic* Frege gives definitions of *zero*, *succeeds* (*"follows directly after"*), and *finite* (*natural*) *number*, and shows, easily enough, that zero is a finite number, that anything that succeeds any finite number is a finite number, that zero succeeds nothing and that if m, m', n, and n' are finite numbers, n succeeds m, and n' succeeds m', then $m = m'$ iff $n = n'$. It is by no means evident, however, that every finite number is succeeded by something, and it was a matter of considerable difficulty for Frege to prove it. The central argument of the *Foundations of Arithmetic* is a fairly complete sketch of a proof that every finite number is succeeded by a finite number; in proving this and the other facts about the numbers, Frege makes use only of Hume's principle and the system of logic set forth in his *Begriffsschrift*. The intricacy of his reasoning is astonishing and repays careful attention. See the appendix for a reconstruction. I do not know whether Frege realized that Hume's principle plus the logic of the *Begriffsschrift* was all he used or needed. Perhaps not; *he* would have had no reason to value the observation. In any event, it is a pity that the derivability of arithmetic from Hume's principle isn't known as *Frege's theorem*.

Frege thus succeeds where Dedekind has failed. He has demonstrated the existence of an infinite system. With hard work, he has proved the analogue of what Dedekind simply assumes to be the case for the system

---

[8] *Treatise*, I, III, I para. 5.
[9] "$F \approx G$" abbreviates the second-order formula: $\exists \varphi (\forall x [Fx \rightarrow \exists y (Gy \wedge \varphi xy)] \wedge \forall y [Gy \rightarrow \exists x (Fx \wedge \varphi xy)] \wedge \forall x \forall x' \forall y \forall y' [\varphi xy \wedge \varphi x'y' \rightarrow (x = x' \leftrightarrow y = y')])$.

of objects x of his thought, that for each x, *there is* a y to which x bears the appropriate relation.

Do not be deceived by the absence of the sort of wallpaper found in the *Begriffsschrift* into thinking that the *Foundations* is not fundamentally a mathematical work.[10] In a letter of September 1882, Carl Stumpf suggested to Frege that it might be "appropriate to explain your line of thought first in ordinary language and then—perhaps separately on another occasion or in the very same book—in conceptual notation. I should think that this would make for a more favorable reception of both accounts."[11] Frege apparently took Stumpf's suggestion, which was that his *mathematical* ideas be first published in ordinary language. At the heart of the *Foundations* there lies a *proof.*

Frege outlines a demonstration in the *Foundations* that arithmetic, i.e., the basic axioms of the second-order arithmetic of zero and successor (from which that of full second-order arithmetic, of addition and multiplication, can be derived, as in *Was sind*; Frege seems never to have been interested in deriving the axioms of addition and multiplication), can be derived from Hume's principle. Second-order arithmetic is consistent, presumably; Frege derives Hume's principle from an inconsistent theory of concepts. Nevertheless, an inconsistent theory may, indeed must, have consistent consequences, and it turns out that in the same sense in which second-order arithmetic may be derived from Hume's principle in the system of logic of the *Begriffsschrift*, Hume's principle may be derived back from second-order arithmetic. (Deriving axioms from theorems has been called "reverse mathematics" by Harvey Friedman.) Hume's principle and second-order arithmetic, which is sometimes called "analysis," are thus equiconsistent, and very effectively so: A proof of an inconsistency from either could easily be turned into a proof of an inconsistency from the other.

In analysis, there are two sorts of variables, one sort ranging over the natural numbers, the other over sets of natural numbers. The axioms are the usual ones: the Peano axioms, together with the usual axioms for addition and multiplication (which, as Dedekind showed, are dispensable), and a comprehension scheme: For any formula of the language of analysis, there is a set of all and only the numbers satisfying the formula.

The *Foundations* of course shows that if analysis is inconsistent, so is Hume's principle. How may the converse be shown? Let α be a set of natural numbers. Call the natural number n the *grumber* belonging to α if either α has infinitely many members and $n = 0$ or α has $n - 1$ members. The grumber of U.S. senators is 101, the grumber of roots of the equation "$x - 5 = 0$" is 2, and the grumber of even numbers is zero, which is also the

---

[10]Cf. (Benacerraf, 1995).
[11](Gabriel et al., 1980), p. 172.

grumber of numbers divisible by four. It is then a *theorem* of analysis that the grumber belonging to α = the grumber belonging to β if and only if the members of α and those of β are in one-one correspondence. Any derivation in second-order logic of a contradiction from Hume's principle could thus be turned into a derivation in analysis of a contradiction from the theorem of analysis about grumbers just cited. Thus if analysis is consistent, so is the result of adjoining Hume's principle to second-order logic.

Some trick like the introduction of grumbers is necessary because "the natural number belonging to" is not defined for all sets of natural numbers, indeed not defined for any set containing infinitely many natural numbers. There is no natural number that is the number of members of the set of evens; but the grumber zero belongs to this set.

Frege, then, gave an intricate and mathematically interesting derivation of arithmetic from a simple, consistent, and trivial-seeming principle. Since the principle is as weak as any from which arithmetic can be derived, Frege's derivation was "best possible."

Dedekind, of course, might well have objected to our suggestion that Hume's principle be used to obtain an infinite system on the ground that the arithmetical notion *the number belonging to* that figures in the principle is undefined, and that arithmetic is therefore not shown to be entirely a part of logic. A weak reply can be made: There is a principle, discussed farther on, that deals only with objects and concepts, which licenses a definition of number, from which Hume's principle can be derived. The principle is that for every concept F there is a unique object x such that for all concepts G, G is in x if and only if F and G are equinumerous. Unlike Hume's, this principle does not explicitly mention numbers. The number belonging to F may of course be defined to be the unique x such that for all G, etc.

Even if the objection that the expression "is in" is not a logical "constant" is waived, this principle cannot be held to be a logical principle for a reason we shall consider at length: It commits us to the existence of too many objects. Here we should note that a truth's being couched in purely logical terms is not sufficient for it to count as truth *of* logic, a logical truth, a truth that is true solely in virtue of logic. A distinction must be drawn between truths of logic and truths expressed in the language of logic. I suspect that failure to draw this distinction was largely responsible for there ever being any thought at all that the axiom of infinity might actually count as a logical truth.

According to Hume's principle, for any concepts F and G, there are certain *objects*, namely, the number x belonging to F and the number y belonging to G, such that x is identical to y if and only if F and G are equinumerous. It is the objecthood of numbers that explains why Hume's principle, despite appearances, cannot be considered to be a truth of logic, a definition,[12] an immediate consequence of a definition, analytic, quasi-analytic, or anything of that sort.

The reason that it may appear so is that it can easily be confused with a principle that has a considerably greater claim to the status of truth of logic. Assume that some version of the theory of types, including axioms of comprehension and extensionality, counts as logic. Then matters are as the theory of types has it: There are individuals, sets of individuals, classes of sets of individuals, etc. (The words "set" and "class" are used here just to keep the types straight: We've got classes of sets of individuals.) According to a comprehension axiom, for any set, there will be a class containing all and only the sets that are equinumerous with that set; by extensionality, there will be at most one such class. We may call classes containing all and only the sets that are equinumerous with any one set *Russellnumbers*, and say that the Russellnumber of a set is the Russellnumber that contains the set. The proposition that Russellnumbers are identical if and only if the sets they are Russellnumbers of are equinumerous is then a theorem of (this version of) the theory of types.

Russellnumbers are classes of sets of individuals. One of the (ineffable?) doctrines of the usual formulation of the theory of types is that the types are disjoint. No set is a class or individual, and no class is an individual. Russellnumbers are not individuals.

The disadvantage of this way of defining numbers, of course, is that arithmetic cannot be derived in the theory of types without postulating that there are infinitely many individuals. With honest toil, however, Frege succeeds in proving from Hume's principle the infinity of the natural numbers.

Observe that although the theory of objects and concepts that is sketched in the *Foundations* is almost certainly inconsistent, there is a consistent fragment of it that is all Frege needs, or uses, to derive arithmetic. According to this fragment, there are objects (or individuals), first level concepts, under which objects may or may not fall, and second level concepts, under which first level concepts may or may not fall. So far, matters look pretty much as they do on the theory of types, if one substitutes "object," "first level concept," and "second level concept" for "individual," "set," and "class." Crucially, though, Frege does not analogously define the numbers as those second level concepts under which fall all and only those first level concepts that are equinumerous with some one concept. Call such second level concepts *numerical*. Instead, he introduces a new primitive relation between objects and concepts ("is the extension of") and then defines a number to be an object that is the extension of some numerical concept. (Numerical concepts, to repeat, are second level.) Thus Frege assumes that

[12]On hearing me say what I meant by "Hume's principle," a *very* famous philosopher exclaimed, "But that's just a *definition!*"

for every first level concept F there is a unique object that is the extension of the second level concept under which fall all and only those first level concepts that are equinumerous with F.

The introduction of second level concepts is not necessary. All that Frege need do is introduce a primitive predicate "is in" for a relation between first level concepts and objects and assume that for any first level concept F there is exactly one object x such that for any first level concept G, G is in x if and only if F and G are equinumerous. Frege may then define the number "belonging to" F as that object x.

Notice the additional step Frege has taken. Unlike Russell, Frege has assumed there is a way of associating objects with numerical concepts so that different objects are associated with different numerical concepts. (Assume that coextensive concepts are identified.) This cannot be done, if there are only finitely many objects; if there are (say) eighteen objects, then there will be nineteen numerical concepts. It is a weighty assumption of Frege's, to put it slightly differently, that the first level concepts can be mapped into objects in such a way that concepts are mapped onto the same object only if they are equinumerous, and *it is a lucky break that the assumption is even consistent.*

The well-known comparison that Frege draws in §§64–69 of the *Foundations* between "the direction of line l" and "the number belonging to concept F" is therefore seriously misleading. We do not suspect that lines are made up of directions, that directions are some of the ingredients of lines. Had Frege appended to the direction principle, "The direction of line l is equal to the direction of line k if and only if l and k are parallel," the claim that directions are points, we would never have regarded the principle as anything like a definition, and would perhaps have wondered whether there are enough points to go around. (In fact, there are: There are continuously many points and continuously many directions.) The principle that directions of lines are identical just in case the lines are parallel looks, and is, trivial only because we suppose that directions are one or more types up from, or at any rate are all distinct from, the things of which lines are made.

The principle that numbers belonging to concepts are identical if and only if the concepts are equinumerous, then, should count as a logical truth only if it is supposed that numbers do not do to concepts or sets the corresponding sort of thing, namely, fall under, or be elements of, them. On the theory of types, matters so fall out. But Frege's proof from Hume's principle that every number has a successor cannot be carried out in the theory of types: the proof cannot succeed unless it is supposed that numbers are objects.

For how does Frege show that the number 0 is not identical with the number 1? Frege defines 0 as the number belonging to the concept *not identical with itself*. He then defines 1 as the number belonging to the concept *identical with* 0. Since no object falls under the former concept, and the object 0 falls under the latter, the two concepts are, by logic, not equinumerous, and hence their numbers 0 and 1 are, by Hume's principle, not identical. Notice that for this argument to work it is crucial that 0 be supposed to be an object that falls under the concept *identical with* 0. 2 arises in like manner: Now that 0 and 1 have been defined and shown different, form the concept *identical with* 0 *or* 1, take its number, call it 2, and observe that the new concept is coextensive with neither of these concepts *because the distinct objects* 0 *and* 1 *fall under it*. Conclude by Hume that 2 is distinct from both 0 and 1.

Frege proves that if n is a finite number, then it is succeeded by the number belonging to *being less than or equal to n*; the proof works because n is an object that can be proved not to fall under *being less than n*.

Thus it is only to one who supposes that numbers are not objects that Hume's principle looks analytic or obvious. Frege's proof that every number has a successor depends vitally on the contrary supposition that numbers are indeed objects.

A sentence is a logical truth only if it is true no matter what objects it speaks of and no matter to which of them its predicates or other non-logical words apply. (The vagueness of the consequent, including that as to which words count as logical, is matched by that of the antecedent.) A sentence is not a logical truth if it is false when interpreted over a domain containing infinitely many things, and it is not a logical truth if, like Hume's principle, it is false when only finitely many things belong to the domain.

It is clear that an account of logical truth that attempts to distinguish Hume's principle as a logical truth will have the hard task of explaining why Hume's principle is a logical truth even though two other similar-looking principles are not. These are the principle about extensions embodied in Frege's Rule (V) and a principle about relation numbers that is strikingly analogous to Hume's principle. They read: Extensions of concepts are identical if and only if those concepts are coextensive; and: Relation numbers of relations are identical if and only if those relations are isomorphic. Russell showed the former inconsistent; Harold Hodes has astutely observed that the latter leads to the Burali–Forti paradox.[13]

It will not do to say: Hume's principle, unlike the other two, and like the principle by which we take ourselves to introduce the two truth-values, is a logical truth because it is consistent.

For say that the concepts F and G *differ evenly* if the number of objects

[13](Hodes, 1984).

falling under F but not G or under G but not F is even (and finite). The relation between concepts expressed by "F and G differ evenly" is an equivalence relation (exercise), and can of course be defined in purely logical (second-order) vocabulary. Now introduce the term "the parity of" for a function from concepts to objects and consider the parity principle: The parity of F is identical with that of G if and only if F and G differ evenly.

The parity principle is evidently consistent. Let X be any finite domain containing the numbers 0 and 1. Let the parity of a subset of X be 0 if it contains an even number of objects and 1 otherwise. Then with "parity" so defined, the parity principle is true in the domain X, and is therefore consistent.

However, the parity principle is true in no infinite domain. Here's a sketch of the proof. Let X be an infinite set. Then, where Y and Z are subsets of X,

$$
\begin{aligned}
&|\{\, Y : Y \text{ differs evenly from } Z \,\}| \\
&\quad = |\{\, (A,B) : A, B \subseteq X,\ A \subseteq Y,\ B \subseteq Z,\ A \cap Z = \emptyset, \\
&\qquad B \cap Y = \emptyset, |A \cup B| \text{ is even, and } Y = (Z \cup A) - B \,\}| \\
&\quad \leq |\{\, (A,B) : A, B \subseteq X \text{ and } |A \cup B| \text{ is even} \,\}| = |X|.
\end{aligned}
$$

Thus $|\{\, Y : Y \text{ differs evenly from } Z \,\}| = |X|$.

Suppose now that $f : PX \to X$ and for all Y, Z, if $fY = fZ$, Y evenly differs from Z. Then for each x in X, $|\{Y : fY = x\}| \leq |X|$, and $|PX| \leq |X| \times |X| = |X|$, contradiction.[14]

Consistent principles of the form: The object associated in some manner with the concept F is identical with that associated in the same manner with G if and only if F and G bear a certain equivalence relation to one another, may therefore be inconsistent with each other. Hume's principle is inconsistent with the parity principle. Which is the logical truth?[15]

---

[14] A less natural example to the same purpose, but one that is less heavily dependent on set theory, is the following. Abbreviate: $(\exists x \exists y[x \neq y \wedge Fx \wedge Fy] \vee \exists x \exists y[x \neq y \wedge Gx \wedge Gy]) \to \forall x(Fx \leftrightarrow Gx)$ as: F Equ G. Equ is an equivalence relation. The principle: $\forall F \forall G (\hat{\ }F = \hat{\ }G$ iff F Equ $G)$ is evidently satisfiable in all domains containing one or two members. It is, however, satisfiable in no domain containing *three* or more members and is therefore inconsistent with Hume's principle. For suppose that $a \neq b \neq c \neq a$. Define Rx as follows: If $x \neq a, b, c$, then Rx iff for some F, $x = \hat{\ }F$ and $\neg Fx$; but if x is one of a, b, c, then let $Ra, Rb, \neg Rc$ hold if none of a, b, c is $\hat{\ }F$ for some F such that $Fa, Fb, \neg Fc$; otherwise let $Ra, \neg Rb, Rc$ hold if none of a, b, c is $\hat{\ }F$, for some F such that $Fa, \neg Fb, Fc$; otherwise let $\neg Ra, Rb, Rc$ hold if none of a, b, c is $\hat{\ }F$ for some F such that $\neg Fa, Fb, Fc$; otherwise let Ra, Rb, Rc hold. Then $\exists x \exists y(x \neq y \wedge Rx \wedge Ry)$. Let $d = \hat{\ }R$. If $d = \hat{\ }F$, then by the principle, $\forall x(Rx \leftrightarrow Fx)$. Thus $d \neq a, b, c$. If Rd, then for some F, $d = \hat{\ }F$ and $\neg Fd$. But then $\forall x(Rx \leftrightarrow Fx)$ and Fd. Thus $\neg Rd$. So $\forall F(d = \hat{\ }F \to Fd)$, whence Rd, contradiction.

[15] Cf. (Hazen, 1985).

Indeed, not only do we have no reason for regarding Hume's principle as a truth of logic, it is doubtful whether it is a truth at all. As the existence of a number, 0, belonging to the concept *not-self-identical* is a consequence of Hume's principle, it also follows that there is a number belonging to the concept *self-identical,* a number that is the number of things that there are. Hume's principle is no less dubious than any of its consequences, one of which is the claim, uncertain at best, that there is such a number.

Crispin Wright claims that "there is a programme for the foundations of number theory recoverable from *Grundlagen.*"[16] He calls the program "number-theoretic logicism" and characterizes it as the view that "it is possible, using the concepts of higher-order logic with identity to explain a genuinely sortal concept of cardinal number; and hence to deduce appropriate statements of the fundamental truths of number-theory, in particular the Peano Axioms, in an appropriate system of higher-order logic with identity to which a statement of that explanation has been added as an axiom."[17] He adds that he thinks that it would "serve Frege's purpose against the Kantian thesis of the *synthetic a priori* character of number-theoretic truths. For the fundamental truths of number theory would be revealed as consequences of an explanation: a statement whose role is to fix the character of a certain concept."[18]

Wright regards Hume's principle as a statement whose role is to fix the character of a certain concept. We need not read any contemporary theories of the a priori into the debate between Frege and Kant. But Frege can be thought to have carried the day against Kant only if it has been shown that Hume's principle is *analytic,* or a truth of logic. This has not been done. Nor has the view Wright describes been shown to deserve the name "(number-theoretic) logicism." It's logicism only if it's claimed that Hume's principle is a principle of logic. Wright quite properly refrains from calling it one.

We have noted that Dedekind would not have been happy with the suggestion that the existence of infinite systems be derived from Hume's principle. Nor, presumably, would Frege have rested content with it as the foundation of arithmetic. Hume's principle may yield a great deal of information about the natural numbers, but it doesn't tell us how they may be viewed as logical objects, nor even which objects they are. Nor, as Frege noted in §66 of the *Foundations,* does it enable us to eliminate the phrase "the number belonging to" from all contexts in which it occurs, in particular not from those of the form "x = the number belonging to F."

Well. Neither Frege nor Dedekind showed arithmetic to be part of logic.

---

[16](Wright, 1983), p. 153.
[17](Wright, 1983).
[18](Wright, 1983).

Nor did Russell. Nor did Zermelo or von Neumann. Nor did the author of *Tractatus* 6.02 or his follower Church. They merely shed light on it.

Appendix: Arithmetic in the *Foundations*

Hume's Principle $\#F = \#G \leftrightarrow F \approx G$.

Definition $0 = \#[x : x \neq x]$. (*Foundations* §74)

1. $\#F = 0 \leftrightarrow \forall x \neg Fx$. (§75)

Proof. Since $0 = \#[x : x \neq x]$, $\#F = 0$ iff $F \approx [x : x \neq x]$. Since $\forall x \neg x \neq x$, $F \approx [x : x \neq x]$ iff $\forall x \neg Fx$. ∎

Definition mPn *iff* $\exists F \exists y (Fy \wedge \#F = n \wedge \#[x : Fx \wedge x \neq y] = m)$. (§76)

2. $mPn \wedge m'Pn' \rightarrow (m = m' \leftrightarrow n = n')$. (§75 para. 5)

Proof. Suppose mPn and $m'Pn'$. Let F, y, F', y' be as in the definition of P. Suppose $m = m'$. Then $\#[x : Fx \wedge x \neq y] = \#[x : F'x \wedge x \neq y']$, whence $[x : F'x \wedge x \neq y'] \approx [x : Fx \wedge x \neq y]$ via some φ. Since Fy and $F'y'$, $F \approx F'$ via $\varphi \cup \{\langle y, y' \rangle\}$ and then $n = \#F = \#F' = n'$. Conversely, suppose $n = n'$. Then since $\#F = \#F'$, $F \approx F'$ via some ψ. For some unique x, $x\psi y'$; for some unique x', $y\psi x'$. Let

$$\varphi = ((\psi - \{\langle x, y' \rangle, \langle y, x' \rangle\}) \cup \{\langle x, x' \rangle\}) - \{\langle y, y' \rangle\}.$$

Then $[x : Fx \wedge x \neq y] \approx [x : F'x \wedge x \neq y']$ via φ and $m = m'$. (Since x and x' might be identical with y and y', it is necessary to include "$-\{\langle y, y' \rangle\}$" in the definition of φ.) ∎

3. $\neg mP0$. (§78 para. 6)

Proof. Otherwise for some y, Fy and $\#F = 0$, contra 1. ∎

Definition xR^*y *iff* $\forall F (\forall a \forall b ([(a = x \vee Fa) \wedge aRb] \rightarrow Fb) \rightarrow Fy)$. (§79)

Thus to show that $xR^*y \rightarrow \ldots y \ldots$, it suffices to let $F = [z : \ldots z \ldots]$, assume $a = x \vee Fa$ and aRb, and show Fb.

4. $xRy \rightarrow xR^*y$. (*Begriffsschrift*, Proposition 91)

Proof. Suppose xRy and $\forall a \forall b ([(a = x \vee Fa) \wedge aRb] \rightarrow Fb)$. Then Fy follows, if we let $a = x$ and $b = y$. ∎

5. $xR^*y \wedge yR^*z \rightarrow xR^*z$. (*Begriffsschrift*, Proposition 98)

Proof. Suppose xR^*y, yR^*z, and $(*)$ $\forall a \forall b([(a = x \lor Fa) \land aRb] \to Fb)$. Show Fz. Since yR^*z, it suffices to show $\forall a \forall b([(a = y \lor Fa) \land aRb] \to Fb)$. Suppose $(a = y \lor Fa)$ and aRb. Show Fb. Since xR^*y, by $(*)$ Fy. We may suppose Fa. But then by $(*)$ we are done. ∎

6. $xP^*n \to \exists m\, mPn \land \forall m(mPn \to [xP^*m \lor x = m])$.

Proof. Let $F = [z : \exists m\, mPz \land \forall m(mPz \to [xP^*m \lor x = m])]$. Suppose $a = x \lor Fa$, aPb. Show Fb. Since aPb, $\exists m\, mPb$. Suppose mPb. By 2, $m = a$. If $a = x$, $x = m$, and we are done. So suppose Fa. Then for some m', $m'Pa$, and xP^*m' or $x = m'$. Since $m'Pa = m$, $m'Pm$. If xP^*m', then $xP^*m'Pm$, whence by 4 and 5, xP^*m; if $x = m'$, xPm, whence by 4, xP^*m. ∎

7. $0P^*n \to \neg nP^*n$. (§83)

Proof. Let $F = [z : \neg zP^*z]$. Suppose $a = 0 \lor Fa$, aPb. Show Fb. Suppose bP^*b. By 6, $bP^*a \lor b = a$, whence by 4 and 5, aP^*a, contra Fa. Thus $a = 0$, $0P^*0$, and by 6, $\exists m\, mP0$, contra 3. ∎

Definition $m \leq n$ iff $mP^*n \lor m = n$. Finite n iff $0 \leq n$.

8. $mPn \land 0P^*n \to \forall x(x \leq m \leftrightarrow x \leq n \land x \neq n)$. (§83)

Proof. Suppose mPn, $0P^*n$. If $xP^*m \lor x = m$, then xP^*n, by 4 and 5; and by 7, $x \neq n$. If $x \leq n$ and $x \neq n$, then xP^*n, and by 6, $x \leq m$. ("$0P^*n$" cannot be dropped: if $n = $ Frege's ∞_1, i.e., $\#[x : 0 \leq x]$, then nPn but $x \leq n$ iff $x = n$.) ∎

9. $mPn \land 0P^*n \to \#[x : x \leq m]P\#[x : x \leq n]$. (§82)

Proof. Suppose mPn, $0P^*n$. By 8, $[x : x \leq m] \approx [x : x \leq n \land x \neq n]$; since $n \leq n$, $\#[x : x \leq m]P\#[x : x \leq n]$. ∎

10. $mPn \to (0 \leq m \land mP\#[x : x \leq m] \to 0 \leq n \land nP\#[x : x \leq n])$. (§82)

Proof. Suppose mPn, $0 \leq m$. By 4 and 5, $0P^*n$. Thus $0 \leq n$. Suppose $mP\#[x : x \leq m]$. By 2, $\#[x : x \leq m] = n$. By 9, $nP\#[x : x \leq n]$. ∎

11. $0P\#[x : x \leq 0]$. (§82)

Proof. Let $F = [x : x \leq 0]$. $F0$; but by 6, if xP^*0, $\exists m\, mP0$, contra 3. Thus $\forall x \neg(Fx \land x \neq 0)$; and by 1, $\#[x : Fx \land x \neq 0] = 0$, whence $0P\#[x : x \leq 0]$. ∎

12. $0 \leq n \to 0 \leq n \land nP\#[x : x \leq n]$.

Proof. If $0 = n$, done by 11. Suppose $0P^*n$. Let $F = [z : 0 \leq z \wedge zP\#[x : x \leq z]]$. Suppose $m = 0 \vee Fm$ and mPn. Show Fn. If $m = 0$, by 11, Fm. And then by 10, Fn. ∎

13.   Finite $n \rightarrow nP\#[x : x \leq n]$. (§83)

Proof. By 12. ∎

14

Whence the Contradiction?

Chapter 17 of Michael Dummett's *Frege: Philosophy of Mathematics*[1] begins with the question: how did the serpent of inconsistency enter Frege's paradise? In the section of that chapter called "How the serpent entered Eden" Dummett says, "The second-order quantifier presents an altogether different problem, and it is to its presence in Frege's language that the contradiction is due." Dummett regards Frege's remarks concerning the second-order quantifier as insufficient to stipulate the references of terms (in particular, the truth-values of sentences) formed by second-order quantification and writes that Frege's "amazing insouciance concerning the second-order quantifier was the primary reason for his falling into inconsistency." But in the very last chapter of the book he says, "We may say that his mistake lay in supposing there to be a totality containing the extension of every concept defined over it; more generally, it lay in his not having the glimmering of a suspicion of the existence of indefinitely extensible concepts." One might wonder whether it was Frege's insouciance or his naïveté that Dummett thinks is to blame for the error. But although it may seem as if Dummett is offering incompatible diagnoses of the contradiction, he has, I believe, a unitary account of its etiology to provide.

There are two pieces of unarguable mathematical fact on which Dummett's explanation of the contradiction rests. The first is that the first-order fragment of the system of Frege's *Grundgesetze* is consistent.[2] The second is that, as one might put it, one cannot assign different members of a set to different subsets of that set so that to every subset at least one member is assigned. (With the aid of the notion of the ordered pair,

From *Aristotelian Society Supplementary Volume* 67 (1993): 213–233. Reprinted by courtesy of the Editor of the Aristotelean Society: ©1993.

[1](Dummett, 1991). Let me say that although I disagree with several of its central contentions, I greatly admire Dummett's book.

[2]A particularly perspicuous description of this system and proof of its consistency have been given by Terence Parsons (Parsons, 1987). A version of Parsons' proof is presented below.

friends of plurals can make the appropriately general statement by saying: no matter what some objects may be, there are some things for which no item is such that the ordered pairs of that item and those things are all and only those objects.) A generalization of the second fact, pointed out by Dummett, is that the set of all functions from and to a given set can be injected into that set if and only if it contains exactly one member.

As the argument of Zermelo shows that $x \cup \{\{y \in x : \neg y \in y\}\}$ is always a proper superset of the set x, and as, on the von Neumann construction of the ordinals, $\bigcup x \cup \{\bigcup x\}$ is always a larger ordinal than any ordinal in the set x, so, according to Dummett, the argument of the Russell paradox shows that $'x \forall F(x = 'F \rightarrow \neg F(x))$—let us call this item r—can never belong to the domain, can never be one of the objects over which the first-order variables range. (I shall modernize notation throughout, often writing "$'F$" instead of "$'yF(y)$" etc.) Note, though, that r is the extension of a concept whose name "$\forall F(\xi = 'F \rightarrow \neg F(\xi))$" begins with a second-order quantifier. Thus the introduction of second-order quantifiers forces an extension of the domain to comprise such new objects as r. If only first-order quantifiers are present, r cannot be defined, and it is possible to satisfy the first-order fragment over the natural numbers, as we shall see. But once second-order quantifiers are added, no domain is large enough to contain all extensions of concepts defined on that domain. Each domain gives rise to a more encompassing one, containing all extensions of concepts defined over the original domain. It was because Frege didn't have a glimmering of a suspicion of the way each domain must give rise to a properly wider one that he could be insouciant about the second-order quantifier.

Dummett's diagnosis is subtle: it is a striking but not particularly well known fact that the first-order fragment of Frege's system is consistent. And his account appears to be attractively commonsensical, for it looks like a reformulation in Fregean terminology of familiar and incontestable set-theoretic facts, such as that the power-set of a set is never injectable into that set. But powerful and unified as it is, I think it is too recherché. There is a much simpler account of what went wrong which lays the blame, not on the stipulations Frege made concerning the truth-conditions for sentences beginning with second-order quantifiers, but on those for identities of value ranges. On the account I want to argue for, the culprit is the obvious one, Basic Law V.

Before discussing Dummett's analysis of Frege's mistake about quantifiers, I want to express some doubt concerning Dummett's idea of an indefinitely extensible concept.[3]

---

[3] A superb discussion of this issue, with which I find myself in complete agreement, is found in (Cartwright, 1994). I myself discuss the matter in Article 2 of this volume, a reply to Charles Parsons' essay "Sets and classes" (Parsons, 1983c).

"What the paradoxes revealed," Dummett writes, "was not the existence of concepts with inconsistent extensions, but of what may be called indefinitely extensible concepts." Dummett defines a definite totality as one "quantification over which always yields a statement determinately true or false."[4] According to Dummett, "No definite totality comprises everything intuitively recognizable as an ordinal number," and the concept of an ordinal number is for that reason a prototypical example of an indefinitely extensible concept. An indefinitely extensible concept has nothing that should be termed its extension, but has, rather, what may be called a sequence of extensions, of indeterminate ordinal length. Governing indefinitely extensible concepts like *ordinal* and *set* are principles of extendibility that take us from one of these extensions to the next in the sequence: the principles for *ordinal* and *set* presumably refer to operations such as $\alpha \mapsto \alpha \cup \{\alpha\}$ and $x \mapsto x \cup \{\{y \in x : \neg y \in y\}\}$.

Dummett's use of the term "totality" rather than "set" or "class" leads me to suspect strongly that he believes that we cannot actually quantify over all the ordinals (or sets) there are; that whenever we quantify over some ordinals, there is at least one ordinal we have failed to quantify over, e.g., $\bigcup W \cup \{\bigcup W\}$, where W is the totality of all those ordinals we take ourselves to be quantifying over. Of course Dummett knows perfectly well that there is no set of all ordinals (an ordinal being a set), no set containing all sets, and no class containing all classes. Nevertheless, it would seem that he does think that there has to be a—what to call it—totality? collection? domain? containing all of the things we take ourselves at any one time to be talking about. He would seem to believe that whenever there are some things under discussion, being talked about, or being quantified over, for example some or all of the ordinals, there is a set-like item, a "totality," to which they all belong. That is, he supposes that whenever we quantify, we quantify not over all the (ordinals or) sets that exist but only over some of them, and that, similarly, whatever sets we do on any occasion quantify over form a totality X which omits the item $\{x \in X : x \notin x\}$. Since $\{x \in X : x \notin x\}$ is a set (or a set-like item, an item "intuitively recognizable as a set"), we have not managed to quantify over all the sets there are.

Benson Mates once stated: "Any thing or things whatever constitute the entire membership of a class; in other words, for any things there are, there

[4]There is a danger here of which Dummett seems unaware. Whether or not a totality is definite will depend, if Dummett's words are taken literally, on what the predicates of the language are. $Th(\langle On, < \rangle)$ is easily decidable (I am grateful to Ehud Hrushovski for explaining the proof of this folkloric result to me). Thus every statement in the language whose predicates are just $<$ and $=$, and in which quantifiers range over what Dummett calls "the intuitive totality of all ordinals," will have a "definite" or "determinate" truth value, and the ordinals under less-than could not be reckoned an indefinite totality.

is exactly one class having just those things as members."[5] And Nelson Goodman has written, "Yet by the logicians' usage, any things whatever make up a class or set."[6] Logicians (like Mates) who think so err. The following things don't constitute the entire membership of any one class: the classes that don't contain themselves.

It may be replied, and I suspect that Dummett would agree, that nevertheless, whenever we use quantifiers, there must always be some domain, some totality of objects, over which our variables of quantification range; so if we take ourselves to be quantifying over all classes, then we must assume that there is a totality or domain containing all classes. And it may be thought that it is part of what we mean by "quantify over" that there must be some such domain. Certain textbooks may reinforce this impression by telling us that to specify an interpretation we must first specify a non-empty set (class, collection, totality), the universe of discourse (or domain), over which our variables range, and then specify subcollections of the domain for each monadic predicate letter, etc.

But not all. If we look at the presentation of class theory found in Kelley's *General Topology*,[7] we find that the theory presented there is a full-fledged theory of classes in which variables range over (pure) classes and in which "set" is defined to mean "member of a class." Kelley writes, "A remark on the use of the term 'class' may clarify matters. The term does not appear in any axiom, definition, or theorem, but the primary interpretation [Kelley's footnote: Presumably other interpretations are also possible.] of these statements is as assertions about classes (aggregates, collections). Consequently the term 'class' is used in the discussion to suggest this interpretation."

Kelley's axiom of extent (extensionality) reads, "For each x and each y it is true that $x = y$ if and only if for each z, $z \in x$ when and only when $z \in y$." His comment: "Thus two classes are identical iff every member of each is a member of the other."

Now it seems to me that insofar as we have any grip at all on the use of the phrase "quantify over," we have to say that Kelley, in laying down his axiom of extent, is quantifying over all classes (aggregates, collections). I take it that when Kelley says "each," he means it. How else are we to understand the axiom of extent, in view of Kelley's comment, except as saying that *any* classes x and y are identical iff x and y have the same members?

But why should we for a moment think that therefore there must be a collection of all the things Kelley was using his variables to range over? If

---

[5](Mates, 1981), p. 43.
[6](Goodman, 1972), p. 155.
[7](Kelley, 1955).

one checks the exposition in *General Topology*, at any rate, one will find no suggestion at all that there must exist some sort of super-class, containing all the classes that the theory talks about. As Kelley's exposition makes plain, we can perfectly well explain or understand a standard formal language without having to suppose that there is a collection or other totality over which the variables of the language range: we can simply say: Our variables range over all classes (and "$\in$" applies to any x and y, in that order, just when x belongs to y).

Or: over all ("absolutely," if you insist, all) objects there ("really") are. If Frege thought his variables could so range, as of course he did, he was not in error. To his credit, Frege did not have the glimmering of a suspicion of the existence of indefinitely extensible concepts.

I've argued elsewhere that by using the plural number in translating sentences that begin with second-order quantifiers we can make *second*-order quantification intelligible, even if we suppose that the first-order variables in our sentences range over all the objects there are. (Hartry Field once pointed out to me that my scheme cannot be used to translate sentences containing dyadic, or, more generally, polyadic, second-order variables.) I won't repeat the argument here, but turn instead to Frege's alleged amazing insouciance.

The account Dummett presents in Chapter 17 of the cause of the inconsistency seems to me to offer a number of interpretive difficulties, some of which arise from the explanations Frege himself gives in the first thirty-three sections of *Grundgesetze*, especially with the notorious "proof" that Frege offers in Section 31.[8] Without being perfectly certain about the matter, I take it that in that chapter, Dummett is attempting to explain the origin of, or primary reason for, the inconsistency of Frege's system, and not merely the failure of Frege's argument in Section 31 for the conclusion that inconsistency cannot occur. If only the latter, then I do not see that he has provided an explanation of the inconsistency. (A failed consistency proof for a consistent system would hardly be a serpent in Eden.) I also take it, though, that Dummett wishes to claim that if one examines the consistency proof and sees why it fails, one will see that the blame for the contradiction is to be ascribed to Frege's carelessness concerning the second-order quantifier. (But I am not sure, since at the end of the chapter Dummett states, correctly in my view, that the argument fails to show the consistency of even the first-order portion of the system.)

A second difficulty concerns causation, responsibility, blame. What caused the avalanche? The cry of the yodeler? Or the presence of snow on the mountainside? If there had been no snow, there certainly would have been

---

[8] In Section 31 Frege attempts to prove that every term of his formal system has a reference [*Bedeutung*], which of course must be unique.

no avalanche. Is second-order quantification the yodel or the snow? I will argue that Dummett has taken what should be thought of as a "background condition" to be the cause of Frege's trouble.

A third worry concerns Dummett's rather surprising claim that Frege's explanation of the second-order quantifier, unlike the one he gives for the first-order quantifier, appears to be substitutional rather than objectual. Dummett says that there is meager evidence for attributing to Frege the classical conception of the totality of all functions from and to the domain.[9] This last worry first.

A rather bad typographical error occurs on p. 218 of Dummett's book, doubtless caused by the difficulty of reproducing Frege's two-dimensional notation. What Frege actually said in the sentence Dummett quotes from Section 20 is "Now we understand by '$\forall F(\forall a\, F(a) \to F(\Gamma))$' the truth value of one's always obtaining a name of the value True whichever function-name one inserts in place of 'F' in '$\forall F(\forall a\, F(a) \to F(\Gamma))$'." Typo or no, one might well think that Dummett is probably right: Frege's talk here of inserting function-names in place of second-order variables certainly accords ill with an objectual interpretation of the second-order quantifier.

However, the evidence for the (standard) attribution of the classical conception strikes me as weighty and that for a substitutional reading of Frege's second-order quantifiers equivocal at best.

It is to be noted that Frege occasionally uses the language of substitutional quantification when discussing *first*-order quantification. In the passage from Section 31 quoted by Dummett on p. 215, Frege says, "Now '$\Phi(\xi)$' has a reference if, for every referential proper name 'Δ', '$\Phi(\Delta)$' refers to something. If so, this reference is either always the value True (whatever 'Δ' refers to) or not always." We certainly should not be willing to ascribe to Frege a substitutional interpretation of the first-order quantifiers on the strength of that remark.

Dummett cites a passage from Section 25 of *Grundgesetze*, where Frege writes, "Let $\Omega_\beta(\varphi(\beta))$ be a second-level function of one argument of the second kind, whose argument place is indicated by φ. Then $\forall F\, \Omega_\beta(F(\beta))$ is the value *True* only when for every suitable argument the value of our second-level function is the value *True*." Dummett says, "this is a comment, not a stipulation, since it is not laid down what $\forall F\, \Omega_\beta(F(\beta))$ is to be when the condition is not fulfilled; and no explanation is given of what constitutes a 'suitable argument'."

"Suitable" is Dummett's translation of *passend* (Furth renders it "fitting"), and at the end of Section 23, where he also defines arguments of the second kind as first-level functions of one argument, Frege does indeed

---

[9]Richard Heck tells me that Dummett is now no longer entirely satisfied with the views he expressed on this matter.

define *passend*. He says, "The objects and functions whose names are suitable for the argument places of the function are *suitable* [itals. Frege's] for the function." The suitability of argument-places is defined syntactically, in the obvious way. Note that Frege says "objects and functions," treating them alike.

Section 25, entitled "Generality with respect to second-level functions. Basic Law IIb," is an explanation of the correctness of Basic Law IIb, "$\forall F \, M_\beta(F(\beta)) \to M_\beta(\mathbf{F}(\beta))$," Universal Instantiation for functions. Frege glosses this Law, "What holds for all first-level functions of one-argument holds for any," a gloss which seems to me to support an objectual reading. The use of "any" here, corresponding to the Roman letter "$\mathbf{F}$", which is an unbound variable in Frege's symbolism, renders rather implausible the suggestion that he is talking only about functions that are definable in the language of the system. His next sentence is, "Consequently $\forall F \, \Omega_\beta(F(\beta)) \to \Omega_\beta(\varphi(\beta))$ is always the True, whatever first-level function of one argument $\varphi(\xi)$ may be ...," which again sounds as if he had an objectual reading of "$\forall F$" in mind. I take it that Frege, like Kelley when he says "each," means it when he says "all," "any," or "whatever." Frege says that $\forall F \, \Omega_\beta(F(\beta))$ is *true* only when a certain condition is met since he is engaged in showing that then $\Omega_\beta(\Phi(\beta))$ will also be true: he is trying to justify IIb. Thus it would not be particularly to the point for him there to give sufficient conditions as well for the truth of $\forall F \, \Omega_\beta(F(\beta))$.

In any event, in Section 24, Frege does give the requisite general account: "If after a concavity with a Gothic function letter there follows a combination of signs composed of the name of a second-level function of one argument and this function letter, which fills the argument-places, then the whole denotes the True if the value of that second-level function is the True for every fitting argument; in all other cases it denotes the False."

To tell whether Frege understands his second-order quantifiers substitutionally or objectually, one must look at how he translates formulas of his symbolism containing those quantifiers into natural language. If he regularly translates "$\forall F$" as "for every function ...," or something similar, then we may conclude that he understands them objectually; if as "for every function name ...," then substitutionally. I think the preponderance of examples favors an objectual interpretation. In addition to those in Sections 24 and 25, three other good objectual examples are found in Sections 20, 34, and 45, where Frege discusses Leibniz's law, Frege's own surrogate for $\in$, and the ancestral, respectively.

I am thus inclined to think it likely that in Section 25 Frege may have been supposing there that every concept is named by some function-name (and hence that he would have to accept that there are uncountably many function-names). Or more likely, lacking the notion of satisfaction, Frege may have simply expressed his meaning imprecisely in the sentence Dummett quotes. Despite his sophistication, care, and rigor, Frege did after all lack the notion of satisfaction (truth of a formula with respect to an assignment of appropriate items to its free variables), which would be introduced by Tarski about 40 years after the first volume of *Grundgesetze* appeared. I find it plausible that Frege would have accepted the natural Tarski-style definition of the satisfaction conditions for formulas of his language as a friendly amendment to his views.

I don't want to discuss the matter at any greater length, though, because no matter how Frege's second-order quantifiers are to be construed, the concept crucial to Dummett's discussion of Frege's explanation of second-order quantification has a concept-name that it is very easy to construct: $\forall F(\xi = {}'F \to F(\xi))$.

If we abbreviate this formula as $H(\xi)$ and then consider the sentence $\forall F({}'H = {}'F \to F({}'H))$, which we shall call TT, we find, according to Dummett, that the "stipulations intended to secure for it a determinate truth-value go round in a circle."

Dummett's argument is that if we try to determine whether or not TT is true according to the stipulations Frege has made, we shall find that we are thrown back to determining whether or not TT is true and therefore cannot suppose that Frege has satisfactorily specified conditions that determine when TT is true. For, according to the specifications, TT is true if and only if the result of substituting each concept name for F in $({}'H = {}'F \to F({}'H))$ is true (or as I prefer to think Frege must have meant, if and only if for every concept F, the extension of the concept H is identical with that of F only if the extension of H falls under F). And if so, then $({}'H = {}'H \to H({}'H))$ is true, and therefore so is $H({}'H)$, which is identical to TT, the very sentence whose truth-value we are trying to determine. Dummett adds that if we had substituted $G(\xi)$, abbreviating $\forall F({}'F = \xi \to \neg F(\xi))$, for $H(\xi)$ "we should, with a little help from Axiom V, have obtained the Russell contradiction."

What this reasoning is supposed to show is unclear to me. At least this much is certain: from certain of Frege's specifications for the second-order quantifier and the signs for equality and the conditional, we cannot, by employing one very obvious line of deduction, determine that $H({}'H)$ is false. (We have after all substituted a concept name for a universal second-order quantifier, an odd way to proceed if we are trying to show $H({}'H)$ true.)

But we haven't shown that we can't show $H({}'H)$ false by some other perhaps not so obvious line of argument, and we haven't shown that there is no way to show $H({}'H)$ true.

The argument, curiously, makes no mention of, or appeal to, Basic Law V. It shows that Frege's specifications tell us that among the conditions necessary for the truth of TT is the condition that TT be true. To show that the specifications tell us that among the conditions necessary for the falsity of TT is the condition that TT be false, it would appear to be necessary to appeal to Basic Law V (which is the obvious license for the passage from $H('H)$ to $\forall F('H = 'F \rightarrow F('H))$. As it stands, however, the argument shows only that if $H('H)$ is true, then Frege's specifications will require that $H('H)$ be true. Supplementation by Basic Law V would seem to be needed to show that if $H('H)$ is false, Frege's specifications will also require that it be false.

However, by means of a non-obvious argument due to Curry, we can in fact deduce that $H('H)$ is false. Let $C(\xi)$ abbreviate

$$\exists F(\xi = 'F \wedge F(\xi)) \rightarrow \neg H('H)$$

Suppose $C('C)$. Then $\exists F('C = 'F \wedge F('C)) \rightarrow \neg H('H)$. But since $'C = 'C$ and $C('C)$, $\exists F('C = 'F \wedge F('C))$, and therefore $\neg H('H)$. Thus we have shown that if $C('C)$ holds, then $H('H)$ is false. We must now show that $C('C)$ holds. Suppose that for some F, $'C = 'F$ and $F('C)$. By Basic Law V, $\forall x(C(x) \leftrightarrow F(x))$. But since $F('C)$, $C('C)$. So if $\exists F('C = 'F \wedge F('C))$, then $C('C)$, i.e., $\exists F('C = 'F \wedge F('C)) \rightarrow [\exists F('C = 'F \wedge F('C)) \rightarrow \neg H('H)]$, and therefore $\exists F('C = 'F \wedge F('C)) \rightarrow \neg H('H)$, i.e., $C('C)$ holds. Of course "$\neg H('H)$" could have been replaced in this argument by "$H('H)$" or "p" or "$\bot$". With the aid of Basic Law V we can prove whatever we please.

In Section 31 of *Grundgesetze*, Frege says, "By our stipulations [*Festsetzungen*], that '$'e\psi(e) = 'e\varphi(e)$' is always to have the same truth-value as '$\forall a(\psi(a) = \varphi(a))$' ..." I find it hard to understand why it is the stipulations Frege gave concerning the second-order quantifier that are to be held responsible for the contradiction *rather* than this stipulation concerning the truth-values of identities between value-range names.

Hume's principle is the statement that no matter which things the Fs and Gs may be, the number of Fs is the same as the number of Gs just in case the Fs and Gs are in one-one correspondence. Dummett calls Hume's principle "the original equivalence"; Crispin Wright calls it $N^=$ (for numerical equality). If we let $F \approx G$ be some standard formula of second-order logic expressing the equinumerosity of the objects assigned to the variables F and G, and let $\#$ be a sign for a function from concepts to objects and read $\#F$ as: the number of Fs, then we may symbolize Hume's principle: $\forall F \forall G(\#F = \#G \leftrightarrow F \approx G)$.

Some years ago I showed that any proof of an inconsistency in the theory obtained by adjoining Hume's principle to second-order logic could be readily converted into the proof of an inconsistency in second-order arithmetic. Charles Parsons seems to have been the first person to realize[10] that the

----

[10]See (Parsons, 1983a).

converse also holds (apart from Frege himself, to whom the ascription of such a recognition would be a charitable anachronism). It is instructive to consider the result of replacing the value-range operator $'$ by the cardinality operator $\#$ in Dummett's discussion. It would seem that if the second-order quantifier were responsible for Frege's difficulty, then by substituting "cardinality," denoted: $\#$, for "value-range" in Dummett's argument, we should be able to show that Frege's specifications do not provide certain sentences about cardinality, analogous to TT, with truth-values. So let $I(x)$ abbreviate $\forall F(x = \#F \to F(x))$ and let us try to determine the truth-value of $I(\#I)$.

We argue: for $I(\#I)$ to be true, $\forall F(\#I = \#F \to F(\#I))$ must be true, thus $\#I = \#I \to I(\#I)$ must be true, and therefore $I(\#I)$ must be true. So we seem to have gone round in a circle and it might occur to us to conclude that the argument shows that had Frege instead stipulated, "$\#e\psi(e) = \#e\varphi(e)$" is always to have the same truth-value as "$\psi \approx \varphi$" [abusing notation], he would not have secured a determinate truth value for $I(\#I)$.

But we have not used all the resources at hand. For, as Basic Law V was available to the real Frege so Hume's principle would have been at the disposal of our imaginary Frege. And Hume's principle enables us to see that $I(\#I)$ is in fact false. For assume $I(\#I)$, i.e., $\forall F(\#I = \#F \to F(\#I))$. Let $G(x)$ be $x \neq x$. Then $\neg I(\#G)$, for otherwise $I(\#G)$, i.e. $\forall F(\#G = \#F \to F(\#G))$, whence $(\#G = \#G \to G(\#G))$ and $G(\#G)$, i.e. $\#G \neq \#G$, impossible. Therefore also $\#I \neq \#G$. And now let $F(x)$ be $(x = \#G \vee [x \neq \#I \wedge I(x)])$. Then since $\#I$ does, but $\#G$ does not, fall under I, $I \approx F$, and by Hume's principle, $\#I = \#F$. Then by the definition of $I(x)$, $F(\#I)$, i.e., $(\#I = \#G \vee [\#I \neq \#I \wedge I(\#I)])$. But that is absurd: $\#I \neq \#G$, as we just saw. And because Hume's principle is consistent, we cannot also show $I(\#I)$ true.

The situation is the opposite for the other truth-teller-like formula similarly obtained from the formula $\exists F(x = {'}F \wedge \neg Fx)$ found in a variant proof of the inconsistency of Basic Law V: if $J(x)$ is $\exists F(x = \#F \wedge Fx)$, then it turns out that we can prove $J(\#J)$ from Hume's principle.

It is certainly not the case that if Basic Law V is replaced with Hume's principle, then the truth-values of all sentences are determined thereby. Each of the four sentences "The number of natural numbers is/is not identical with the number of numbers" and "the number of numbers is/is not identical with the number of (self-identical) objects" is consistent with the result of replacing V by Hume. Thus apart from the inescapable Gödelian incompleteness, there are some very fundamental questions about cardinality that Hume fails to resolve.

To recapitulate: Dummett's argument shows only that if one does not appeal to Basic Law V, one cannot readily deduce from Frege's stipulations concerning the second-order quantifier and the other usual logical symbols, what the truth-value of $H('H)$ is. Similarly for Hume's principle and $I(\#I)$. If we supplement Frege's explanations of the quantifiers with Hume's principle, we can show that $I(\#I)$ is true (and its companion $J(\#J)$ false); we cannot show $I(\#I)$ false, since second-order logic plus Hume's principle is consistent. However, assuming Hume's principle hardly settles all questions, even all elementary questions. If we supplement Frege's explanations with Basic Law V, we find that we can deduce that $H('H)$ is true and also deduce that it is false.

It is, in my view, not so much Frege's insouciance concerning second-order quantifiers that was responsible for his downfall as his adoption of a theory about a function from second- to first-order objects that could not possibly be true, facilitated by a lingering attachment to the idea that "contextual definitions" like Hume's principle and Basic Law V, are, if not logically true, then near enough as could make no difference. If the difficulty were where Dummett takes it to be, the introduction of the cardinality operator should be as uncertain and dangerous as that of the operator assigning value-ranges to concepts. Some uncertainty there indeed is; but danger is not now conceivable, for the result of adjoining Hume's principle to second-order logic and second-order arithmetic ("analysis") are equiconsistent.

It is of interest to examine Terence Parsons' construction of a model for the first-order fragment of Frege's system, which works by inductively assigning denotations to value-range terms. Seeing why it cannot be extended to the full system brings out that it is not any ill-foundedness in Frege's stipulations concerning the second-order quantifier that is to blame for the contradiction.

In the model, the variables, which are all first-order, are to range over the natural numbers. Inductively define the *rank* of a value-range term $'\alpha A(\alpha)$, which may contain free variables, to be the least natural number greater than the rank of any value-range term contained in the formula $A(\alpha)$. Now add to the language of the system a constant $\mathbf{i}$ denoting i, for each natural number i. Order all *closed* value-range terms of the expanded language in an ω^2-sequence, with those of lower rank preceding those of higher. Let J be a standard pairing function. Now inductively assign a natural number as denotation to each closed value-range term $'\alpha A(\alpha)$, as follows: let m be the rank of $'\alpha A(\alpha)$. If for some term $'\beta B(\beta)$ that is earlier in the ω^2-sequence than $'\alpha A(\alpha)$, $A(\mathbf{i})$ and $B(\mathbf{i})$ have the same value (perhaps a truth-value) for all natural numbers i, then let $'\alpha A(\alpha)$ denote the same number as $'\beta B(\beta)$. (The definition is OK, since if there are two such terms $'\beta B(\beta)$ and $'\gamma C(\gamma)$, we may inductively assume that they have been assigned the same denotation.) Otherwise, let $'\alpha A(\alpha)$ denote the least number $J(m, n)$ not yet assigned to any closed term of rank m. (Again, the definition is OK, since there are only finitely many terms of *rank m* that precede $'\alpha A(\alpha)$ in the sequence.)

The notion *the value of A*(**i**) is unproblematic, for the value of $A(\mathbf{i})$ will depend only on the denotations of closed terms of *lower rank,* which may be assumed inductively to have been fixed, and a universal quantification $\forall x\, D(x)$ is true if and only if for every i, $D(\mathbf{i})$ is true.

Then since there cannot be an earliest closed value-range term $'\alpha A(\alpha)$ such that for some earliest earlier closed value-range term $'\beta B(\beta)$, $'\alpha A(\alpha)$ and $'\beta B(\beta)$ have the same denotation iff for some i, $(A(\mathbf{i}) \leftrightarrow B(\mathbf{i}))$ is false, Basic Law V holds in the model.

As rereading it will make plain, Frege's argument in Section 31 can in no way be regarded as an anticipation of this proof of Parsons'. It is a natural question to ask where Parsons' proof fails, as it must, if second-order quantifiers are added to the language. Although, following Dummett, one might guess that an ill-foundedness in the truth-conditions will turn out to be responsible for the failure, that guess would be mistaken.

The trouble is that if second-order variables are added, the language will then contain the open terms $'\alpha F(\alpha)$, F a second-order variable. The rank of these terms is 0. We handled first-order quantifiers by adding constants **i** denoting members of the domain to the language. Were we to try to handle (monadic) second-order quantifiers similarly, we should have to add to the language a new monadic predicate constant **S** for each set S of members of the domain. But then whenever S and T are different subsets of the domain, we should have to assign $'\alpha \mathbf{S}(\alpha)$ and $'\alpha \mathbf{T}(\alpha)$ different members of the domain as their denotations, which cannot be done. Thus it is the Cantor–Russell *aporia* that screws up the attempt to construct a model for Basic Law V in the full second-order language, and not any ill-foundedness in the truth-conditions for the second-quantifier.

A recent paper by Richard Heck[11] sheds considerable light on Basic Law V and the origin of the contradiction. His article seems to me to do in, once and for all, the idea that "contextual definitions" like Hume's principle or Basic Law V, have, in general, any privileged logical status.

Heck observes that every sentence is equivalent to the existential quantification (over %) of some "contextual definition," by which we here mean a sentence of form $\forall F \forall G(\%F = \%G \leftrightarrow E(F, G))$, where F and G are variables of the same type, usually not that of individuals, % is a sign for a function from the entities over which F and G range to individuals, and $E(F, G)$ defines an equivalence relation on those entities. Basic Law V and Hume's principle are contextual definitions, with $\forall x(F(x) \leftrightarrow G(x))$ and

---

[11](Heck, 1992).

$F \approx G$ playing the role of $E(F, G)$.

Heck's observation is proved thus: let a sentence φ be given. Then the relation $E(F, G)$ on concepts F and G expressed by: $\varphi \vee \forall x(F(x) \leftrightarrow G(x))$ is an equivalence relation on concepts. For either φ is false, in which case $E(F, G)$ holds if and only if the same objects fall under F and G, or φ is true, in which case $E(F, G)$ is the universal relation on concepts, i.e. holds no matter what F and G might be. Either way, E is an equivalence relation. Now suppose φ holds. Let a be some object in the domain. For all F, let $\%F = a$. Then, no matter what F and G may be, $\%F = \%G$ holds; but also, by our supposition, so does $\varphi \vee \forall x(Fx \leftrightarrow Gx)$. Conversely, suppose $\forall F \forall G(\%F = \%G \leftrightarrow (\varphi \vee \forall x(Fx \leftrightarrow Gx)))$ holds, but φ does not. Then $\forall F \forall G(\%F = \%G \leftrightarrow \forall x(Fx \leftrightarrow Gx))$ holds. Russell's argument then yields a contradiction. (Substitute % for $'$.)

Thus there is a contextual definition that implies any given non-logical truth φ; moreover φ implies the existential quantification of the definition. How to demarcate those contextual definitions that should count as logical truths in some extended sense of the expression from those that should not would seem to be a philosophical problem we have no hope of solving at present.

An observation quite analogous to Heck's, but concerning sentences of form $\mathrm{Tr}(\ulcorner \psi \urcorner) \leftrightarrow \psi$ ("T-sentences," or "Tarski biconditionals"), has been made by Vann McGee.[12] McGee shows that in any formal theory capable of proving the diagonal lemma (not a stringent requirement), any sentence whatsoever will be equivalent to some Tarski biconditional. (His argument is independent of the choice of the truth-predicate Tr( ) for the language of the theory.) The argument is brief. Let φ be an arbitrary sentence of the language. Let $A(x)$ be the formula $\mathrm{Tr}(x) \leftrightarrow \varphi$. By the diagonal lemma, there is a sentence ψ such that the theory proves $\psi \leftrightarrow A(\ulcorner \psi \urcorner)$, i.e., $\psi \leftrightarrow (\mathrm{Tr}(\ulcorner \psi \urcorner) \leftrightarrow \varphi)$. Therefore, by the associativity (!) and commutativity of the biconditional, φ is equivalent in the theory to the Tarski biconditional $\mathrm{Tr}(\ulcorner \psi \urcorner) \leftrightarrow \psi$.

McGee's observation renders thoroughly implausible the assumption, natural enough, that T-sentences, apart from a few special, well known, and easily isolable problem cases (e.g. $\mathrm{Tr}(\ulcorner \psi \urcorner) \leftrightarrow \psi$, ψ having been obtained by applying the diagonal lemma to $\neg \mathrm{Tr}($)), are "partial definitions" of truth or possess only minimal content. One currently fashionable view of truth holds that all there is to the concept of truth is the infinite set of all Tarski biconditionals (minus a few trouble-makers), each of which makes almost no claim at all, but which taken together make a significant claim about truth. McGee's result suggests that the problem of purging the undesirables

---

[12](McGee, 1992).

is likely to be difficult at best.

With McGee's and Heck's examples in mind, it doesn't take much to come up with an instance of the comprehension scheme of set theory equivalent (in the presence of the null set axiom) to any sentence φ whatsoever: $\exists y \forall x (x \in y \leftrightarrow (\neg \varphi \wedge x \notin x))$. Fortunately, we are past considering comprehension statements as having the slightest claim to our credence. Contextual definitions deserve the same regard.

Sentences $\forall F \forall G(\% F = \% G \leftrightarrow E(F, G))$ assert that the function denoted by %, which standard notational conventions insist be a function of "mixed" type defined on all concepts, maps concepts into individuals in a way that respects the equivalence relation on those concepts defined by $E(F, G)$. Whether there is any such function will depend typically only on how many individuals there are. If $E(F, G)$ is $\forall x(F(x) \leftrightarrow G(x))$, as in V, there will be no suitable function for % to denote regardless of what the domain is; if $E(F, G)$ is $F \approx G$, as in Hume, there will be a suitable function iff there are (Dedekind) infinitely many individuals; if $E(F, G)$ asserts that (the values of) F is finitely different from G, then a suitable function will exist iff there are only finitely many individuals; if $E(F, G)$ is $\forall x((F(x) \leftrightarrow F(x)) \wedge (G(x) \leftrightarrow G(x)))$, then no matter what the individuals may be, there will always exist a suitable function.

Of course, for his derivation of arithmetic from Hume to work, Frege had to understand # as a function from concepts to objects (and not, say, as one from concepts to second-level concepts). Two difficulties then beset the thought that Hume is analytic: there must be infinitely many individuals and there must be a biggest number, the number of all the things there are. The second may be surmountable; the first, I have elsewhere argued, is not. Properly understood, Hume only seems to be analytic; seeing how it can be put to work reveals it as synthetic if either analytic or synthetic.

And Basic Law V is not only not analytic, it is simply a (higher-order) logical falsehood. For contraposing and making obvious abbreviations, we may write the bad half of Basic Law V: $\forall F \forall G(F \neq G \rightarrow \, 'F \neq \, 'G)$. The smooth breathing sign $'$ is then seen to be a sign for a function allegedly mapping concepts one-one into objects, the non-existence of which was carefully proved by Frege in the appendix to *Grundgesetze*, after his derivation of the Russell paradox.

Here is a suggestion about the genesis of Frege's error in putting forth Basic Law V. Seeing no way to "solve the Julius Caesar problem," Frege plumps in the *Grundlagen* for extensions and the principle "extensions of concepts are the same iff the same objects fall under the concepts." His qualms are assuaged by the evident structural similarity of this principle to Hume's principle, which both appears to be *thoroughly* obvious to Frege and strikes him as the preeminent analytic principle governing the notion of a number. He takes it that the principle about extensions is, similarly, the preeminent analytic truth governing the notion of an extension. After *Grundlagen*, he is struck by another analogy: concepts-as-functions. And if concepts are functions, there must be objects, value-ranges of functions, corresponding to extensions, and a corresponding principle, (the full) Basic Law V, governing functions and their value-ranges. Frege is now hooked: the principle about extensions strikes him as a *confirming instance* of Basic Law V, indeed as its single most important confirming instance.

But although we may guess at Frege's trains of thought, I think we cannot explain how the serpent entered Eden except to say: it is a brute fact that you cannot inject the power set of a set into that set (although you can, if the set is infinite, map its power set into it in such a way that equinumerous subsets are always taken to the same element). Perhaps hoodwinked by an analogy with an analytic-looking principle about numbers, Frege simply failed to notice that he was trying to do precisely that.

Confronted with the Russell paradox, Frege responded with a patch, which failed. It is not clear that had the patch worked, it would have counted as vindicating logicism, for it is by no means certain that Frege's revised notion of an extension and the new version of Basic Law V governing it could have counted as any more a logical notion and a logical truth than the notion of number itself and Hume's principle.

But there is a patch that works, and one not too distant in character from Frege's idea of concepts as "two-sided" entities.[13] And it seems fair to say that it has as much claim as Frege's would have had (had it worked) to vindicate logicism about the natural numbers. But like Frege's revision of Basic Law V, and unlike Hume's principle and the original Basic Law V, its drawback is that there is no notion, and certainly no unquestionably logical notion, that the patched axiom can be thought to be analytic *of.*

So let $V(x)$ be $x = x$. Call a concept F *bad* if $F \approx V \approx [x : \neg Fx]$, i.e., if there are just as many F objects, as non-F objects, as *objects*. Let $E(F, G)$ hold iff either the same objects fall under F and G or F and G are both bad. $E(F, G)$ is easily seen to define an equivalence relation.

Then the arithmetic of the natural numbers can be derived in the result of adjoining $\forall F \forall G(\%F = \%G \leftrightarrow E(F, G))$ to second-order logic; we may define zero as $\%[x : x \neq x]$ and successor in either the Zermelo or the von Neumann manner. Supplementation is required to obtain the theory of the real numbers. And it is not difficult to show $\forall F \forall G(\%F = \%G \leftrightarrow E(F, G))$ satisfiable.[14]

---

[13]Terry Parsons suggested to me this two-sided improvement of the repair given in my "Saving Frege from contradiction" (Article 11 in this volume), and inquired whether it is consistent.

[14]Thus: $\emptyset$ is (von Neumann) 0; no ordered pair is a von Neumann ordinal. Let

The example shows that Frege's stipulations concerning the conditions under which an equality between value-range terms is true may be revised so that they assign each sentence at most one truth-value, without altering the class of well-formed expressions and without preventing the derivation of arithmetic or otherwise "paralyzing" the system, in a way that preserves a reasonable amount of the naive understanding of the notion of an extension. (Of course we have drawn on an understanding of sets sophisticated by decades of experience and theory in devising the repair.)

Dummett sketches a consistency proof for the first-order fragment of the system. It would be interesting to know exactly what the strength of the fragment is. Zero and the unordered pair can be defined, as $'x(x \neq x)$ and $'z(z = x \lor z = y)$. Hence so can each particular hereditarily finite set and successor (à la Zermelo): $\{x\}$, i.e. $'y(y = x)$. And one can define complements of these: V, i.e. $'x(x = x)$; $V - x, y$, i.e. $'z(z \neq x \land z \neq y)$, etc. On the other hand, one can certainly not define $\in$ (otherwise one could form the term $'x(\neg x \in x)$), and hence not $\subseteq$ (for $x \in y$ iff $\{x\} \subseteq y$) and hence neither $\cup$ nor $\cap$ (for $x \subseteq y$ iff $x \cup y = y$, iff $x \cap y = x$). Thus Dummett is certainly right to call the system "paralyzed." But it seems to me that the greater the paralysis, the less tenable his opinion that it is second-order quantification that is the primary reason for the inconsistency; only if the first-order fragment had been strong enough to yield arithmetic or an interesting portion of it, would it be tempting to ascribe the inconsistency to the second-order quantifiers. For example, if some constructive or predicative second-order version of Frege's system could be defined and shown to be consistent and adequate for arithmetic, then Dummett's claim would acquire a force it now lacks.

One may compare the question why the *Grundgesetze* system is inconsistent with the somewhat curious question why (formal) arithmetic is incomplete. Quite evidently, one cannot either say that it is the presence of variables or that it is the presence of addition (and successor) even though variable-free arithmetic and arithmetic with multiplication alone are complete. The reason, I think, is that the two theories are mutilated, or at least very much weakened, fragments of the original. On the other hand,

---

$D_0 = \{\emptyset\}$, $D_{n+1} = \{\langle z, i \rangle : z \subseteq D_n \land i = 0, 1\}$, and $D = \bigcup D_n$. D is infinite. If F is a finite subset of D, then for some n $F \subseteq D_n$, and $\langle F, 1 \rangle \in D_{n+1} \subseteq D$; if F is a cofinite subset of D, then for some n, $D - F \subseteq D_n$, and $\langle D - F, 0 \rangle \in D_{n+1} \subseteq D$. Otherwise F is infinite coinfinite and therefore satisfies "bad." Set $\%F = \langle F, 1 \rangle, \langle D - F, 0 \rangle$, or $\emptyset$ according as F is finite, cofinite, or infinite coinfinite. And then $\langle D, \% \rangle \models \forall F \forall G (\%F = \%G \leftrightarrow (\forall x(Fx \leftrightarrow Gx) \lor F$ and G are bad$))$. For let F and G be arbitrary subsets of D. If F is cofinite, then $\%F = \langle D - F, 0 \rangle \in D$. Moreover, since F is cofinite, F and G are not both bad. And then $\%F = \%G$ iff $\%G = \langle D - F, 0 \rangle$ iff G is cofinite and $D - G = D - F$, iff $\forall x(Fx \leftrightarrow Gx)$. Similarly, if F is finite. Finally, if F is infinite coinfinite, then $\%F = \emptyset$. Then $\%F = \%G$ iff $\%G = \emptyset$, iff G is infinite coinfinite, iff G satisfies "bad," iff F and G satisfy "bad." And if $\forall x(Fx \leftrightarrow Gx)$, then F and G satisfy "bad."

it is perhaps slightly tempting to say, "You know, surprisingly, it's the presence of multiplication in the language," the idea being that addition without multiplication, as opposed to variable-free arithmetic or arithmetic without addition, is an interesting and natural fragment of the whole arithmetic. (Since addition "comes before" multiplication, arithmetic without addition is not "natural.") Were there some minor and salient feature of formal arithmetic whose removal would yield completeness (multiplication is hardly "minor"), we might be tempted to ascribe incompleteness to the presence of that feature. I conclude that since there are revisions to the system of *Grundgesetze* that restore consistency, permit the development of arithmetic, and are significantly less drastic than the wholesale elimination of second-order quantification, it is not the presence of second-order quantifiers or Frege's specifications concerning them that are to be held responsible for the contradiction.

15

1879?

In "Peirce the Logician," a paper in his recent collection of articles called *Realism with a Human Face*,[1] Hilary Putnam takes exception to a remark of Quine's, "Logic is an old subject, and since 1879 it has been a great one," the first sentence of the preface to each of the first two editions (1950, 1959) of Quine's textbook *Methods of Logic*.[2] (Putnam used *Methods of Logic* as a textbook in his logic courses in the late 1950s.) Putnam justifiably considers the statement a slight to Boole. But the remark is dropped from the prefaces to the third and fourth editions and I have not been able to find it anywhere else in either later edition.[3]

In any case, I am grateful to Putnam for recalling the excised remark and thereby prompting me to rethink a view about the history of logic that I had held for a long time: that 1879 was a *watershed* year for logic. 1879, of course, was the year in which Frege's *Begriffsschrift* was published. There is no question that with its publication logic took a gigantic step forward. I want to suggest here that there is a respect in which the advance represented by the *Begriffsschrift* may not have been so great as some (myself certainly included) have supposed. Not the advance wasn't great; just not *so* great.

Harvard University Press publishes an anthology entitled *From Frege to Gödel: a Source Book in Mathematical Logic 1879-1931*.[4] Among those whose help the editor, Jean van Heijenoort, acknowledges are Dreben, Parsons, Quine, and Wang, all of whom have at one time been on the faculty of the university whose press publishes that work. It is interesting to see the Putnam of "Peirce the Logician" standing in opposition to the Frege-centrism that has hitherto prevailed in his home university. Much of this paper consists of ruminations on the history of logic which support the

This paper was first published in *Reading Putnam*, edited by Peter Clark and Robert Hale, Oxford: Basil Blackwell, 1994.

[1](Putnam, 1990). The essay "Peirce the Logician" is found on pp. 252–260.

[2]The quoted sentence is on p. vii of both editions.

[3](Quine, 1972) and (Quine, 1982).

[4](van Heijenoort, 1967).

iconoclastic tendency of Putnam's article.

The historical note to Chapter 46 of the fourth edition of *Methods of Logic,* called "Classes," reads, "The construction illustrated in the definition of ancestor was introduced by Frege in 1879 for application to number. It was rediscovered independently a few years later by Peirce, and again by Dedekind, who propounded it in 1887 under the name of the method of *chains.*" [5] In his *Mathematical Logic,* Quine writes, "The line of reasoning used in D30 was first set forth by Frege in 1879 (*Begriffsschrift,* p. 60) in defining what I have called the proper ancestral." [6] Later we shall later take a look at this historical commonplace.

It is well-known that Quine's later, post-*Mathematical Logic* view is that the discovery of the ancestral was not an advance in logic at all, but only an advance in the theory of classes, a portion of "mathematics." Although the ancestral is described in his textbook on logic, Quine's account of it is contained in Part IV, "Glimpses Beyond"—beyond logic, as the term, according to the later Quine, is properly used, of course. Putnam has views about the scope of logic that differ interestingly from Quine's and from my own; before discussing the year 1879, I want to lay out Putnam's views about the scope of logic and what I take to be our differences on this question.

In *Philosophy of Logic,* Putnam argues that "(a) it is rather arbitrary to say that 'second-order' logic is not 'logic'; and (b) even if we do say this, the natural understanding of first-order logic is that in writing down first-order schemata we are implicitly asserting their validity, that is, making second-order assertions." [7]

He suggests that it is one quite natural choice to take statements like "For all classes, S, M, P, if all S are M and all M are P, then all S are P," which refer explicitly to classes, as statements of logic. He holds that this statement expresses the validity of the inference: $\forall x(Sx \rightarrow Mx)$, $\forall x(Mx \rightarrow Px)$ $\therefore \forall x(Sx \rightarrow Px)$. At least some statements expressing the validity of certain valid inferences should thus be counted as logical truths.

He writes, "The decision of the great founders of modern logic ... was unhesitatingly to count such expressions as $\exists F$ as part of logic, and even to allow such expressions as $\exists F^2$, with the meaning *for every class of classes.*" [8] He continues, "Suppose, however, we decide to draw the line at 'first-order' logic ('quantification theory') and to count such expressions as '$\exists F$', '$\exists F^2$', etc. as belonging to mathematics." [9] But, one might wonder, if one draws

---

[5] (Quine, 1982), p. 294.
[6] (Quine, 1955), p. 221.
[7] (Putnam, 1971), p. 32.
[8] (Putnam, 1971), p. 30.
[9] (Putnam, 1971), p. 31.

the line here, what happens to expressions like "$\forall F$" and "$\forall F^2$"? Are these still expressions of logic? Putnam here seems to be hinting at a position on which truths of the form $\forall F_1 \forall \ldots \forall F_n \varphi$, φ a valid first-order formula, should be counted as logical truths, while truths $\exists F \varphi$, which on his view assert the existence of classes with certain properties, should not.

To object that $\exists$ is definable in terms of $\forall$ and negation is to miss the point of this view, which is that those "Π_1" statements obtained from valid formulae of first-order logic by prefixing strings either of universal second-order quantifiers $\forall F$ or of their definitional equivalents $\neg \exists \neg F$ are more justifiably counted as logical truths than class-existence assertions and other true "Σ_1" statements.

In "Peirce the Logician," Putnam writes:

> (1) Where to draw the line between logic and set theory ... is not an easy question. The statement that a syllogism is valid, for example is a statement of second-order logic. (Barbara is valid just in case
>
> $$\forall F \forall G \forall H (\forall x (Fx \to Gx) \land \forall x (Gx \to Hx) \to \forall x (Fx \to Hx)),$$
>
> for example). If second-order logic is "set theory," then most of traditional logic becomes set theory. (2) The full intuitive principle of mathematical induction is definitely second-order in anybody's view. Thus there is a higher-order element in arithmetic whether or not one chooses to "identify numbers with sets" just as Frege realized.[10]

And at the Quine conference in San Marino, Putnam explicitly said he wanted to "split the difference" between Quine and me, and to count second-order universal quantifications of valid first-order schemata, but only those, as logical truths. Some fretting over a few uncomfortable aspects of this position is called for.

First of all, Putnam writes, "Barbara is valid just in case

$$\forall F \forall G \forall H (\forall x (Fx \to Gx) \land \forall x (Gx \to Hx) \to \forall x (Fx \to Hx))."$$

This is what R. C. Jeffrey calls Loglish. What's not perfectly clear to me is what the formula "$\forall F \ldots$" is supposed to mean. What does the first-order variable x range over? One naturally supposes over (absolutely) all the things there are. Otherwise, the second-order statement would not give the full force of the validity; one can, one supposes, make a Barbara syllogism about any things whatsoever. But if so, then what do the

---

[10](Putnam, 1990), p. 259. There are some first-order quantifiers missing from the text, which I have here restored.

second-order variables range over? Classes? Well O.K.; but then it had better not be that all classes are things. But then can one or can't one speak about everything—and by everything, I mean everything including all classes there might be—with one's first-order variables? Putnam would be the last person to want to say that one could use the word "everything" to quantify over everything but one could not use a universal quantifier to do so.

But if it means "If $F, G,$ and H are classes, everything that belongs to F belongs to G, and everything that belongs to G belongs to H, then everything that belongs to F belongs to H," how is one to symbolize:

> If each thing that is a class containing all ordinals is a class that is not a member of anything and each thing that is a class that is not a member of anything is a class that is the same size as the universal class, then each thing that is a class containing all ordinals is a class that is the same size as the universal class

a valid, indeed syllogistically valid, statement, each of whose propositional constituents is true, according to the currently standard theory of classes? If one symbolizes it as $\forall F \forall G \forall H(\forall x(Fx \rightarrow Gx) \wedge \forall x(Gx \rightarrow Hx) \rightarrow \forall x(Fx \rightarrow Hx))$, with "$x$" ranging over all classes, and "Fx" abbreviating "x is a class containing all ordinals," etc., then how is this statement supposed to be a consequence of the second-order assertion about classes? The problem, of course, is that according to the standard theory of classes, there aren't any classes that contain all classes that contain all ordinals, etc.

I don't wish to suggest that these difficulties can't somehow be overcome. In a pair of articles published some years ago,[11] I suggested that the plural number can be used in explaining what validity of Barbara comes to. It is admittedly rather taxing to pronounce the explanation, and to do so risks the Hilarious response: "*That's* clearer than introducing classes?" But here goes:

> No matter what certain things—call them F things—may be, no matter what certain things—call them G things—may be, and no matter what certain things—call them H things—may be, if everything that is an F thing is a G thing, and everything that is a G thing is an H thing, then everything that is an F thing is an H thing.

Not so bad after all, and even if appreciably more awkward, certainly less involved with classes and the serious theoretical difficulties that attend their

---

[11] "To Be is To Be a Value of a Variable (or To Be Some Values of Some Variables)," and "Nominalist Platonism," reprinted as Articles 4 and 5 in this volume.

introduction than the formulation offered by Putnam. Whether use of the plural number frees one from these worries or not, it remains the case that talk about classes won't get one out of the sorts of difficulty that prompted their introduction in the first place.

Other possible responses, of course, are to admit a hierarchy of classes, superclasses, superduperclasses, etc., to claim that, actually, we can't talk about everything at once, to invoke some doctrine of typical or systematic ambiguity, or to mutter something about a ladder. I prefer the tongue-twisting to the mystical or the nonsensical, and particularly to the nonsensical that advertises itself as such, "strictly speaking."

Putnam's remark that mathematical induction is second-order and that there is a higher-order element in arithmetic is perplexing. The position of the remark in Putnam's paper makes it seem as if he were arguing *against* a Quinean. But the remark is one that Quine could happily accept.

I suspect—hope—that the view Putnam wanted to put forth is that just as statements like the second-order statement expressing the validity of Barbara should be regarded as logical truths, so the ancestral should be counted as a logical notion. The (weak) ancestral R_* of a relation R may be defined by a formula that is a universal second-order quantification of a first-order formula: xR_*y iff

$$\forall F(Fy \land \forall w \forall z(Fz \land wRz \to Fw) \to Fx).$$

Thus a good deal of the content of arithmetic, expressible with the aid of the ancestral, would count as logic.

If Putnam wishes to count the ancestral as a notion of logic, then one wants to know about inferences involving the ancestral. Are any of them to be counted as logically valid? What about inferences in which a statement to the effect that one person is an ancestor of another is a *premiss*? Here's a favorite example of mine, a near-relative of an inference made by Frege in deriving (83) of the *Begriffsschrift*:

Xavier is an ancestor of Yolanda.	xP_*y
Yolanda is blue.	By
Any parent of anyone blue is red.	$\forall w \forall z(Bw \land zPw \to Rz)$
Any parent of anyone red is blue.	$\forall w \forall z(Rw \land zPw \to Bz)$
$\therefore$ Xavier is either red or blue.	$\therefore Rx \lor Bx$

Of course "xP_*y" here abbreviates its second-order definition.

There is a way to express, in English, an argument that shows the validity of this inference without explicitly introducing the notion of a class. By the first premiss, no matter who certain people may be, if Yolanda is one of them and every parent of any one of them is also one of them, then Xavier is one of them too. Now consider the people who are either red or blue.

By the second premiss, Yolanda is one of them. By the third and fourth premisses, every parent of any one of them is also one of them. Thus Xavier is one of them, and is therefore either red or blue.

I find it an uncomfortable position to want to admit the ancestral as a logical notion, but not to admit as (logically) valid inferences such as the one just given which involve the ancestral in at least a moderately interesting way. Does one want to accept a doctrine on which the foregoing argument is not a piece of logic, in the fullest sense of the word?

But then if the inference is logically valid, then either the statement that it is valid is not itself valid or some true Π_2 statements are logical truths. The reason is that the statement that the inference is valid will be a universal quantification (with respect to P, R, B) of a formula in which a universal second-order quantifier occurs in the *antecedent* of a conditional; its prenex equivalent will thus begin: $\forall P \forall R \forall B \exists F \ldots$ I would have supposed that the desire to count the statement that Barbara is valid as itself valid would have arisen from a more general, and perfectly reasonable, wish: to count a true statement to the effect that any given inference is valid as itself valid.

Thus Putnam's position on the scope of logic strikes me as unstable: either certain plainly logical modes of reasoning fail to count as such or more than just the valid Π_1 sentences are going to have to come out as logical truths.

In the articles I mentioned, I offered a scheme of translation from the notation of second-order logic into natural language augmented with devices for cross-reference. (Such devices are a necessary addition to natural language if one wishes to translate sentences in *first*-order notation with even a moderately complicated quantificational structure into natural language.) The key feature of the scheme was the clause for the translation of the second-order existential quantifier "$\exists X$," which was to be rendered roughly as "There are some things that$_x$ are such that $\ldots$"[12] Under the assumption that the first-order variables range over all the things there are, the translation of $\exists X \forall x (Xx \leftrightarrow \neg x \in x)$ into English is equivalent, not to a contradiction, but to the trivial truth: if there is at least one thing that is not a member of itself, then there are some things that are such that each thing is one of them if and only if it is not a member of itself.

As was mentioned in Article 4 above, Charles Parsons claimed that there appeared to be no non-artificial way to translate "$\forall X$" into natural language, and that any translation would seem to have to proceed via the equivalence of "$\forall X$" with "$\neg \exists X \neg$." David Lewis later suggested, however, that the construction "No matter what some (or certain) things X may be $\ldots$" is a perfectly adequate way to render second-order monadic universal

---

[12] A small qualification must be made to account for the "null class."

quantifiers into familiar English.

Lewis's suggestion was particularly striking to one who had been taught[13] that it was not perfectly correct to read "$\forall x(Fx \rightarrow Gx)$" as "for every x, if x is an F, x is a G," since "every" cannot be followed by a linguistic item that also functions as a pronoun, and that the proper way to read it was: "No matter what a thing may be, if it is an F, then it is a G." Pluralizing "a" to "some" (and pronounced: [sm], not [sum], as Helen Cartwright points out), or to "certain," yields the desired correct and natural version of "$\forall X$."

A more serious drawback, first pointed out to me by Hartry Field, is that the scheme provides no way to translate into natural language second-order dyadic, or, more generally, polyadic, quantification. In favorable cases, of course, pairing functions will be definable in the language and higher-degree second-order quantification reducible to monadic. But the availability of pairing functions cannot be considered to be guaranteed by logic: any domain closed under an ordered pair function contains (Dedekind) infinitely many objects if it contains at least two. Field's observation appears to be unassailable.

In the absence of a way to translate sentences of second-order logic containing second-order polyadic quantifiers into natural language, must we regard second-order polyadic logic as capable of being made intelligible only via quantification over polyadic relations, and thus as legitimate only when first-order quantifiers range over a set?

I am not sure, but I don't see that we must. We need not regard translatability of a notation into language we already understand as a necessary condition of the intelligibility of that notation. To be sure, our ability to translate second-order monadic statements into English enables us to take them as having a sense—the one given by the translation. (The issue, of course, is whether they have a sense when the first-order variables range over objects that do not together constitute a set.) But the provision of a translation scheme into an antecedently understood language need not be the only way to confer sense upon statements in some notation; we didn't learn our mother tongue that way, for sure. And after all, we understand the basic formal machinery of second-order logic rather well; the syntax (including the devices of quantification and predication, as well as elementary proof theory) of polyadic second-order notation can be understood by one who understands that of polyadic first-order logic and monadic second-order logic. Moreover, we can imagine sufficiently well how enough additional resources—"pro-verbs"—might have been present in natural language for us to be able to translate into it the entire formalism of polyadic second-order logic. Thus I incline to think that our understand-

---

[13]As I had been, by C. G. Hempel.

ing of both the syntax of polyadic second-order logic and the semantics of polyadic first-order and monadic second-order logic has combined with our ability to envisage an extension (possibly a quite radical one) of the language we speak to afford us an understanding of the semantics of polyadic second-order logic, even though we cannot express that understanding in the language we presently speak.

I shall not speculate on the question whether we can imagine extensions of English that would make possible a translation scheme for such third-order statements as "$\exists a \exists F \exists x (aF \wedge Fx)$," "$x$" again being understood to range over all objects. Instead, hoping that I have opened the way for Putnam to move closer to my position, and aware that I have sufficiently nudged him in that direction, I shall now turn to one of the historical issues raised by Putnam's paper, the extent to which 1879 was an "epochal"[14] year in the history of logic.

The main point of "Peirce the Logician" seems to me entirely correct: great accomplishments in logic *were* made before 1879. I want to begin to sharpen the point by describing an observation about propositional logic that Boole made towards the end of his 82-page monograph *The Mathematical Analysis of Logic*,[15] which was published in 1847.

In the next-to-last section, called "Properties of Elective Functions" (elective functions are truth functions, or in current parlance, Boolean functions), Boole notes that, as we would now put the matter, any formula $\varphi(x)$ of propositional logic containing the propositional variable x is equivalent to the formula $(\varphi(1) \wedge x) \vee (\varphi(0) \wedge \neg x)$. Here 1 and 0 are constants of propositional logic for truth and falsity. The disjuncts are incompatible, and nothing is lost by replacing the Boolean exclusive disjunction with the inclusive $\vee$.

What Boole realized was that iterating this operation shows that an arbitrary propositional formula $\varphi(x_1, \ldots, x_m)$ is equivalent to the disjunction of the 2^m formulas $(\varphi(i_1, \ldots, i_m) \wedge \pm x_1 \wedge \ldots \wedge \pm x_m)$, where each i_j is either 1 or 0 and $\pm x_j$ is x_j or $\neg x_j$ according as i_j is 1 or 0. Boole termed the equivalence the law of development, and called (his analogues of) the constant formulae $\varphi(i_1, \ldots, i_m)$ the moduli of $\varphi(x_1, \ldots, x_m)$. Since each modulus is equal to 1 or 0 (and it can be easily calculated which), every propositional formula $\varphi(x_1, \ldots, x_m)$ is, as Boole saw, equivalent to the disjunction of those formulae $x_1 \wedge \ldots \wedge \pm x_m$ for which the moduli $\varphi(i_1, \ldots, i_m)$ do not vanish (are not $= 0$). Thus Boole knew that every formula of propositional

---

[14]Thus van Heijenoort: "A great epoch in the history of logic did open in 1879 when Gottlob Frege's *Begriffsschrift* was published." (van Heijenoort, 1967), p. vi.

[15](Boole, 1847). The monograph is much less well known than Boole's *Laws of Thought* (Boole, 1916), probably because it is too short for a proper book, and too long to be included in a collection of articles.

logic is equivalent to one in what we now call perfect disjunctive normal form.

Boole clearly had the idea of all possible distributions of truth-values: "It is evident that if the number of elective symbols is m, the number of different moduli will be 2^m, and that their separate values will be obtained by interchanging in every possible way the values 1 and 0 in the places of the elective symbols of the given function." Thus one main feature of the method of truth-tables, usually credited to Post and Wittgenstein, was on prominent display in Boole's early monograph.[16] Another feature was not: the now familiar manner in which the truth-values of compound sentences are inherited from those of their components; for that reason, it would be injudicious, I think, to try to credit Boole with the discovery of truth-tables.

No edition of *Methods of Logic* provides a natural deduction system for Boolean notions. It may be that Quine was inclined to minimize the significance of the propositional calculus simply because truth-functional validity is decidable. Although quantifiers are important in logic—very important— it is not strictly true that "their importance cannot be overemphasized." It may have taken the rise of the computer for us to see the interest and importance of the propositional calculus, but it is not now possible to forget that a problem about the propositional calculus, the $P = NP$ problem (which Putnam worked on with Martin Davis in the '50s, by the way), is generally considered to be one of the ten most important unsolved problems in the whole of mathematics. We now see the decidable as a realm with an interesting structure in which Boolean notions are anything but trivial. Truth-functional validity is certainly decidable, but it is also, in the apt technical term, hard. *Logic is an old subject, and since 1847 it has been a hard one.* Quine himself has of course made important contributions to propositional logic, which, he has noted with apparent pride, contributors to engineering journals have frequently cited. The offending sentence having been removed from later editions of *Methods* and with "Boolean" a term now known to every student of programming, amends should be thought made.

Along with the publication of the *Begriffsschrift,* another event of logical note occurred in 1879. A passage in the introduction to Volume 4 of the *Writings of Charles S. Peirce*[17] relates that according to lecture notes taken by a student of Peirce's, Allan Marquand, Peirce gave a lecture in December of 1879 in which he presented the following axiomatization of arithmetic:

[16](Kneale and Kneale, 1984) contains an extensive and very useful account of Boole's *Mathematical Analysis of Logic* and *Laws of Thought.*

[17]The passage, to which Joe Ullian called my attention, is in (Frisch et al., 1982), vol. 4, p. xliv.

1. Every number by process of increase by 1 produces a number.

2. The number 1 is not so produced.

3. Every other number is so produced.

4. The producing and produced numbers are different.

5. In whatever transitive relation every number so produced stands to that which produces it, in that relation every number other than 1 stands to 1.

6. What is so produced from any individual number is an individual number.

7. What so produces any individual number is an individual number.

Letting "Pxy" mean "x by process of increase by 1 produces y," we may symbolize these:

1. $\forall x(Nx \rightarrow \exists y(Pxy \wedge Ny))$

2. $N1; \forall x(Nx \rightarrow \neg Px1)$

3. $\forall y(Ny \wedge y \neq 1 \rightarrow \exists x\, Pxy)$

4. $\forall x \forall y(Nx \wedge Ny \wedge Pxy \rightarrow x \neq y)$

5. $\forall R(\text{Trans}(R) \wedge \forall x \forall y(Nx \wedge Ny \wedge Pxy \rightarrow Ryx) \rightarrow \forall x(Nx \wedge x \neq 1 \rightarrow Rx1))$

6. $\forall x \forall y(Nx \wedge Pxy \rightarrow Ny)$

7. $\forall x \forall y(Ny \wedge Pxy \rightarrow Nx)$

"Trans(R)" abbreviates: $\forall x \forall y \forall z(Rxy \wedge Ryz \rightarrow Rxz)$, of course.

The editors write that Peirce went on to define the relation "greater than." Not having seen Marquand's notes, I can only guess that the definition ran, more or less: a number is greater than another if it stands to that other in every transitive relation in which every number so produced stands to that which produces it. If so, then Axiom 5 would be equivalent to the statement that every number other than 1 is greater than one.

There are at least one and possibly two serious omissions from Peirce's axiomatization. It is not clear from Peirce's language that at most one number is produced by process of increase by 1 from any number. What he says allows, I think, that some number produces two distinct numbers, and that the tree of finite sequences of zeros and ones or even the full infinitary tree of finite sequences of objects in any one set whatsoever might be a model of his axioms. Perhaps, though, the phrase "the producing and the produced numbers" in Axiom 4 is meant to imply uniqueness of the produced number; we might then adjoin the conjunct $\forall x \forall y \forall z (Nx \land Pxy \land Pxz \rightarrow y = z)$ to the symbolization of 4.

The second omission is that Peirce's axioms, even with the emendation of Axiom 4 just given, do not guarantee that production is one-one, that is, that different numbers produce different numbers. Dedekind's condition δ in *The Nature and Meaning of Numbers* [*Was sind und was sollen die Zahlen?*] explicitly provided just such a guarantee. The omission from Peirce's list is serious: Peirce's axioms are true in the three-element model in which 1 produces 2, 2 produces 3, and 3 produces 2:

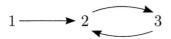

But we can easily supply Peirce with the necessary axiom:

$$\forall x \forall y \forall z \forall w (Pxy \land Pzw \rightarrow (x = z \leftrightarrow y = w)).$$

What is worth dwelling on in Peirce's list of axioms is not what it leaves out, but what it contains. The fifth axiom, which says that whenever R is a transitive relation which includes the relation on numbers *is produced by*, then every number other than 1 bears R to 1, is remarkable. It is, audibly, a second-order axiom, universally quantifying over all relations of a certain sort: "in whatever transitive relation ... in that relation ... " What the axiom says, and how it is supposed to work, though, may not be immediately apparent.

However, translating "is produced (by process of increase by 1) by" as "succeeds" enables us see that it implies the principle of mathematical induction. For suppose that 1 has a certain property F and that every number that succeeds a number with property F also has property F. [$F1 \land \forall x(Nx \land Ny \land Fx \land Pxy \rightarrow Fy)$.] We are to show that every number has F. Let R be the relation that holds between w and z if and only if w and z are numbers and if z has F then w has F. [Rwz iff $Nw \land Nz \land (Fz \rightarrow Fw)$.] R is transitive, by the transitivity of material implication. Moreover, every number produced by a number stands in R to that number, for if y is produced by x, i.e. y succeeds x, then by our supposition, if x has F, y has F, that is to say, y stands in R to x. [$Pxy \rightarrow (Fx \rightarrow Fy)$, and so $Pxy \rightarrow Ryx$.] By Axiom 5, every number other than 1 stands in R to 1.

That is, if x is a number other than 1, if 1 has F, then x has F; since 1 does have F, x has F. Thus every number x has F.

In the presence of the other axioms, mathematical induction implies Peirce's Axiom 5; we may safely leave this derivation to the reader.

It is at least moderately plausible to conjecture that Peirce recognized that mathematical induction thus followed from Axiom 5 and conversely. But whether or not he did, it is certain that the idea of applying the logic of relations to the "primitive" relation of one number's succeeding another in order to characterize the natural number series was in the air over Baltimore, far from that over Jena, the year the *Begriffsschrift* was published.

The date of the preface of the Begriffsschrift is 18 December 1878, and there is absolutely no question that Frege's achievements were thus much in advance of those of Peirce, even if one ignores Peirce's omissions. The opening paragraph of van Heijenoort's (1967) introduction to the *Begriffsschrift* enumerates several of the excellencies of Frege's book. I note with pleasure, by the way, that van Heijenoort refers to the definition of the ancestral as "a *logical* [my italics] definition of sequence."[18] I am not now concerned with the excellencies of Frege's work but want instead to raise the question of whether Frege was actually the first to define the ancestral. Peirce came close, we have seen, but whatever he may be thought to have done, Frege had him beat by at least a year.

x bears the (strong) ancestral of the relation R to y if y belongs to every class containing all objects to which x or some member of the class bears R: xR^*y iff $\forall K(\forall z(xRz \lor \exists w(Kw \land wRz)) \rightarrow Kz) \rightarrow Ky)$, i.e. $\forall K(\forall z(xRz \rightarrow Kz) \land \forall w \forall z(Kw \land wRz \rightarrow Kz) \rightarrow Ky)$. The class of all objects to which x bears the ancestral of R is itself a class containing all objects to which x or some member bears R. Was Frege the first person to define a class in this manner, as the class of objects belonging to all classes satisfying a certain condition, where that class is itself one of the classes that satisfy the condition? Frege gives the definition of the ancestral of a relation in Section 26 of the *Begriffsschrift*. His "elucidation" of his (symbolic) definition is:

> If from the two propositions that every result of an application
> of the procedure f to x has property F and that property F
> is hereditary in the f-sequence it can be inferred, whatever F
> may be, that y has property F, then I say: "y follows x in the
> f-sequence," or "x precedes y in the f-sequence."

Did anyone give such a definition before Frege? Of course it must be somewhat vague what "such" denotes here.

[18] (van Heijenoort, 1967), p. 1.

The preface to the second edition of Richard Dedekind's monograph, *The Nature and Meaning of Numbers*, dated 24 August 1893, contains the statement,

> About a year after the publication of my memoir I became acquainted with G. Frege's *Grundlagen der Arithmetik*, which had already appeared in the year 1884. However different the view of the essence of number adopted in that work is from my own, yet it contains, particularly from §79 on, points of very close contact with my paper, especially with my definition (44).[19]

Definition (44) runs: "If A is an arbitrary part [subset] of S, then we will denote by A_0 the intersection of all those chains (e.g. S) of which A is a part; this intersection A_0 exists (cf. 17) because A itself is a common subset of all these chains. Since by 43, A_0 is moreover a chain, we will call A_0 the chain of the system A, or for short the chain of A."[20] K is a *chain* with respect to a function φ, it may be recalled, if every image of a member of K is a member of K, if as we would say, K is closed under φ. Thus according to Dedekind's definition, $y \in A_0$ iff $\forall K(\forall z(z \in A \rightarrow z \in K) \wedge \forall w(w \in K \rightarrow \varphi(w) \in K) \rightarrow y \in K)$. It is clear that Dedekind's definition of A_0 is strikingly like Frege's of "y follows x in the f-sequence," and is "such" a definition.

Was sind was first published in 1888, nine years after the *Begriffsschrift*. In the preface to the first edition, dated 5 October 1887, Dedekind wrote,

> The design of such a presentation I had formed before the publication of my paper on *Continuity*, but only after its appearance and with many interruptions occasioned by increased official duties and other necessary labors, was I able in the years 1872 to 1878 to write out a first draft on a few sheets of paper, which several mathematicians examined and in part discussed with me.[21]

The question thus arises whether those few sheets of paper contained anything like the definition of A_0 Dedekind would later give in Section 44 of his published monograph.

The answer may have come to light only as recently as 1976, with the publication of Pierre Dugac's *Richard Dedekind et les Fondements des Mathématiques*,[22] which contains a large number of unedited texts, includ-

---

[19](Dedekind, 1901), p. 42.

[20](Dedekind, 1901), pp. 57–58.

[21](Dedekind, 1901), p. 32.

[22](Dugac, 1976). I am extremely grateful to Jan Sebestik, of the Institut d'Histoire des Sciences, Université de Paris I, for calling my attention to Dugac's book and its appendix LVI, which contains Dedekind's 1872–1878 draft of *Was sind und was sollen die Zahlen?*

ing the first draft of *Was sind* mentioned by Dedekind. It turns out that
the draft contains most of the ideas and proofs found in the later version,
including the definition of a chain, that of a principal element (*Hauptele-
ment*), which is an element of the given set S that is contained in *all* chains
(Dedekind emphasizes "all"), and the argument that the system of principal elements is itself a chain. These all appear near the beginning of the
draft. Dedekind's first definition of "chain" reads: "A part K of S shall
be called a *chain* (or any other name) if K' is a part of K."[23] It strongly
appears that different parts of the draft were composed at different times,
for Dedekind later defines a group in the same way: "(with respect to this
mapping) G is called a group if G' is a part of G."[24] B is then called
dependent on A "if B is a part of every group of which A is a part."[25] A
theorem immediately follows: "The system of all things dependent on A is
a group, which shall be designated A_0."[26] Dedekind was evidently dither-
ing over which term to use. Later in the manuscript he settles on "chain,"
defines it as before, defines a thing, b, in S to be dependent on a thing, a,
in S if "every chain that contains a contains b," introduces the notation:
(a) to denote the system of all things dependent on a, and proves that (a)
is a chain.[27] Towards the end of the draft there is even a fourth series of
similar definitions and theorems.

Dedekind and Dugac give only the span of years 1872–1878 as the period
during which the draft was written. But since it appears that different
parts of the draft were composed at different times (for it seems rather
unlikely that Dedekind would have written down much the same thing four
times over within the space of a year or two), it is quite possible that
Dedekind formulated the definition of A_0 for the first time several years
before 1878, possibly towards 1872, and quite possibly before Frege arrived
at the definition of "y follows x in the f-sequence." It may be that we shall
never know who came first, and perhaps that is all to the good. Reading the
draft convinces me, however, that it is at least as probable that Dedekind
had the definition of A_0 before Frege had that of the ancestral as that he
had it after.

The draft contains much else that is familiar from the monograph: definitions of one-one mapping, (Dedekind!) infinity, the connection between
the principle of complete induction[28] and the notion of a principal element,
the recursion equations for addition (not multiplication), and the assertion

---

[23](Dugac, 1976), p. 295.
[24](Dugac, 1976), p. 296.
[25](Dugac, 1976).
[26](Dugac, 1976).
[27](Dugac, 1976), p. 298.
[28]Early on in the draft one finds the parenthetical phrase "(Schluß von n auf $n + 1$)";
in the *Grundlagen* Frege refers to induction as "die Schlussweise von n auf $n + 1$."

that numbers are creations of the human mind (*Geist*). Notable is the elaborateness of the development of the theory of chains that Dedekind gives towards the end of the draft, after the fourth definition of "chain" (counting that of "group" as the second). Another remarkable feature, as Dugac notes, is the *absence* of an analogue of Theorem 66 of *Was sind*, "There are infinite systems," and hence of any dubious or non-mathematical proof of that theorem.

Dedekind, of course, did not quite define the ancestral of a relation. But the difference between his definition of A_0 (with respect to a mapping φ) and Frege's of the ancestral is small. y is an element of A_0 iff $\forall K(\forall z(z \in A \to z \in K) \land \forall w(w \in K \to \varphi w \in K) \to y \in K)$; xR^*y iff $\forall K(\forall z(xRz \to Kz) \land \forall w \forall z(Kw \land Rwz \to Kz) \to Ky)$. Dedekind's notion is less general than Frege's in what is perhaps a significant respect: it covers only functional, rather than arbitrary, relations. For Dedekind to have had a notion fully as general as Frege's, however, he need only have changed the definition of K' from: $\{\varphi w : Kw\}$, i.e. $\{z : \exists w(Kw \land \varphi w = z)\}$, to: $\{z : \exists w(Kw \land Rwz)\}$. The difference between the definitions of A_0 and R^* is insignificant in comparison with the idea of making definitions in this manner at all.

Whether or not Dedekind anticipated by a few years one of Frege's greatest discoveries, there is another year before 1879, 1858, which it would not be implausible to take as the one in which logic became a "great" subject. Indeed, logic as we now know it might be said to have arisen precisely on 24 November 1858. For on that date, according to the preface to *Continuity and Irrational Numbers*, Dedekind succeeded in discovering the "true origin in the elements of arithmetic" of the theorem that every bounded increasing function on the reals approaches a limit.[29] The key idea needed to prove the theorem rigorously was one of the most celebrated of all logical constructions, Dedekind's definition of the real numbers via cuts in the rational numbers. Although exceedingly familiar, the construction seems to me to be of possible philosophical interest in a way that has not been much remarked upon. Let me review it and describe the aspect I find noteworthy.

Take as given the set Q of rationals and the less-than relation on them. In Section 4 of *Continuity*, called "Creation of the Irrational Numbers [*Schöpfung der irrationalen Zahlen*]," Dedekind defines a cut as a pair (A_1, B_1) of non-empty classes of rationals such that every element of A_1 is less than every element of B_1. Every rational number r *produces* two cuts, viz. $([q : q \leq r], [q : q > r])$ and $([q : q < r], [q : q \geq r])$, which are identified, regarded as "inessentially different." Some cuts, however, e.g. $([q : q < \sqrt{2}], [q > \sqrt{2}])$, are not produced by any rational number.

[29](Dedekind, 1901), p. 2.

"Now anytime a cut (A_1, A_2) occurs that is produced by no rational number, we then *create* [*erschaffen*,[30] emphasis in the original] a new, irrational number α, which we regard as completely defined by this cut (A_1, A_2); we shall say that the number α corresponds to this cut, or that it produces this cut."[31] If α is produced by (A_1, A_2) and β by the essentially different cut (B_1, B_2), then α is said to be less than β if and only if A_1 is properly included in B_1 (in which case B_2 is also properly included in A_2).

In *The Nature and Meaning of Numbers* Dedekind argues that the term "free creation [*freie Schöpfung*]" is an appropriate one to apply to the natural numbers. His thought is that one may come to recognize the numbers by considering an arbitrary system satisfying the "Dedekind–Peano" axioms while "disregarding" all non-structural aspects of that system. Despite Dedekind's use of "*erschaffen*" in *Continuity*, it is plausible that he regarded the creation of the irrationals and that of the natural numbers as two instances of the same phenomenon. On Frege's view, of course, nothing at all is created by inattention or postulation: one merely recognizes what there is. Frege, though, would certainly have accepted the mathematical part of Dedekind's claim, which can be symbolized:[32]

$$\forall X [X \subseteq Q \wedge \Lambda \neq X \neq Q \wedge \forall q(Xq \to [\forall r(r < q \to Xr) \wedge$$
$$\exists r(q < r \wedge Xr)]) \to \exists! \alpha (\forall r(Xr \to r < \alpha) \wedge \forall r(\neg Xr \to \alpha \leq r))]$$

From our point of view, the crucial aspect of Dedekind's construction is that iterating it produces nothing new, as he proves in Section 5 of *Continuity*. Unlike the rationals, every cut in the reals (a well-defined notion, since less-than on the reals has been defined) is produced by a real. (Indeed the cut (A_1, A_2) in the reals is produced by the real corresponding to the cut $(\{r : \exists \alpha(A_1 \alpha \wedge r < \alpha)\}, \{r : \exists \alpha(A_2 \alpha \wedge \alpha \leq r)\})$ in the rationals.) Thus after taking the rationals, making all possible cuts in them, and then introducing in Dedekind's manner numbers corresponding to these cuts, one obtains no *new* numbers by taking the numbers one has so far gotten, making all possible cuts in them, and then introducing new numbers corresponding to the cuts made the second time around.

Cuts are two-sided. Dedekind defined them as pairs (A_1, A_2) of sets of certain sorts of numbers in which A_2 is the set of all numbers of the

[30]My colleague Irene Heim informs me that the etymologies of "erschaffen" and "Schöpfung" converge, but only at a rather remote date.

[31](Dedekind, 1901), p. 15.

[32]Q is the set of rationals, "q" and "r" are variables ranging over the rationals, and "α" a variable ranging over the reals. In words, "for every downward closed concept X under which only rational numbers fall, under which some but not all rational numbers fall, and under which no greatest rational number falls—cuts determined by rationals necessitate this proviso—there is a unique real number greater than every rational falling under X and less than or equal to every rational number not falling under X."

relevant sort not in A_1. Cuts, however, are not numbers; instead, they "are produced by" numbers, "correspond to" them, or "define" them. It is thus natural to think of a cut not as a first-order object, but either as a pair of second-order entities (A_1, A_2), as *two* second order entities A_1 and A_2, or, perhaps most naturally, as a second-order entity A together with another one N, a "sort of number," such that $\forall x(Ax \rightarrow Nx)$, $\exists x\, Ax$ and $\exists x \neg Ax$. It is equally natural to formalize Dedekind's account of the reals in second-order logic with the second-order variables ranging over cuts (or their left-hand halves), and relativized first-order variables ranging over the numbers, which are the objects Dedekind is primarily interested in.

Like cuts, Fregean concepts are also two-sided; they are functions from objects to the two truth-values. There is a striking analogy between Dedekind's definition of the reals as objects to which cuts in the rationals correspond uniquely (if one ignores inessentially different cuts) and Frege's ill-fated attempt in Basic Law (V) of *Grundgesetze* to introduce extensions as objects corresponding uniquely to concepts. Both begin with a domain of objects, take all possible second-order two-sided entities of an appropriate sort over that domain, and then introduce (recognize, Frege would say) certain objects in the domain in one-one correspondence with those second-order entities. Repeating the operation yields (or in the case of Frege, is supposed to yield) no objects not obtained (or recognized) after the operation has been performed only once.

Frege, as is well-known, was not altogether confident about Basic Law (V). "I have never concealed from myself its lack of the self-evidence which the others possess, and which must properly be demanded of a law of logic" he wrote in the Appendix to *Grundgesetze*; earlier, as he notes there, he had said, "A dispute can arise, so far as I can see, only with regard to my Basic Law concerning courses-of-values (V)." Saul Kripke once wondered aloud why Frege did not make the experiment of seeing whether or not Cantor's paradox could be derived in the formal system of *Grundgesetze*. It is of course conceivable that he simply did not know of it; it is not conceivable that he did not know of Cantor's proof that the power class of a class is not equinumerous with it, for Section 164 of Volume II of *Grundgesetze* contains the following noteworthy paragraph:

> We thus require a class of objects, which stand to one another in the relations of our domain of quantities, and this class must certainly contain infinitely many objects. Now to the concept *finite number* there belongs an infinite number, which we have called "endless" [*endlos*]; but this infinity still does not suffice. If we call the extension of a concept that is subordinate to the concept *finite number*, a CLASS OF FINITE NUMBERS, then

to the concept *class of finite numbers* there belongs an infinite
number, which is bigger than endless; i.e. the concept *finite
number* can be mapped into the concept *class of finite numbers*,
but the latter cannot be mapped into the former.[33]

Frege cites Dedekind's work in his own. It strikes me as a quite plausible speculation that the success of Dedekind's well-known construction
may have given Frege confidence that a similar procedure could be used to
introduce extensions in the manner he wanted and needed.

Dedekind's construction, of entities of one sort out of entities of another
sort, is certainly not the first "logical construction" in mathematics. Hamilton's definition of the complex numbers as ordered pairs of reals, as good a
logical construction as any ever made, antedates it by twenty-five years.[34]
I shall not attempt to say whether I think Hamilton's philosophically brilliant construction counts as a contribution to *logic*, however.[35] Of special
interest in Dedekind's work, as opposed to Hamilton's, is the use of what
Quine would regard as set theory and what I, and I hope Putnam, would
call logic. I wonder whether any piece of mathematics remotely comparable
in logical sophistication antedates it.

In thus speculating upon the causes of Gottlob Frege's acceptance of
Grundgesetze's deadly Basic Law (V), I have strayed rather far from the
theme of *Reading Putnam*. But I recall from undergraduate days a certain
professor of mine who once remarked that the way to seduce good students
into philosophy was to teach them the Frege–Russell definition of number.
I had been thus led astray, and if I am now to be faulted for dallying still
in the early history of logic, I simply propose to transfer the blame to the
author of that remark, Hilary Putnam.

---

[33](Frege, 1903), vol. II, p. 161.

[34]Thirty-nine, if one considers publication dates.

[35]Hamilton's treatment is a piece of philosophy if ever there was one. The insight was
not merely to recognize that a problematic sum of a real and a product of a real with
the square root of −1 could be explicated as an unproblematic pair of reals, but also to
understand that the new problem the explication seems to give rise to—whether one can
really *add* and *multiply pairs* of reals—is irrelevant to mathematics, since it is obvious
which definitions of the various operations on pairs of reals have to be given.

16

The Advantages of Honest Toil over Theft

> He [Russell] had a secret craving to have proved *some* straight mathematical theorem. As a matter of fact there *is* one: "$2^{2^a} > \aleph_0$ if a is infinite." Perfectly good mathematics.

> — J. R. Littlewood[1]

In the section of his and Martha Kneale's *Development of Logic* called "Russell's Theory of Logical Types," William Kneale writes,

> It is essential for mathematics that there should be no end to the sequence of natural numbers, and so Russell finds himself driven to introduce a special Axiom of Infinity, according to which there is some type with an infinity of instances, and that presumably the type of individuals, which comes lowest in the hierarchy. Without this axiom, he tells us, we should have no guarantee against the disastrous possibility that the supply of numbers would give out at some highest number, i.e., the number of members in the largest admissible set.

> There is something profoundly unsatisfactory about the axiom of infinity. It cannot be described as a truth of logic in any reasonable use of that phrase, and so the introduction of it as a

From *Mathematics and Mind*, Alexander George, ed., Oxford: Oxford University Press, 1994, pp. 27–44. Copyright ©1994 by Oxford University Press, Inc. Used by permission of Oxford University Press, Inc.

I am grateful to Tony Anderson, David Auerbach, Richard Cartwright, Philippe de Rouilhan, Michael Hallett, Elliott Mendelson, Michael Resnik and Linda Wetzel for helpful comments. Research for this paper was carried out under grant no. SES–8808755 from the National Science Foundation.

[1](Littlewood, 1986), p. 128.

primitive proposition of logic amounts in effect to abandonment
of Frege's project of exhibiting arithmetic as a development of
logic ... But even if we abandon all hope of carrying out Frege's
programme in full and say boldly that Russell's axiom is required as an extra-logical premiss for mathematics, how can we
justify our acceptance of it? What are the individuals of which
Russell speaks, and how can we tell whether there are infinitely
many of them? ... [H]e even suggests that there may be [no in-
dividuals] because everything which appears to be an individual
is in fact a class or complex of some kind. With regard to [this]
possibility, which seems very mysterious, he adds cheerfully that
if it is realized, the axiom of infinity must obviously be true for
the types which there are in the world. But he does not profess
to know for certain what the situation is, and he ends by saying
that there is no known method of discovering whether the axiom of infinity is true or false. [Footnote in Kneale and Kneale:
Introduction to Mathematical Philosophy, p. 143.][2]

The irritated tone of Kneale's commentary is noticeable; but one might
well think that something more like utter exasperation with Russell's procedure is called for: In *Principia Mathematica*,[3] a work supposedly intended
to show arithmetic a part of logic, more than *nine hundred and fifty pages* of
text[4] precede the official introduction of the axiom of infinity. Just once in
Volume I is the axiom mentioned, in the introduction to the second edition,
on p. xxiv. On p. 335, Russell states,

> We might, of course have included among our primitive propo-
> sitions the assumption that more than one individual exists, or
> some assumption from which this would follow, such as
>
> $$(\exists \phi, x, y).\phi!x. \sim \phi!y.$$
>
> But very few of the propositions which we might wish to prove
> depend upon this assumption, and we have therefore excluded
> it. It should be observed that many philosophers, being monists,
> deny this assumption.

The wisecrack may distract the reader from the outrageous claim that few
of the propositions we might wish to prove depend on the assumption that
there are at least two individuals.

[2](Kneale and Kneale, 1984), pp. 657–672, esp. p. 669.
[3](Whitehead and Russell, 1927).
[4]More than 800 in the first edition.

Perhaps there are only a few propositions that depend *just* on that assumption and on nothing stronger; but the existence of the cardinal number 2, equivalent in *Principia* to the existence of at least two individuals is one of those, and without its truth the development of arithmetic is impossible. The importance of the propositions depending on this axiom that we might wish to prove may offset the smallness of their number.

And of course a much stronger statement is needed than that of the existence of at least two individuals. The first two Peano postulates, in the order given them by Russell in *Introduction to Mathematical Philosophy,* assert that zero is a (natural) number and that the successor of a number is a number; the fourth states that zero is not the successor of a number; the fifth is the principle of mathematical induction. These are very easily proved in *Principia* without the assumption of any special axiom. The third, however, states that different numbers have different successors; together with the first three and Russell's definitions of *zero, successor,* and *natural number,* it implies the truth of the axiom of infinity, which asserts there are infinitely many individuals. The first four Peano postulates are theorems of every formal system for arithmetic that I know of; it is hard to see how any development of *arithmetic* could fail to deliver them.

Three axioms of *Principia* have struck commentators as having diminished claims to *logical* truth: those of reducibility, choice, and infinity. (Russell calls the axiom of choice the "multiplicative axiom.") Of these only the axiom of infinity is required for a *Principia*-style development of the arithmetic of the natural numbers, basic to all mathematics, but it is the only one of the three of which no mention is made in the first edition of Volume I, where indeed not a word is spoken of the need to assume a special axiom guaranteeing the truth of the third Peano postulate.

In order to determine whether Russell has unjustifiably minimized the role of the axiom of infinity by thus tucking it away, to raise certain further worries, to point out certain perhaps underappreciated virtues of his procedure, and to compare his with the sublime (and therefore consistent) account of number found in Frege's *Grundlagen der Arithmetik,* we shall have to race over some all too familiar material: the development of arithmetic in the modernized theory of types TT, which, for the sake of simplicity and ignoring Russell's own strenuous efforts to dispense with classes, we shall pretend was the theory Russell was expounding. The version we shall explain is essentially the one given in Gödel's "On formally undecidable propositions of *Principia Mathematica* etc.," but without symbols for zero and successor, and without the assumption that the natural numbers are individuals.

In TT, the objects of type 0 are the *individuals,* whatever they are; those of type $n + 1$ are the classes of objects of type n, n a natural number.

Objects of types 1, 2, and 3 we'll call *sets*, *classes*, and *class-classes*, respectively. Variables $x_n, y_n, z_n, \ldots$ range over objects of type n; for every natural number n, there is an axiom

$$\forall x_{n+1} \forall y_{n+1} (\forall z_n (z \in x \leftrightarrow z \in y) \to x = y)$$

of extensionality and infinitely many comprehension axioms

$$\exists y_{n+1} \forall x_n (x \in y \leftrightarrow \varphi)$$

φ a formula not containing y_{n+1} free.[5]

We shall frequently use $a, b, c, \ldots$ as variables ranging over individuals (in addition to $x_0, y_0, z_0, \ldots$); $x, y, z, \ldots$, over sets; $m, n, A, B, C, \ldots$, over classes, and $X, Y, Z, \ldots$, over class-classes.

Λ is the null set; V is the universal set, that is, the set of all individuals. $\emptyset$ is the null class; 0, alias zero, is $\{\Lambda\}$. Like those that follow, these sets and classes all exist by comprehension and are unique by extensionality. $x - a$ is $\{b : b \in x \land b \neq a\}$, $y + a$ is $\{b : b \in y \lor b = a\}$, and sA, alias the successor of A, is $\{x : \exists a (a \in x \land x - a \in A)\}$.

À la Frege and Russell, n is a number if and only if

$$\forall X (0 \in X \land \forall A (A \in X \to sA \in X) \to n \in X),$$

that is, iff n belongs to every class-class to which zero and the successor of every member belong. $m, n, \ldots$ range over (natural) numbers.

The first, second, and fifth Peano postulates are trivial to prove. (*Applications* of induction of course require comprehension.) And it is very easy to prove the fourth, that 0 is not the successor of a number: every member x of sn is non-empty but 0 has an empty member. The difficulty is to see that different numbers have different successors. This will turn out to be the case iff $\emptyset$ is not a number.

Infin ax, introduced in Section 120 of *Principia Mathematica*, reads

$$\alpha \in \text{NC induct.} \supset_\alpha . \exists! \alpha.$$

In our terminology, for all n, n has at least one member; equivalently, $\emptyset$ is not a number.

Not only is it more than dubious whether any version of the axiom of infinity can be regarded as a logical truth, this formulation disguises what is being asserted more than need be. As usual, define a set x to be finite if and only if

$$\forall A (\Lambda \in A \land \forall y \forall a (y \in A \to y + a \in A) \to x \in A)$$

[5]We will often drop type subscripts when the type of a variable is clear form the context.

A less ad hoc formulation of the axiom of infinity is: V, the set of individuals, is not finite. Of course, the two versions are fairly easily interderivable. Thus it might be thought a matter of "taste" which one assumes. Perhaps so, but it would be absurd to claim that "∅ is not a number" expresses the statement that there are infinitely many individuals as directly as does "*V* is not finite."

However that may be, I shall want to argue that this lapse from perspicuity is the only charge against Russell mentioned in this essay that can be made to stick and that in *Principia Mathematica* Russell has in no way given us grounds for complaint that he has disguised, obscured or minimized the role of the axiom of infinity.

If, following Russell, we say that x sm y if and only if there is a one-one function with domain x and range y, then with the aid of a lemma provable by induction on n, and asserting that if $x \in n$, then x sm y iff $y \in n$, it is easy enough to show that ∅ is not a number if and only if the third Peano postulate holds, i.e., iff different numbers have different successors.

The proofs, found in Appendix I, are short and routine. They show how short the logical distance is between the axiom of infinity and the third Peano postulate. One could well think it not much less of a cheat for Russell to have assumed the axiom of infinity and then derived the third Peano postulate from it than it would have been for him to proclaim the postulate a truth of logic outright.

Russell once wrote, sarcastically, I believe, that "The method of 'postulating' what we want has many advantages; they are the same as the advantages of theft over honest toil. Let us leave them to others and proceed with our honest toil."[6]

Russell's procedure may seem to suffer further when compared with the account of number found in Frege's *Grundlagen der Arithmetik*. It will be recalled that in Sections 74–83 of that work, Frege outlines a derivation of (second-order) arithmetic in the logical system given in his *Begriffsschrift* from the principle that the number belonging to the concept F is the same as that belonging to the concept G if and only if the objects falling under F are in one-one correspondence with those falling under G. Frege derives this principle, sometimes called the number principle, or Hume's principle, from an inconsistent theory of objects, extensions (a species of object), and concepts of various levels. A number is defined as the extension of some second-level concept under which falls some first level concept along with all and only those first-level concepts that are equinumerous with it. Being extensions, numbers are objects. Frege's criterion for the identity of extensions, that extensions of concepts (of the same level) are identical

---

[6](Russell, 1919), p. 71.

if and only if the same entities fall under them, is inconsistent, not only
with respect to extensions of first-level concepts, as Russell showed, but
also with respect to extensions of concepts of any higher level. Thus it is
clear that the theory Frege implicitly employed in the *Grundlagen* to define
number is inconsistent.

Suitably formalized, however, Hume's principle can be shown to be equi-
consistent with the arithmetic that Frege wished to derive from it: a proof of
a contradiction in the system that results when Hume's principle is adjoined
to the logic of the *Begriffsschrift* can (easily) be turned into a contradiction
in second-order arithmetic, and, as Frege in effect showed, vice versa.[7] The
derivation of arithmetic from Hume's principle that Frege sketched can
be elaborated into formal deductions of the (infinitely many) axioms of
second-order arithmetic. The most remarkable part of Frege's argument
is his proof that every natural number has a successor. It utilizes a much
more interesting mathematical idea than any found in Russell's derivation
of the Peano postulates: *zero, successor of,* and *natural number* having
been defined, and *less than* being defined as the ancestral of the relation
an object bears to any of its successors, the number of objects less than or
equal to any given natural number *a* can be shown to be a successor of *a*.

Recall also that Frege wished to show how numbers could be "conceived as
logical objects." It is clear enough that before Russell's communication to
him of the Contradiction, Frege supposed that the identification of numbers
with certain sorts of *extensions* expressed a recognition of numbers as logical
objects, and that the mere recognition of the truth of Hume's principle
did not. As many commentators have noted, what is perhaps not clear
is why Frege should have supposed this. Questions of consistency aside
(!), what is there about extensions that makes *them*, and not numbers,
logical objects in the absence of an account such as Frege tried to give?
Extensions of concepts are supposed to be the same if and only if the objects
falling under one of the concepts are identical with those falling under the
other. To say when numbers are the same, simply change "extensions of"
to "numbers belonging to" and "identical" to "in one-one correspondence"
in the foregoing sentence. Although it certainly requires a somewhat more
complex formula to express that some objects are in one-one correspondence
with others than to express that some objects are identical with others, one
may reasonably doubt whether that difference entitles us to conclude that
extensions are logical objects, but numbers are not.

Frege, it is also well known, failed to find a way out: his proposed solution
to the difficulty turned out to be inconsistent with the assertion that there
are at least two numbers. There is a modification of the notion of an

---

[7]Cf. Articles 12 and 13 in this volume.

extension that works, however. Say that a concept F is *small* iff the objects falling under F cannot be put in one-one correspondence with all the things there are. Say that F equiv G if and only if, if either F or G is small then the same objects fall under both. Equiv is an equivalence relation. Introduce subtensions by assuming that the subtension $*F$ of the concept F is identical with $*G$ if and only if F equiv G. This assumption can be shown consistent relative to second-order arithmetic, and can be used to define numbers: let $0 = *[x : x \neq x]$, i.e., let $0 = *F$, where $\forall x(Fx \leftrightarrow x \neq x)$; let $sy = *[x : x = y]$; and let x be a number iff, as usual, $\forall F(F0 \wedge \forall y(Fy \rightarrow Fsy) \rightarrow Fx)$. The development of arithmetic then proceeds smoothly enough. (Peano three is no problem since $\exists x\, x = x$; thus $0 \neq *[x : x = x]$; thus there are at least two objects; thus for every y, $[x : x = y]$ is small.)

If subtensions are logical objects, then we have a way of recognizing numbers as logical objects; if not, despite their resemblance to extensions and the consistency of the axiom governing them, then we have even less reason than before to agree with the view that extensions, "had Rule V been consistent," would be logical objects.

Whether extensions, subtensions, or numbers are logical objects or not, it may seem, from a Fregean point of view, that Russell's definition of the numbers as certain sorts of class fails in two respects: invoking the axiom of infinity invalidates a claim to have shown numbers to be *logical* objects; defining them as certain classes (of sets of individuals) forbids him from thinking he has shown them to be logical *objects*. To show numbers to be objects, Russell would have had to show which individuals they are.

My aim so far has been to depict Russell's account of number in the worst possible light, as a series of tedious definitions and deductions in an inadequate theory to which an inelegantly formulated axiom has been surreptitiously adjoined with no justification other than to derive an indispensable but otherwise unobtainable theorem, and in which the definitions, moreover, obviously fail to satisfy one basic requirement of the enterprise of setting up a theory of number at all.

What then did Russell achieve? The answer may be found by reflecting on the "perfectly good" piece of mathematics mentioned in Littlewood's remark. This proposition and its proof, found in Vol. II at $*124.57$, constitute, I want to claim, the mathematical core of the theory of natural numbers given in *Principia Mathematica*.[8]

Never forget that the natural numbers form not merely an infinite totality, but one that is *Dedekind* infinite. Assuming now some theory of sets such as ZF, we say that a set is *finite* if and only if (as in the definition given above)

---

[8]The theorem is erroneously ascribed to Tarski in one well-known excellent text: (Lévy, 1979), p. 80.

it belongs to all classes (here = sets) that contain the null set and contain all results of adjoining to any member any one object. Equivalently, a set x is finite if and only if there is a natural number i such that x can be put into one-one correspondence with the set of natural numbers less than i. A set is *Dedekind infinite* if and only if it can be put in one-one correspondence with a proper subset of itself. Equivalently, a set x is Dedekind infinite if there is a one-one correspondence between the set of all natural numbers and a subset of x (not necessarily a proper subset). The set of natural numbers is, trivially, Dedekind infinite according to the either of these equivalent definitions. A set is infinite if and only if it is not finite, Dedekind finite if and only if not Dedekind infinite. It is easy to show that no finite set is Dedekind infinite; it requires some assumption that is not a theorem of ZF such as the axiom of choice to show that no infinite set is Dedekind finite. Russell, who was admirably clear on the distinction, called the finite sets "inductive" and the Dedekind infinite sets "reflexive"; it is a pity that this excellent terminology has not become standard.

According to the theorem Littlewood ascribed to Russell, if a is an infinite number, then $2^{2^a} > \aleph_0$. What does the theorem mean? Theorems about cardinal numbers are often best understood as encrypted theorems about one-one correspondences. After decoding, the theorem states that if x is an infinite set (with cardinal number a), then the set of natural numbers (which has cardinal number $\aleph_0$) can be mapped one-one into the power set $\mathcal{PP}x$ of the power set $\mathcal{P}x$ of x (which thus has cardinal number 2^{2^a}; thus $\aleph_0 \leq 2^{2^a}$), that is, that $\mathcal{PP}x$ is Dedekind infinite; but that there is no one-one correspondence between the set of natural numbers and the power set of the power set of x (thus $\aleph_0 \neq 2^{2^a}$, and so $2^{2^a} > \aleph_0$). The more interesting half of the theorem is thus that if x is an infinite set, then $\mathcal{PP}x$ is Dedekind infinite.

How, then, may this half of theorem be proved? Let x be an infinite set. The null set is a subset of x of cardinality 0. If y is a subset of x of cardinality n, then since y is a finite subset of the infinite set x, y is not identical with x; thus there is some element a of x not in y and $y \cup \{a\}$ is a subset of x of cardinality $n + 1$. By mathematical induction, for every natural number n, there is a subset of x of cardinality n. Thus for each finite n the set S_n of subsets of x of cardinality n is nonempty, and if $m \neq n$, S_m and S_n are disjoint and hence distinct. Each S_n is a subset of the power set $\mathcal{P}x$ of x. Thus $n \mapsto S_n$ is a one-one function from the set of natural numbers into $\mathcal{PP}x$.

(The other half of the theorem, according to which $2^{2^a} \neq \aleph_0$, is immediate: if $2^{2^a} = \aleph_0$, then since $a < 2^a < 2^{2^a}$ by Cantor's theorem, $a < \aleph_0$, a is finite, and then so are 2^a, 2^{2^a}, and $\aleph_0$, impossible. I am not sure whether Littlewood had this, the "*strictly*-less-than," half of the theorem in mind when he made his remark.)

Thus although one can point to a specific place in *Principia* where Russell proved the theorem ascribed to him by Littlewood, it would not be unreasonable to give: "*PM*, passim" as a citation for the theorem. To belabor the obvious: call the members of the infinite set x *individuals*. Then $\mathcal{PP}x$ comes to type 2 and S_n to the Russellian version of n; the Dedekind infinity of $\mathcal{PP}x$ is witnessed by S_0 and the function, which assigns S_{n+1} to each S_n and A itself to each member A of $\mathcal{PP}x$ not of the form S_n. Put in Russellian terminology, the point is that Russell did not assume the type of individuals to be reflexive. He supposed it non-inductive and showed that it follows from that weaker supposition that type 2 is reflexive, and thus includes a subcollection similar to the set of natural numbers.

Not only can it not be proved in set theory without choice that there are no infinite Dedekind finite sets, it cannot even be proved that there do not exist infinite sets *whose power set* is Dedekind finite. And by adapting to the theory of types the Fraenkel–Mostowski method for showing the independence of various forms of the axiom of choice from set theory with individuals it can be shown that it is consistent with the theory of types supplemented with the axiom of infinity that the type of all individuals is infinite while the type of sets, i.e., all classes of individuals, and hence the type of all individuals as well, is Dedekind finite.[9]

The idea of the proof is simple. Working in the theory of types, we shall build a model $\{T_0, T_1, T_2, \ldots\}$ of the theory of types in which T_0 is infinite, but in which there is no one-one mapping of the Russell numbers into T_1.

Begin with an infinite (Dedekind infinite if you like) set T_0, of individuals. Define a permutation π to be a one-one function whose domain and range are T_0. Say that π fixes a set x of individuals if for every $a \in x$, $\pi a = a$. Now suppose T_n defined, and $\pi\alpha$ defined for all α in T_n. If β is a subset of T_n, let $\pi\beta = \{\pi\alpha : \alpha \in \beta\}$, and let T_{n+1} be the set of those subsets β of T_n such that for some *finite* set x of individuals, $\pi\beta = \beta$ for all π that fix x. Thus each T_n is a subset, in general a proper subset, of type n.

It is easy to see that T_1 consists of the sets of individuals that are either finite or have a finite complement (relative to T_0). If n is a Russell number, then $\pi n = n$, for *every* π, and thus n is in T_2 (take $x = \Lambda$); similarly for the set N of Russell numbers: $\pi N = N$ for every π, and therefore N is in T_3.

The sets T_n, together with the sets belonging to them, turn out to form a model m of the theory of types and the statements that there are infinitely many individuals but Dedekind finitely many classes of individuals. The details of the proof are given in Appendix II.

---

[9]Cf. (Jech, 1973), ch. 4, and (Felgner, 1971), ch. 3.

Russell showed that there are Dedekind infinitely many classes of classes of individuals from the assumption that there are infinitely many individuals. But, as we have just observed, Dedekind infinity could not have been found any lower: without the aid of some such assumption as the axiom of choice it cannot be proved from the axiom of infinity that the individuals or the classes of them form a Dedekind infinite totality.

Of course there is a simpler reason why the numbers must first appear two types up if only the axiom of infinity is assumed. In the theory of types there is no way to define the numbers as sets of individuals and hence no way to define them as individuals. More precisely, for every formula $\varphi(x)$ containing exactly the (set) variable x free, the sentence $\exists!3x\,\varphi(x)$[10] expressing the existence of exactly three sets satisfying $\varphi(x)$ is not a theorem of the theory of types. Thus there are no formulae $0(x)$, $1(x)$, and $2(x)$ such that $\imath x0(x)$, $\imath x1(x)$, and $\imath x2(x)$ can be proved to exist and differ from one another; otherwise $\exists!3x\,(0(x) \vee 1(x) \vee 2(x))$ would be provable.

In fact, it can be shown more generally that for any formula $\varphi(x)$ of TT and any integer $i > 2$, the sentence

$$[\exists!ix\,\varphi(x) \to \bigvee\{\exists!na\,a = a : n \leq i \text{ and for some } \sigma \subseteq \{0,\ldots,n\},$$
$$i = \textstyle\sum\{nCr : r \in \sigma\}\}]$$

is provable in TT. (nCr is the binomial coefficient.) As the only rows of Pascal's triangle from which 3 can be obtained by summing entries are 121 and 1331, for any formula $\varphi(x)$, $\exists!3x\,\varphi(x) \to \exists!2a\,a = a \vee \exists!3a\,a = a$ is provable in TT. Since $\exists!2a\,a = a \vee \exists!3a\,a = a$ is not a theorem, neither is $\exists!3x\,\varphi(x)$. Thus, if our resources are confined to those of the theory of types with the axiom of infinity, the natural numbers can't be classes of individuals. (The mod 2 numbers could be, however.)

In his first proof that every set can be well-ordered, Zermelo in effect showed how to extend the theory of types plus the axiom of infinity to make it possible to define the numbers as individuals. It will be instructive to examine the extension and definition, which it is perhaps not too farfetched to take to formalize the theory of arithmetic of Frege's interlocutor at the beginning of *Die Grundlagen der Arithmetik*, who, according to Frege, will likely invite us to "*select* something for ourselves—anything we please—to call one."

Let us add to the language of the theory a symbol ϑ for a function f whose values for arguments of type 1 are of type 0. And now let us take as a new axiom a strengthened version of the axiom of choice for type 1:[11]

$(*)$ $\exists a\,a \in x \to \vartheta x \in x$

---

[10]That is, $\exists x \exists y \exists z(\varphi(x) \wedge \varphi(y) \wedge \varphi(z) \wedge x \neq y \wedge x \neq z \wedge y \neq z \wedge \forall w(\varphi(w) \to w = x \vee w = y \vee w = z))$.

[11]By replacing items with their singletons, we can see that "choice drops down"; thus

We can now define 0 as fT_0, 1 as $f(T_0 - 0)$, 2 as $f(T_0 - 0 - 1)$, etc. (Had we asked Frege's man on the street to tell us what two was, he would surely have invited us to select something *else*—anything *else* we please—and call it two.)

By the argument of Zermelo's proof, there is a unique well-ordering R of T_0 in which $f(T_0 - A)$ is the R-least element of $T_0 - A$, for any proper initial segment A of R. We may then define b to be the successor of a if aRb and for no c, $aRcRb$, and a to be a natural number if every b such that bRa or $b = a$ is zero or a successor. The axiom of infinity here guarantees that every natural number has a successor.

Thus simply by adding a new function symbol to the language of the theory of types and a suitable axiom governing the function denoted by it, we have found a way to "recognize" the numbers as individuals. Of course, there was no need to bring in a *function* symbol; we could have adhered more closely to the syntactic style of the theory of types by introducing a constant $\mathcal{C}$ of type 4, along with the axiom

$$\exists a\, a \in x \rightarrow \exists! a(a \in x \wedge \{\{x\}, \{x, \{a\}\}\} \in \mathcal{C}).$$

"But," it may be objected, "isn't that cheating? We are trying to find individuals with which to identify the natural numbers. However, not any old means of finding them is allowed. We have to use means that are recognizably logical. I don't see that the importation of a brand-new function sign, designating who knows what function (or the use of a higher-type constant: there's no difference), counts as a logical means of finding individuals that can serve as the natural numbers. We don't know which function ϑ denotes; you've just pulled something out of thin air to do the work you wanted to have done."

Let us note this objection for now and examine another means of recognizing the numbers as individuals.

Suppose that we add to the theory of types a function sign $\#$ whose values for arguments of type 1 are of type 0 and take as a new axiom:

$$\#x = \#y \leftrightarrow x \text{ sm } y$$

("sm" abbreviates "is similar to," defined as usual).

Then, as Frege showed in *Grundlagen*, if, *working without axioms of extensionality, the axiom of infinity or any version of the axiom of choice,* we define 0 as

$$\imath a\, \exists y (\forall c\, c \notin y \wedge a = \#y)$$

if we had introduced a function sign ρ and a strengthened version of choice for type $n+1$:

$$\exists x_n\, x_n \in x_{n+1} \rightarrow \rho x_{n+1} \in x_{n+1},$$

we could then have defined a suitable ϑ and proved $(*)$.

define c to succeed b iff

$$\exists a \exists y \exists z(z = y + a \wedge a \notin y \wedge b = \#y \wedge c = \#z)$$

and define a to be a natural number iff

$$\forall x(0 \in x \wedge \forall b \forall c(b \in x \wedge c \text{ succeeds } b \rightarrow c \in x) \rightarrow a \in x)$$

then we can prove the Peano postulates, together with all necessary existence and uniqueness assumptions. It is an immediate consequence that the individuals form a Dedekind infinite totality, and that the axiom of infinity therefore holds after all. Moreover, the numbers have indeed been defined as individuals.

For all its excellences, this method of obtaining the natural numbers at the lowest level of the type hierarchy is as much subject to the objection that we have no idea which function the new symbol refers to as was the postulation described above of a *particular* choice function f for type 1. (E.g., if π is a permutation of T_0, then where a is the value of $\#x$ and b that of $\#y$, $\pi a = \pi b$ iff x sm y holds.[12]) It can be said with equal justice in both cases that nothing establishes, determines, fixes the function to which the newly introduced function symbol refers. No one struggled harder than Frege to overcome the apparent lack of fixity of the function referred to by "the number of (belonging to)." But it has often been remarked that whatever other problems may have beset Rule V of *Grundgesetze,* for Frege to use that axiom to introduce extensions and then to define a number as a certain sort of extension, is to advance little if at all in settling the question to which items number words refer: if we are uncertain whether numbers are conquerors, we are not going to be helped out of the slough by being told that numbers are extensions. (I think Michael Dummett pointed this out to me more than twenty-five years ago.)

It may be thought that we know what it is for one item to bear the relation indicated by "$\in$" to another better than we know which particular function is designated by "the number of," and certainly better than we know which function is designated by ϑ. To the extent that this is so, or supposed so, Russell's treatment of the numbers will be, or seem, *ideologically* superior to Frege's in the sense of Quine, superior in respect of the clarity or determinacy of the notions of which it avails itself. Russell may assume as an axiom a statement that Frege can prove, but Frege utilizes a notion that can neither be expressed in Russell's language, a sublanguage of Frege's, nor, apparently, freed from a very familiar sort of indeterminacy.

Of course, there is indeterminacy aplenty in the theory of types. As in the theory of complex numbers, i and $-i$ are indiscernible—any truth

[12]The "Irving Caesar" problem.

remains true in which "i" and "$-i$" are everywhere interchanged—so in the theory of types "$\in$" and "$\notin$" may be uniformly interchanged at any one type (thanks to the existence of a unique complement in its type for every item not of the type of individuals). More exactly, for any n, if φ is a theorem of the theory of types, then so is the result of replacing every atomic formula of the form $x_n \in y_{n+1}$ in φ with its negation. (In set theory we cannot perform this sort of switch: $\exists y \forall x \, \neg x \in y$ is, but $\exists y \forall x \, x \in y$ is not, a theorem of set theory.) Moreover, such interchange can be performed at any one type independently of whether it is performed at any others. Thus the theory of types is indeterminate in at least $2^{\aleph_0}$ ways.

But this sort of indeterminacy also infects the theory of objects and first- and higher-level concepts that was employed by Frege: we are free to interpret the predication Fx as asserting that the value of x fails to fall under the concept denoted by F. Thus in any event a *new* sort of indeterminacy arises with the introduction of either $\#$ or ϑ.

Of course the axiom: $\#x = \#y \leftrightarrow x$ sm y (Hume's principle) is not to be regarded as a *definition* of number; it is merely a consistent principle whose addition to a suitable higher-order (indeed, second-order) logic yields a system in which the basic notions of the arithmetic of natural numbers can be defined and their most familiar properties proved. Thus with the aid of a familiar-seeming principle, Frege has given a remarkably simple axiomatization of arithmetic whose consistency is not at present subject to doubt. (The tragedy of Russell's paradox was to obscure from Frege and from us the great interest of his actual positive accomplishment.) It has been my aim these last few pages to point out a number of respects in which Russell's account of arithmetic stands comparison with the one Frege is now known to have provided.

The construction of the numbers with the aid of a choice function, which was sketched above, shows, I think, that Hume's principle cannot be thought to be *the* foundation of arithmetic. One of zero's properties, and a very important one too, is that it is the number of things there are that are not self-identical; but, as our discussion of Frege's man in the street showed, there is also a perfectly sensical alternative development of the idea that zero, or one (if you prefer to begin the number series there), is the "typical object." It is also to be noted that there is no trace in the construction of the idea that 2, for example, is the class of all couples; nor is use made in the construction of a function injecting Russell numbers into the individuals.

Moreover, by the trick of reserving 0 for the number of things that are self-identical and "pushing each natural number up one," we can define $\#$ so as to prove Hume's principle in the theory of types plus the axiom of infinity and our strengthened version $(*)$ of choice.

I now want to take up the question whether Russell's introduction of the axiom of infinity in Volume II of *Principia Mathematica* amounts, as Kneale put it, "to abandonment of Frege's project of exhibiting arithmetic as a development of logic." Of course the axiom of infinity cannot be counted as a truth of logic, and no one was clearer on that score than Russell himself.

> From the fact that the infinite is not self-contradictory, but is also not demonstrable logically, we must conclude that nothing can be known a priori as to whether the number of things in the world is finite or infinite. The conclusion is, therefore, to adopt a Leibnizian phraseology, that some of the possible worlds are finite, some infinite, and we have no means of knowing to which of these two kinds our actual world belongs. The axiom of infinity will be true in some possible worlds and false in others; whether it is true or false in this world we cannot tell.[13]
>
> We may take the axiom of infinity as an example of a proposition which, though it can be enunciated in logical terms, cannot be asserted by logic to be true. ... We are left to empirical observation to determine whether there are as many as n individuals in the world. ... There does not even seem any logical necessity why there should be even one individual [Footnote in original:The primitive propositions in *Principia Mathematica* are such as to allow the inference that at least one individual exists. But I now view this as a defect in logical purity.]— why in fact there should be any world at all.[14]

In *Principia Mathematica*, Whitehead and Russell say,

> If, for example, Nc'Indiv $= \nu$, then this proposition is false for any higher type; but this proposition, Nc'Indiv $= \nu$, is one which cannot be proved logically; in fact it is only ascertainable by a census, not by logic. Thus among the propositions which can be proved by logic, there are some which can only be proved for higher types, but none which can only be proved for lower types.
>
> "Infin ax," like "Mult ax," is an arithmetical hypothesis which some will consider self-evident, but which we prefer to keep as a hypothesis, and to adduce in that form whenever it is relevant. Like "Mult ax," it states an existence theorem ...
>
> It seems plain that there is nothing in logic to necessitate its

[13](Russell, 1919), p. 141.
[14](Russell, 1919), pp. 202–203.

truth or falsehood, and that it can only be legitimately believed or disbelieved on empirical grounds.[15]

And, in Volume III:

Great difficulties are caused, in this section ["Generalization of number"], by the existence-theorems and the question of types. These difficulties disappear if the axiom of infinity is assumed, but it seems improper to make the theory of (say) 2/3 depend upon the assumption that the number of objects in the universe is not finite. We have, accordingly, taken pains not to make this assumption, except where, as in the theory of real numbers, it is really essential, and not merely convenient. When the axiom of infinity is required, it is always explicitly stated in the hypothesis, so that our propositions, as enunciated, are true even if the axiom of infinity is false.[16]

But if Russell made it plain that he did not consider the axiom of infinity to be a truth of logic, "asserted by logic to be true," what becomes of the project of showing arithmetic to be a development of logic, of logicism? Russell was a logicist, wasn't he?

To determine whether or not he was one, it might just be advisable to consult his writings instead of common opinion. It turns out that Russell was rather more cautious in certain works than others in proclaiming that mathematics can be reduced to logic, or is identical with it. The question whether Russell was or was not a logicist cannot, I think, be given a direct answer, and ought to be replaced with questions of the form, "Was Russell a logicist in work X?" What can be said is that he expressed logicist views in certain works and refrained—significantly, it seems to me—from expressing them in others, notably *Principia Mathematica,* in which, as it happens, there are rather few remarks on the relation between logic and mathematics; perhaps Whitehead and Russell considered it unnecessary to supply many, for the work is, after all, an extended disquisition upon just that subject. Those there are, however, make it doubtful that the authors should be considered logicists, i.e., defenders of the view that mathematics, or arithmetic, or at least the Peano postulates, can be derived by logical means alone from statements true solely in virtue of logic and appropriate definitions of mathematical notions. *Principia* is not quite 2000 pages long, and it is hard to be perfectly certain that one has not overlooked a significant remark or failed to put together separated comments that would

[15]Quotations from (Whitehead and Russell, 1927), vol. II, x, pp. 203 and 183, respectively.
[16](Whitehead and Russell, 1927), vol. III, p. 234.

make it plain that its authors do after all count as logicists. However, there appears to be only one section of *Principia* that explicitly deals with the relation between logic and mathematics, at the beginning of the introduction to the first edition. There Russell and Whitehead list three aims of the logic which occupies Part I of *Principia*. They are, in reverse order, the avoidance of the contradictions, the precise symbolic expression of mathematical propositions, and the one that concerns us:

> effecting the greatest possible analysis of the ideas with which it deals and of the processes by which it conducts demonstrations, and ... diminishing to the utmost the number of the undefined ideas and undemonstrated propositions (called respectively *primitive ideas* and *primitive propositions*) from which it starts.[17]

Later, the first aim is described, rather differently, as "the complete enumeration of all the ideas and steps in reasoning employed in mathematics."[18]

It is evident that one who claims to have enumerated all the ideas and steps involved in mathematical reasoning need not imply that that reasoning is logical reasoning, or even that the third Peano postulate is a truth of logic. However justly, it might well be said that Zermelo–Fraenkel set theory provides such an enumeration; to say so is, obviously, not to be committed to the view that its axioms are logical truths. Russell's second description of his first aim provides no reason to take him to be committed to the central thesis of logicism.

Nor does his first description. The most thorough analysis possible of mathematical ideas and argumentation might well have as its outcome that the third Peano postulate is equivalent to the axiom of infinity, but leave entirely open the question whether the latter is a truth of logic. Russell repeatedly states that it is not one, and he did not take it to be a primitive proposition; moreover, he claimed to have proved from primitive propositions only the conditional with consequent Peano three and antecedent Infin ax.

One may distinguish, as Carnap has usefully done,[19] two theses of logicism, the first of which states that the concepts of mathematics can be explicitly defined from logical concepts; the second, that the theorems of mathematics can be deduced from logical axioms by logical means alone. We may call these the *definability thesis* and the *provability thesis* of logicism.

---

[17](Whitehead and Russell, 1927), vol. I, p. 1.
[18](Whitehead and Russell, 1927), vol. I, pp. 2–3.
[19](Carnap, 1931).

Establishing the definability thesis will show that all truths of mathematics can be expressed in the vocabulary of pure logic. But it is important to distinguish truths expressed in the vocabulary of pure logic from truths that are true "by virtue of logic alone," i.e., *logical truths* or *truths of logic* properly so called. Russell's way of making this distinction was between propositions that are "enunciated in logical terms" and those that are "asserted by logic to be true."[20] "$\exists x \exists y\, x \neq y$" is a truth, and expressed in logical vocabulary, which Russell, correctly in my view, did not regard as a logical truth. One who accepts the theory of types will almost surely regard Infin ax as true and in logical vocabulary, but one who so regards it need not therefore take it to be a logical truth. Establishing both theses would certainly show the truths of mathematics to be logical truths, but establishing the definability thesis alone does not suffice to do this, and hence certainly does not establish the provability thesis. No one, I take it, counts as a full-fledged logicist who does not endorse the provability thesis as well as the definability thesis.

It seems fair to take Russell's aim in *Principia* to have been the systematic exposition of a sufficiently large portion of mathematics to enable the reader to see that, and how, the whole of the subject could be treated in its system, in the sense that every concept of mathematics could be defined in terms of the primitive ideas of the system and every theorem of mathematics either proved from its primitive propositions, *or suitably related* to other propositions of mathematics. In *Principia* then, Russell was an advocate of the definability thesis, but not of the provability thesis of logicism. It was never part of his aim there to show that (say) Peano three, as opposed to "If Infin ax then Peano three," could be derived from the primitive propositions of the system. Whitehead and Russell might have paraphrased Boole and called their work *The Logical Analysis of Mathematics.* To provide such an *analysis,* however, it is not requisite to derive from logic the whole of elementary mathematics.

Once the idea is abandoned that the aim of *Principia* is to vindicate full-fledged logicism, to exhibit arithmetic as a development of logic, there is little to object to in Russell's *modus operandi.* The axiom of infinity is introduced at the appropriate point: in Subsection ∗120 "Inductive cardinals," of Section C, "Finite and Infinite," of Part III, "Cardinal Arithmetic," the part with which Volume II begins. Part I of *Principia* is entitled "Mathematical Logic," Part II, "Prolegomena to Cardinal Arithmetic." Where else should the axiom of infinity have been introduced?

When pronouncing on the relation of logic to mathematics, Russell is significantly less circumspect in the exoteric *Introduction to Mathematical*

---

[20](Russell, 1919), pp. 202–203.

Philosophy than he is in *Principia:*

> Pure logic, and pure mathematics (which is the same thing),
> aims at being true, in Leibnizian phraseology, in all possible
> worlds, not only in this higgledy-piggledy job-lot of a world in
> which chance has imprisoned us.

> The consequence is that it has now become wholly impossible
> to draw a line between the two; in fact, the two are one ... The
> proof of their identity is, of course, a matter of detail ... If there
> are still those who do not admit the identity of logic and mathe-
> matics, we may challenge them to indicate at what point in the
> successive definitions and deductions of *Principia Mathematica*
> they consider that logic ends and mathematics begins. It will
> then be obvious that any answer must be arbitrary ...

> Assuming that the number of individuals in the universe is not
> finite, we have now succeeded not only in defining Peano's three
> primitive ideas, but in seeing how to prove his five primitive
> propositions, by means of primitive ideas and propositions be-
> longing to logic.[21]

These remarks and others that might be cited might well lead one to take
Russell to be advocating a position he himself has given the best of reasons
for rejecting, since he has elsewhere been as explicit as possible that he does
not regard the axiom of infinity as a logical truth. To the challenge Russell
lays down, one may respond that every proposition deduced in *Principia*
is indeed a truth of logic, but Peano three, a proposition of mathematics if
any is, has not been deduced there.

The last quotation, however, suggests a more charitable reading of *Introduction to Mathematical Philosophy,* under which one may interpret Russell
to be claiming the identity of the concepts of mathematics with those of
logic, the derivability of all the Peano axioms but the third, and the provability of "if Infin ax then Peano three." On this reading, the frequent omissions of an important qualification of the logicist thesis must be thought
careless, if not propagandistic. In *Introduction to Mathematical Philosophy,*
then, Russell can perhaps be considered to espouse the definability thesis
of logicism, but to hedge significantly on the question whether the provability thesis holds. It is therefore arguable that Russell does not significantly
change his mind between the writing of *Principia Mathematica* and *Introduction to Mathematical Philosophy,* and that in neither work should he be
seen as fully committed to logicism.

---

[21]Quotations from (Russell, 1919), p. 192, pp. 194–195, and pp. 24–25, respectively.

Appendix I

Lemma *If $x \in n$, then x sm y iff $y \in n$.*

Proof. Induction. The lemma is trivial if $n = 0$, since Λ sm y iff $y = \Lambda$. Suppose $x \in sn$. Then for some $a \in x, x - a \in n$. Suppose x sm y via f. Then $fa \in y, x - a$ sm $y - fa, y - fa \in n$ by the i.h., and $y \in sn$. Conversely, suppose $y \in sn$. Then for some $b \in y, y - b \in n$. By the i.h., $x - a$ sm $y - b$ via some f, and thus x sm y via g, where domain$(g) = x$, $gc = fc$ for $c \in x - a$, and $ga = b$. ∎

Theorem $\emptyset$ *is not a number iff different numbers have different successors.*

Proof. Suppose that $\emptyset$ is a number. $\emptyset$ is empty. 0 is not empty. By induction we may assume that for some number n, n is not empty, but sn, which is a number, is empty. Thus $n \neq sn$. Since sn is empty and $ssn = \{x : \exists a(a \in x \wedge x - a \in sn)\}, ssn$ is also empty, and by Ext, $sn = ssn$. Since ssn is also a number, n and sn are different numbers with the same successor. Conversely, assume that $\emptyset$ is not a number, m, n are numbers and $sm = sn$. Since sm is a number, $sm \neq \emptyset$, and for some $x, x \in sm = sn$. Then for some $a, b, a \in x, b \in x, x - a \in m$ and $x - b \in n$. Then $x - a$ sm $x - b$ via f, where domain$(f) = x - a, fb = a$ if $b \in x - a$, and $fc = c$ for $c \in x - a, c \neq b$. If $z \in m, z$ sm $x - a$ by the lemma, whence z sm $x - b$, and $z \in n$ by the lemma again. Similarly, if $z \in n, z \in m$. By Ext, $m = n$. ∎

Appendix II

In $\mathcal{M}$, T_0 satisfies the formula "x is infinite": since there is in fact no one-one function from any finite set of natural numbers onto T_0, no function in $\mathcal{M}$ witnesses the finitude of T_0.

We now show that T_1 does not satisfy "x is Dedekind infinite" in $\mathcal{M}$.

Suppose that $f, \in \mathcal{M}$, witnesses the Dedekind infinity of T_1. Abbreviate "$\imath z \{\{n\}, \{n, \{z\}\}\} \in f$" by "$fn$." Then f is a one-one function with domain N such that for every n in N, $fn \in T_1$. Since $f \in \mathcal{M}$, there is some finite $x \subseteq T_0$, such that $\pi f = f$ for every π that fixes x. There are only finitely many y such that $y \subseteq x$ or $T_0 - y \subseteq x$. Thus for some n in N and some finite $y \subseteq T_0$, y is not a subset of x, and either $fn = y$ or $fn = T_0 - y$. Let a be an individual in $y - x$, and let b be an individual in neither y nor x (some such b exists since x and y are finite and there are infinitely many individuals). Let π permute a and b but do nothing else. π fixes x; so $\pi f = f$. Then if $fn = y$, $\pi y = \pi fn = \pi f \pi n = fn = y$, and if $fn = T_0 - y$, then $y = T_0 - fn$, and since $\pi T_0 = T_0, \pi y = \pi(T_0 - fn) = \pi T_0 - \pi fn = T_0 - \pi f \pi n = T_0 - fn = y$. In either case $\pi y = y, a \in y$, whence $b = \pi a \in \pi y = y$. But $b \notin y$, contradiction.

We now show that $\mathcal{M}$ is a model of the theory of types.

That extensionality holds in $\mathcal{M}$ is clear: if $x, y \in T_{n+1}, x \neq y$, then for some $z, z \in x$ or $z \in y$. But then $z \in T_n$.

As for comprehension, let $x^1, \ldots, x^n$ be a list containing all variables free in a formula φ; each x^i ranges over some one type or other. By induction on φ, for any π, any objects $o^1, \ldots, o^n$ of the appropriate types, $\mathcal{M} \models \varphi(o^1, \ldots, o^n)$ iff $\mathcal{M} \models \varphi(\pi o^1, \ldots, \pi o^n)$.

Now let x_n be a variable ranging over type $n, x_n, x^1, \ldots, x^m$ be a list containing all variables free in a formula φ. We must see that for any objects $o^1, \ldots, o^m$ in $\mathcal{M}$ of the appropriate types $\{o \in T_n : \mathcal{M} \models \varphi(o, o^1, \ldots, o^m)\}$ $\in T_{n+1}$. Notice that for each n, T_n is a definable subset of type n, and therefore for each formula φ, "$\mathcal{M} \models \varphi(o, o^1, \ldots, o^m)$" defines a definable relation. It thus suffices to show that if $q = \{o \in T_n : \mathcal{M} \models \varphi(o, o^1, \ldots, o^m)\}$, then for some finite $z \subseteq T_0, \pi q = q$ for every π that fixes z.

For each $i, 1 \leq i \leq m$, let z_i be a finite subset of T_0 such that $\pi o^i = o^i$ for every π that fixes z_i. Let $z = z_1 \cup \ldots \cup z_m$. Suppose π fixes z. We show that $\pi q = q$. π fixes $z_1, \ldots, z_m$, and so $\pi o^i = o^i, 1 \leq i \leq m, \pi T_n = T_n$. Suppose $o \in \pi q$. Then $\pi^{-1} o \in q$; $\pi^{-1} o \in T_n$ and $\mathcal{M} \models \varphi(\pi^{-1} o, o^1, \ldots, o^m)$; $o \in \pi T_n$ and $\mathcal{M} \models \varphi(o, \pi o^1, \ldots, \pi o^m)$; $o \in T_n$ and $\mathcal{M} \models \varphi(o, o^1, \ldots, o^m)$; and so $o \in q$. Thus $\pi q \subseteq q$. And if $o \in q$, then $\pi^{-1} o \in \pi^{-1} q$, whence, similarly, $\pi^{-1} o \in q$, and $o \in \pi q$. So $q \subseteq \pi q \subseteq q$, done.

17

On the Proof of Frege's Theorem

In his doctoral dissertation, "Logicism, Some Considerations" (Princeton University, 1960), Paul Benacerraf defended a two-part thesis: (1) what arithmetic can be reduced to should not count as logic; and (2) the reductions of arithmetic to [whatever] do not provide the epistemological support for arithmetic which it had been thought a reduction to logic would supply.

Thirty-five years have gone by since Paul submitted his dissertation. Reread today, it remains remarkably persuasive, its argumentation shrewd and common-sensical. The main claim of the dissertation has passed into the folklore of the philosophy of mathematics, in part because of Paul's teaching, but primarily, I believe, because claims similar to those of the dissertation were advanced and defended by Quine in subsequent publications. To say so is certainly not to suggest that Quine was influenced by Paul, but rather to voice regret that Paul never saw his way to publishing, early on, a book derived from his dissertation. Such a book would have amplified and strengthened the reasons given by Quine for rejecting logicism. (Or rather, Quine's argumentation would have amplified and strengthened the reasoning that would have appeared in Paul's book.)

In the spring of 1961, Paul presented his dissertation in a class he gave on the philosophy of mathematics. He read it to us. I remember being irritated by what I was hearing. It seemed perverse, contrary to doctrines that had been put forth by Russell (hadn't he discovered what the number two really was?), by Hempel (hadn't he explained in a famous article that the work of Frege, Russell and others had reduced arithmetic to logic?), and by Quine himself (hadn't he shown how to derive arithmetic in the system of a book called *Mathematical Logic*?). My sense that something was wrong

First published in Adam Morton and Stephen P. Stich, eds., *Benacerraf and his Critics*, Cambridge, Mass.: Blackwell, 1996, pp. 143–59. Reprinted by kind permission of Blackwell Publishers.

with Paul's line of argument took a long time to disappear, perhaps twenty or twenty-five years, and was in part responsible for a number of papers. I hope that some of these are not completely in error.

I have come to agree with Paul, or Paul as he was in 1961, but my curiosity about *sub*-programs of logicism, about reductions of arithmetic or fragments thereof to interesting theories, which Paul's course also stimulated, persists. It gave rise to the present paper.

We shall consider the concept expressed by "the number of" from a logical, indeed a Fregean, point of view. With Frege, we shall suppose that "the number of" denotes a kind of function that takes concepts as arguments and yields objects as values; we shall further suppose that the function is total, i.e., defined on every concept, and further still, that the function is *extensional* in the sense that it assigns the same object to concepts F and G whenever the same objects fall under F and G.

We do not make any further assumptions about the function denoted by "the number of." In particular, we do not assume that it assigns the same object to F and G whenever and only whenever F and G are equinumerous, that is, we do not assume the truth of Hume's principle, explained below. So we may call ours a logical investigation of the concept of number, or at least a logical investigation of that which is expressed by "the number of."

Of course, we shall carry out our considerations in a formal setting: we use the sign "#", sometimes called "octothorpe,"[1] to symbolize "the number of" and take as our background system of logic standard axiomatic second-order logic, together with a principle expressing the extensionality of the function denoted by "#", Thus octothorpe, when attached to a monadic second-order variable, yields a term of the type of individual variables and the formula $x = \#F$, F a monadic second-order variable and x an individual variable, is well formed. It would therefore be appropriate to call # a concept-object function sign. Under standard semantics, according to which function signs denote *total* functions, $\forall F \exists! x\, x = \#F$ counts as a logical truth.

In *the language* L with which we shall be concerned, # is the sole non-logical constant. A *model* (for L) will be an ordered pair $\langle D, f \rangle$, D a non-empty set and f a function from the power set of D into D. Of course, $\forall F \exists! x\, x = \#F$ is true in any model.

Hume's principle is the sentence: $\forall F \forall G (\#F = \#G \leftrightarrow F \approx G)$; $F \approx G$ is some standard second-order formula expressing the existence of a one-one correspondence between the objects falling under the denotation of F and those falling under that of G (all such "standard" formulas are provably equivalent in axiomatic second-order logic, indeed in predicative axiomatic

---

[1] Cf. p. 42 of *The New Hacker's Dictionary*, second edition (Raymond, 1993), where the term is called "rare." On p. 39 of the first edition (1991) it was classified as "common."

second-order logic). Hume's principle is not valid; as is well known and as will readily follow from what is shown below, it is false in any model in which D is finite. It follows from the axiom of choice that for every infinite D, there is a function f such that Hume's principle is true in $\langle D, f \rangle$. Indeed, Hume's principle is true in an arbitrary model $\langle D, f \rangle$ if and only if for all subsets A, B of D, $fA = fB$ iff there is a one-one function from A onto B.

Although Hume's principle is false in all models in which the domain is finite, it does have models $\langle N, f \rangle$ where N is the set of natural numbers.

> Define f as follows: for any subset A of N, let fA
> $= n + 1$ if A is a finite set containing n members, and
> $= 0$ if A is infinite.

Then Hume's principle holds in $\langle N, f \rangle$. For let $A, B \subseteq N$. If A is finite, and has, say, n members, then $fA = n + 1$, and so $fA = fB$ iff $fB = n + 1$, iff B has n members, iff there is a one-one function from A onto B. But if A is infinite, then $fA = 0$ and so $fA = fB$ iff $fB = 0$, iff B is infinite, iff there is a one-one function from A onto B. (Any two infinite sets of natural numbers are equinumerous.)

Frege arithmetic is the system obtained by adjoining Hume's principle to axiomatic second-order logic. The proof given above that Hume's principle has a model in the natural numbers can easily be adapted to show the relative consistency of Frege arithmetic to standard second-order arithmetic ("analysis").

Frege's theorem is the result, sketched by Frege in *The Foundations of Arithmetic* and correctly proved in detail by him in *Basic Laws of Arithmetic*, that "zero," "(immediately) precedes,"[2] and "natural number" can be so defined in L that the Peano postulates can be proved in Frege Arithmetic.

The suggestion[3] that, even if not logically true, Hume's principle can be regarded as an "explanation" of the concept of a natural number, might lead us to inquire where in the proof of Frege's theorem Hume's principle is actually used and how much of the proof can be carried out by means that we *should* regard as purely logical. Our inquiry will shed light on the way in which Hume's principle acquires its strength. It will also reveal the surprising fact that on Frege's definitions of "zero," "precedes," and "natural (= finite) number," the Peano postulates contain a redundancy.

We shall take the Peano postulates to be the following seven statements:

---

[2]Terminology is awkward here. We take it that "precedes" and "succeeds" are converses and that a number is succeeded only by its successor. Thus 2 precedes 3 but not 4.

[3]Advanced by Crispin Wright in (Wright, 1983), p. 153.

(a) zero is a natural number;

(b) If m is a natural number and m precedes n, then n is a natural number;

(c) If m is a natural number and precedes n and n', then $n = n'$, so that the restriction of the relation precedes to the natural numbers is a many-one, i.e., a functional, relation;

(d) If m is a natural number, then for some n, m precedes n, so that the restriction of precedes to the natural numbers is a "serial" relation, and hence a relation that is the graph of a total function from the natural numbers to the natural numbers;

(e) No natural number precedes zero, so that zero is not in the range of that function;

(f) If natural numbers m and m' precede n, then $m = m'$, so that the restriction of precedes to the natural numbers is a one-many relation and the function is an injection; and

(g) Mathematical induction holds for the natural numbers: For all classes X, if 0 is in X and whenever a natural number m is in X and m precedes n, n is in X, then all natural numbers are in X.

Frege's theorem then states that the Peano postulates can be proved in Frege arithmetic: that is, there exist definitions of "zero," "precedes" and "natural number" under which the translations into L of (a)–(g) can be proved in Frege arithmetic.

We shall frequently use the term "finite" as a synonym of "[a] natural number."

Before beginning our discussion of Hume's principle and the proof of Frege's theorem we should discuss the logical properties of concept-object function signs like octothorpe.

In standard first-order logic a function sign, when interpreted over a model, denotes a function that assigns to every object in the domain of a model a unique object in the domain. We have similarly supposed that # is a sign for a total function, thus # will assign a unique object in the domain of a model to every subset of that domain.

Though nothing in the sequel depends upon it, we also make a supposition that is not completely routine, viz. that the formula $\forall x(Fx \leftrightarrow Gx) \rightarrow \#F = \#G$ is a logical truth. Let us refer to the formula $\forall x(Fx \leftrightarrow Gx) \rightarrow \#F = \#G$ as "FE," for "functional extensionality (for octothorpe)." FE will be a logical truth on any one of three sorts of extensional semantics.

Most importantly, on the usual model-theoretic account, the axiom of extensionality guarantees that FE will be true in any model under any assignment of subclasses of the domain of the model to its free variables F and G. For if $\forall x(Fx \leftrightarrow Gx)$ is true, then by extensionality, the same subclass will be assigned to F and G, and $\#F$ will denote the same element of the domain as $\#G$.

If one interprets second-order formulas using plurals, then if the antecedent $\forall x(Fx \leftrightarrow Gx)$ is true, then the objects assigned to F will be the same objects as those assigned to G, in brief and *par abus de langage*, the Fs and Gs will be the same. Then, since the Fs are the Gs, whatever "the number of" may mean, the number of Fs will be the same as the number of Gs, and therefore $\#F = \#G$ will be true. Thus again FE will be a logical truth.

Finally, if one understands FE in a manner like that in which Frege himself would have understood it, it is also always true. For on the way of taking it we are envisaging, the antecedent will say, approximately, that the same objects fall under F as fall under G ("approximately," because of the "concept horse" difficulty: "fall under" requires completion with names or variables of the type of objects). But then if the antecedent $\forall x(Fx \leftrightarrow Gx)$ is true, the concepts F and G bear to each other the relation that Frege calls the analogue of identity for concepts.[4] It then follows, or would certainly seem to follow, that F and G fall under the same higher-level concepts. And then G will fall under the higher-level concept $[H : \#F = \#H]$, since F falls under it, and therefore the consequent $\#F = \#G$ will be true.

Another way of expressing the last argument is this: On Fregean semantics, $\forall x(Fx \leftrightarrow Gx) \rightarrow \forall \chi(\chi F \leftrightarrow \chi G)$ is a logical truth (χ a second-level concept variable), from which it follows by instantiating $\chi(\ )$ as $\#F = \#(\ )$, that FE is also a logical truth, since $\#F = \#F$ certainly is one.

So let us take it that since $\#$ is a total concept-object function sign that "acts extensionally," FE counts as a logical truth. Hoping that Frege's proof will be recognized as a major accomplishment of the standard logical tradition, I emphasize that on the obvious and natural extension of the usual model-theoretic semantics to function signs of concept-object type, FE is *valid*: its universal closure is true in all models.

Of course FE is directly derivable from Hume's principle in second-order logic, for $F \approx G$ immediately follows from $\forall x(Fx \leftrightarrow Gx)$.

It is also to be noted therefore that on these three systems of semantics, Frege's Basic Law (Vb), which may be symbolized, nearly enough, as $\forall x(Fx \leftrightarrow Gx) \rightarrow\ 'F =\ 'G$, will also count as a logical truth. It was the converse (Va) that got Frege into trouble. Notice that (Va) may be

---

[4]See his "Comments on Sense and Meaning" in (Hermes et al., 1979).

written: $\neg\forall x(Fx \leftrightarrow Gx) \to {}'F \neq {}'G$, and read: if concepts bear the analogue of difference to each other, then their extensions are different. (One can't *inject* all subdomains of a domain into that domain.) Of course, $\neg\forall F\forall G({}'F = {}'G \to \forall x(Fx \leftrightarrow Gx))$ is without doubt a logical truth.

Log is the system in the language L extending axiomatic second-order logic to which FE has been added as the sole new axiom. Totality of the function denoted by $\#$ is expressed in the rules of inference of Log, which permit free instantiation and generalization with terms $\#F$. The comprehension-scheme

$$\exists R^n \forall x_1 \ldots \forall x_n (Rx_1 \ldots x_n \leftrightarrow \varphi),$$

φ an arbitrary formula of L and R not free in φ, is certainly assumed to be an axiom-scheme of second-order logic and hence also of Log. We also suppose that all ordinary rules of classical logic are available in Log. Every theorem of Log is therefore valid.

Since the set of its theorems is recursively enumerable, Log is incomplete with respect to standard, i.e., full, semantics for second-order logic. But it is of course sound with respect to that semantics, and the usual sorts of model-theoretic arguments can be used to show non-derivability in Log. So Hume's principle is not a theorem of Log.

Since every occurrence of a monadic second-order variable in a formula of Log is either before an object variable, after $\forall$ or $\exists$, or after $\#$, the provability of FE in Log entails that of the schema

$$\forall x(Fx \leftrightarrow Gx) \to (\ldots F \ldots \leftrightarrow \ldots G \ldots),$$

as the obvious induction on complexity of formulas $\ldots F \ldots$ shows.

We now note that for any formula φ and any variable x, we may introduce the term $\#[x : \varphi]$ by defining the formula $y = \#[x : \varphi]$ as

$$\exists F(y = \#F \land \forall x(Fx \leftrightarrow \varphi)) \qquad (F \text{ not free in } \varphi).$$

Comprehension alone, in the presence of the rules of inference, yields

$$\exists F(\exists! y\, y = \#F \land \forall x(Fx \leftrightarrow \varphi));$$

comprehension and FE then yield

$$\exists! y \exists F(y = \#F \land \forall x(Fx \leftrightarrow \varphi)),$$

since $\forall x(Fx \leftrightarrow \varphi)$ and $\forall x(Gx \leftrightarrow \varphi)$ imply $\forall x(Fx \leftrightarrow Gx)$. Thus for any variable x and formula φ, comprehension and FE license the introduction of the term $\#[x : \varphi]$.[5] For any formulas φ and ψ, we can then prove $\forall x(\varphi \leftrightarrow \psi) \to \#[x : \varphi] = \#[x : \psi]$.

[5]It is appropriate to comment here on slight differences in the present notation from

These preliminaries completed, let us now see what portions of Frege's argumentation can be carried out in Log. There are six definitions to be made, of "number," $\mathbf{0}$,[6] P (precedes, i.e., immediately precedes), $*$ (the strong ancestral), $*=$ (the weak ancestral), and "natural number."

"x is a number" is defined: $\exists F \, x = \#F$.

$\mathbf{0}$ is defined as the term $\#[x : x \neq x]$, licensed by FE and comprehension as usual. There are many other constant terms that can be formed with the aid of $\#$. Frege defines $\mathbf{1}$ as $\#[x : x = \mathbf{0}]$. We may also define $\mathbf{2}$, $\mathbf{3}$, etc. as $\#[x : x = \mathbf{0} \vee x = \mathbf{1}]$, $\#[x : x = \mathbf{0} \vee x = \mathbf{1} \vee x = \mathbf{2}]$, etc. The existence of 0, of 1, of 2, of 3, ... is thus provable in Log; what is not is the distinctness of 0 and 1, of 0 and 2, of 1 and 2, ... The existence of "anti-zero," the number $\#[x : x = x]$ of things there are, is also provable, but not its distinctness from 0. We might also introduce the number $\#[x : x$ is a number] of numbers there are, and once we define "finite," the number $\#[x : x$ is finite] of finite numbers there are. (It will be easy to show that anything finite is a number.) Even with the aid of Hume's principle, all that can be proved about the identity and distinctness of these three numbers is that if antizero is the number of finite numbers then antizero is also the number of numbers.

Frege defined "m immediately precedes n" as $\exists F \exists x (Fx \wedge \#F = n \wedge \#[y : Fy \wedge y \neq x] = m)$. His phrase for this relation (in Austin's translation) was "n follows in the series of natural numbers immediately after m." It will be convenient to use a slight variant of Frege's definition. So we define mPn:

$$\exists F \exists G \exists y (Gy \wedge \forall x (Fx \leftrightarrow Gx \wedge x \neq y) \wedge m = \#F \wedge n = \#G).$$

It is easy to prove $\mathbf{0}P\mathbf{1}$ in Log: take F as $[x : x \neq x]$, G as $[x : x = \mathbf{0}]$, and y as $\mathbf{0}$. It is something of an exercise to prove $\mathbf{1}P\mathbf{2}$; one way to do it is by cases: $\mathbf{0} = \mathbf{1}$ vs. $\mathbf{0} \neq \mathbf{1}$. However, matters are somewhat delicate: $\mathbf{2}P\mathbf{3}$ cannot be proved. For let the domain consist of two objects, a and b, and, where $\#$ denotes f, let $f\emptyset = f\{a,b\} = a$ and $f\{a\} = f\{b\} = b$. Then $\mathbf{0}$, $\mathbf{2}$, and $\mathbf{3}$ denote a, $\mathbf{1}$ denotes b, and there are no subsets A and B of the domain such that $fA = fB = a$, but A is obtained by removing one element from B. Thus $\mathbf{0}P\mathbf{1}$ and $\mathbf{1}P\mathbf{2}$, but not $\mathbf{2}P\mathbf{3}$, count as truths of logic.

that of Wright, Dummett, and Heck. Where I write $\#F$ and $\#[x : \varphi]$, these authors would all write: $\#xFx$ and $\#x\varphi$, respectively. Thus these authors all take $\#$ (which they write: N) to be a variable-binding operator, attaching to an object variable and a formula, while I take it to be a functor, attaching to a monadic concept variable. I prefer $\#F$ because I do not find it necessary to indicate at every turn that concepts may or may not apply to objects and $\#[x : \varphi]$ because it is on occasion useful to pretend that it is composed of $\#$ and a notation $[x : \varphi]$ for a concept. But it is to be emphasized that each of us has the resources for defining the other's notation (see three sentences back) and that these differences are merely stylistic.

[6]We use boldface for terms of *Log* and lightface for ordinary decimal numerals.

$\forall m \neg mP\mathbf{0}$ is derivable in Log from Hume's principle: Assume $mP\mathbf{0}$. Then for some F, G, y, Gy, $\forall x(Fx \leftrightarrow Gx \wedge x \neq y)$, $m = \#F$, and $\mathbf{0} = \#G$, i.e., $\#[x : x \neq x] = \#G$. By Hume's principle, $[x : x \neq x] \approx G$, whence $\forall x \neg Gx$, contra Gy.

But we certainly can't prove $\forall m \neg mP\mathbf{0}$ in Log alone, for it is consistent that there is only one object, and thus consistent that $\mathbf{0} = \mathbf{1}$; hence since we can prove $\mathbf{0}P\mathbf{1}$, $\mathbf{0}P\mathbf{0}$ is consistent; thus we can't even prove $\neg\mathbf{0}P\mathbf{0}$. (The attempt to deduce a contradiction from $(Gy \wedge \ldots \wedge \mathbf{0} = \#G)$ fails for lack of Hume's principle, which is needed to infer $[x : x \neq x] \approx G$ or its equivalent $\forall x \neg Gx$ from the supposition $\mathbf{0} = \#G$.)

$\forall m \neg mP\mathbf{0}$ is not explicitly mentioned in *Die Grundlagen*, but Frege explicitly proves it in *Grundgesetze*, as Proposition 108. Proposition 78.6 of *Die Grundlagen* is that every number except 0 is preceded by some number. §44 of *Grundgesetze* makes it plain that Frege did not intend 78.6 to imply its converse, $\forall m \neg mP\mathbf{0}$. Unlike its converse, 78.6 is in fact provable in Log. For on disabbreviating and slightly reformulating the natural first symbolization:

$\quad \forall n(n$ is a number $\wedge\ n \neq \mathbf{0} \to \exists m(m$ is a number $\wedge mPn))$,

one obtains:

$\quad \forall n(\exists G\, n = \#G \wedge n \neq \#[x : x \neq x] \to \exists m \exists F \exists G \exists y(Gy \wedge$
$\quad\quad \forall x(Fx \leftrightarrow Gx \wedge x \neq y) \wedge m = \#F \wedge n = \#G))$.

But if $n = \#G$ and $n \neq \#[x : x \neq x]$, then by FE, for some y, Gy. Then take $F = [x : Gx \wedge x \neq y]$ and $m = \#F$.

The natural proof of 78.5 of *Die Grundlagen*, which states that if mPn and $m'Pn'$, then $m = m'$ iff $n = n'$, i.e., that P is a one-one relation ("beiderseits eindeutig"), appeals to Hume's principle at several points. In the natural proof of this crucial proposition, one freely passes back and forth between the existence of one-one correspondences between objects falling under concepts with certain numbers and the identity of those numbers. Since the natural derivation of 78.5 from Hume's principle is given in a paper by Heck and the author,[7] we omit it here.

And like $\forall m \neg mP\mathbf{0}$, the one-one-ness of P is not provable in Log. As before, let the domain consist of two objects, a and b, but this time let $f\emptyset = f\{b\} = f\{a, b\} = a$ and $f\{a\} = b$. Then $\mathbf{0}$ denotes a, $\mathbf{1}$ denotes b, $\mathbf{2}$ denotes a, and so $\mathbf{1} \neq \mathbf{2}$ is true; but $\mathbf{0}P\mathbf{1}$ is (always) true and $\mathbf{0}P\mathbf{2}$ is also true: take $\{b\}$ as G.

Let us give the designations "NPZ," "PM1," "P1M," and "P11" to the sentences

$\quad \forall m \neg mP\mathbf{0}$,

---

[7] Article 20 in this volume.

$$\forall m \forall n \forall n'(mPn \land mPn' \to n = n'),$$

$$\forall m \forall m' \forall n(mPn \land m'Pn \to m = m'), \text{and}$$

$$\forall m \forall m' \forall n \forall n'(mPn \land m'Pn' \to (m = m' \leftrightarrow n = n')).$$

These express, respectively, that nothing precedes zero, that *precedes* is a many-one relation, that *precedes* is a one-many relation, and that *precedes* is a one-one relation. Of course P11 is equivalent to the conjunction of PM1 and P1M. Thus although we cannot prove either NPZ or P11 in Log, we can, as we have seen, prove them in Log *from* Hume's principle.

We now wish to show that the conjunction of NPZ and P11, although implied by Hume's principle, does not imply it. It suffices to construct a model $\langle N, g \rangle$, where N is the set of natural numbers, in which NPZ and P11 are true and Hume's principle is false.

For any $A \subseteq N$, let gA

$= \quad n + 2$ if A is a finite set containing n members,

$= \quad 0$ if A is infinite and cofinite, and

$= \quad 1$ if A is infinite and coinfinite.

Then the term **0** designates $f\emptyset$, which is the natural number 2; the term **1** designates $f\{f\emptyset\}$, which is the natural number 3; the term **2** designates $f\{f\emptyset, f\{f\emptyset\}\}$, etc.[8]

It is clear that Hume's principle fails in $\langle N, g \rangle$. Let A be N and B be the set of even natural numbers. Then $F \approx G$ is true when A and B are assigned to F and G, but $\#F$ and $\#G$ respectively denote 0 and 1 under this assignment.

To see that NPZ and P11 are true in $\langle N, g \rangle$ we shall investigate the conditions under which they are true in an arbitrary model $\langle D, f \rangle$.

The corollary to our first lemma asserts that NPZ holds in a model $\langle D, f \rangle$ iff the object assigned by f to the null set is assigned to no other object.

Lemma 1 $\vdash \text{NPZ} \leftrightarrow \forall F(\#F = \#[x : x \neq x] \to \forall x \neg Fx)$

Proof. When decoded, NPZ, i.e., $\forall m \neg mP\mathbf{0}$, is $\neg \exists m \exists F \exists G \exists y (Gy \land \forall x (Fx \leftrightarrow Gx \land x \neq y) \land m = \#F \land \#[x : x \neq x] = \#G)$, which is equivalent to: $\neg \exists G (\exists y Gy \land \#[x : x \neq x] = \#G)$. (The latter clearly implies the former, and a counterexample to the former can easily be constructed from one to the latter.) ∎

Corollary $\langle D, f \rangle \models \text{NPZ}$ *iff for all* $A \subseteq D$, *if* $fA = f\emptyset$ *then* $A = \emptyset$.

It is clear from the Corollary that NPZ is true in $\langle N, g \rangle$, for if $gA = g\emptyset$, $gA = 2$, A contains no members and $A = \emptyset$.

---

[8]What do you get if you drop all "f"s?

Let us abbreviate "$\exists y(Gy \wedge \forall x(Fx \leftrightarrow Gx \wedge x \neq y))$" by "$F \subseteq_1 G$." The notation is intended to suggest: F falls short of being G by just one object. Now call A and B *f-equivalent* iff $fA = fB$. And write: $A \subseteq_1 B$ to mean: for some x in D, $x \notin A$, and $B = A \cup \{x\}$. The corollary to the next lemma says that PM1 holds in $\langle D, f \rangle$ if and only if, whenever $A \subseteq_1 B$, $A' \subseteq_1 B'$, and A and A' are f-equivalent, then B and B' are f-equivalent.

Lemma 2 $\vdash$ PM1 $\leftrightarrow \forall F \forall F' \forall G \forall G'(F \subseteq_1 G \wedge F' \subseteq_1 G' \wedge \#F = \#F' \rightarrow \#G = \#G')$.

Proof. When decoded, PM1 reads: $\forall m \forall n \forall n' \forall F \forall F' \forall G \forall G' \forall y \forall y'(m = \#F \wedge m = \#F' \wedge n = \#G \wedge n' = \#G' \wedge Gy \wedge G'y' \wedge \forall x(Fx \leftrightarrow Gx \wedge x \neq y) \wedge \forall x(F'x \leftrightarrow G'x \wedge x \neq y') \rightarrow n = n')$, i.e., $\forall F \forall F' \forall G \forall G' \forall y \forall y'(\#F = \#F' \wedge Gy \wedge G'y' \wedge \forall x(Fx \leftrightarrow Gx \wedge x \neq y) \wedge \forall x(F'x \leftrightarrow G'x \wedge x \neq y') \rightarrow \#G = \#G')$. ∎

Corollary $\langle D, f \rangle \models$ PM1 *iff for all* A, A', B, B', *if* $A \subseteq_1 B$ *and* $A' \subseteq_1 B'$, *then if* $fA = fA'$, *then* $fB = fB'$.

The following corollaries supply similar conditions for P1M and P11.

Lemma 3 $\vdash$ P1M $\leftrightarrow \forall F \forall F' \forall G \forall G'(F \subseteq_1 G \wedge F' \subseteq_1 G' \wedge \#G = \#G' \rightarrow \#F = \#F')$.

Proof. Like that of Lemma 2. ∎

Corollary $\langle D, f \rangle \models$ P1M *iff for all* A, A', B, B', *if* $A \subseteq_1 B$ *and* $A' \subseteq_1 B'$, *then if* $fB = fB'$, *then* $fA = fA'$.

Lemma 4 $\vdash$ P11 $\leftrightarrow \forall F \forall F' \forall G \forall G'(F \subseteq_1 G \wedge F' \subseteq_1 G' \rightarrow (\#F = \#F' \leftrightarrow \#G = \#G'))$.

Proof. By Lemmas 2 and 3. ∎

Corollary $\langle D, f \rangle \models$ P11 *iff for all* A, A', B, B', *if* $A \subseteq_1 B$ *and* $A' \subseteq_1 B'$, *then* $fA = fA'$ *iff* $fB = fB'$.

We can now see that P11 also holds in $\langle N, g \rangle$, in brief because any set that differs by one element from an infinite–cofinite or infinite–coinfinite set is itself infinite–cofinite or infinite–coinfinite, respectively.

In more detail: Suppose $A \subseteq_1 B$ and $A' \subseteq_1 B'$. If A is finite, with, say, n members, $gA = n + 2$ and $gB = n + 3$; then $gA = gA'$ iff $gA' = n + 2$, iff A' has $n + 2$ members, iff B' has $n + 3$ members, iff $gB' = n + 3$, iff $gB = gB'$. If A is infinite and cofinite, $gA = 0$ and therefore $gB = 0$, since a superset by one element of an infinite–cofinite set is also infinite–cofinite;

but then $gA = gA'$ iff $gA' = 0$, iff A' is infinite–cofinite, iff B' is infinite–cofinite, iff $gB' = 0$, iff $gB = gB'$. And, similarly, if A is infinite–coinfinite, $gA = 1$ and $gB = 1$, whence $gA = gA'$ iff A' is infinite–coinfinite, iff B' is infinite–coinfinite, iff $gB = gB'$.

We have now concluded the proof that Hume's principle does not follow from the conjunction of NPZ and P11. The reason, in a nutshell, is that Hume looks only at the "insides" of concepts and not at their "outsides," and thus if Hume holds in a model, then the same object must be assigned to any two equinumerous subsets of the domain, regardless of whether their complements are equinumerous or not.

We can now see more clearly how Hume's principle exerts its influence through NPZ and P11. Assume that NPZ and P11 hold in a model $\langle D, f \rangle$. Suppose now that there are natural numbers i and i' and subsets B and B' of D, such that $i < i'$, B contains $i+1$ elements, B' contains $i'+1$ elements, and $fB = fB'$. Then there are subsets A, A' of B, B', respectively, that contain i and i' elements, respectively, and so by P1M, $fA = fA'$. By induction, there is a non-empty subset A' of D such that $f\emptyset = fA'$, contra NPZ. Thus if two finite subsets of D are f-equivalent, they are equinumerous. And in like manner, if A is finite and A' is infinite, $fA \neq fA'$.

Suppose that for some natural number i, any two i-element subsets of D are f-equivalent. Then by PM1, any two $(i+1)$-element subsets of D are also f-equivalent. Thus finite subsets of D are f-equivalent if and only if they are equinumerous.

We now define a sequence $\{a_i\}_{i \in N}$ of elements of D: $a_0 = f\emptyset$; $a_{i+1} = f\{a_0, \dots, a_i\}$. Suppose that for some least j, some $i < j$, $a_i = a_j$. By NPZ, $i \neq 0$, and therefore for some q, r, $q < r$, $i = q+1$ and $j = r+1$. Then $f\{a_0, \dots, a_q\} = a_{q+1} = a_i = a_j = a_{r+1} = f\{a_0, \dots, a_r\}$. So $\{a_0, \dots, a_q\}$ and $\{a_0, \dots, a_r\}$ are equinumerous. But since $q < r < j$, the cardinality of $\{a_0, \dots, a_q\}$ is $q+1$ and the cardinality of $\{a_0, \dots, a_r\}$ is $r+1$. Thus $i = j$, contradiction. So if $i \neq j$, $a_i \neq a_j$, and $\{a_i : i \in N\}$ is a countably infinite subset of D. Moreover, if A is any infinite subset of D, $fA \neq a_i$.

Thus a restricted version of Hume's principle holds in $\langle D, f \rangle$: if A is finite, then $fA = fA'$ iff A and A' are equinumerous.[9]

Notice that we could not have concluded that if A and A' are countably infinite subsets of D, then $fA = fA'$: N and {the evens} are countably infinite subsets of N, $gN = g\{\text{the evens}\}$, but NPZ and P11 are true in $\langle N, g \rangle$.

We turn now to the definition of the natural numbers and the deduction

---

[9]Thanks here to Richard Heck. Heck has observed that this weakened form of Hume's principle is equivalent to the conjunction of NPZ, (c), and $\forall n(0P^{*=}n \to \forall m \forall m'(mPn \wedge m'Pn \to m = m'))$, which is a consequence of P1M. Here one can define "is finite" as "possesses a well-ordering whose converse is a well-ordering."

of the Peano postulates from NPZ and P11. It is apparent at the outset that, however we define "natural number," (c), (e) and (f) will follow, for these are relativizations of PM1, NPZ and P1M to the predicate "natural number." (Note that there are no function signs in our formulation of the Peano postulates.)

If R is an arbitrary relation, then the strong ancestral R^* is defined: $xR^*y \equiv \forall F(\forall d(xRd \rightarrow Fd) \wedge \forall d\forall a(Fd \wedge dRa \rightarrow Fa) \rightarrow Fy)$. Two important facts about the strong ancestral, proved in Frege's *Begriffsschrift*, are that if xRy, then xR^*y and that if xR^*y and yR^*z, then xR^*z.

The weak ancestral $xR^{*=}y$ may be defined: $xR^*y \vee y = x$. It is an exercise to prove this definition equivalent to: $\forall F(Fx \wedge \forall d\forall a(Fd \wedge dRa \rightarrow Fa) \rightarrow Fy)$.

Important for our purpose is the following lemma connecting the strong and weak ancestral:

Lemma 5 $\vdash xR^*y \rightarrow \exists z(zRy \wedge xR^{*=}z)$.

Proof. Let $Fa \equiv \exists z(zRa \wedge xR^{*=}z)$. Then $xRa \rightarrow Fa$, for if xRa, then certainly Fa: take $z = x$. And $Fd \wedge dRa \rightarrow Fa$: Suppose Fd and dRa. Then for some z, zRd and $xR^{*=}z$, and therefore $xR^{*=}d$. But since dRa, Fa. The lemma follows by the definition of the strong ancestral. ■

Finally, "n is finite," i.e., "n is a natural number" is defined, simply, as: $0P^{*=}n$. It is trivial to prove (a) that 0 is finite and easy to prove (b) that if m is finite and precedes n, then n is finite and that anything finite is a number.

It is also easy to prove—should this be surprising?—mathematical induction in the form:

$$F0 \wedge \forall m\forall n(Fm \wedge mPn \rightarrow Fn) \rightarrow \forall n(n \text{ is finite} \rightarrow Fn).$$

We can also prove induction in the more useful and stronger-looking form:

$$F0 \wedge \forall m\forall n(m \text{ is finite} \wedge Fm \wedge mPn \rightarrow Fn) \rightarrow \forall n(n \text{ is finite} \rightarrow Fn).$$

To prove the stronger-looking form from the weaker-looking, define Gx as: $Fx \wedge x$ is finite; then, using (a) and (b), apply the weaker-looking form to G. (g) is nothing other than a reformulation of the stronger-looking form.

We have just seen that there is at least one clear and correct interpretation of the claim that mathematical induction is a truth of logic plus appropriate definitions. Moreover, the proof of induction in Log simply formalizes the usual trivial argument for taking it to be a logical truth.

To sum up thus far: Mathematical induction, in either of the two forms, can be proved in Log; (a) is a trivial consequence of the definition of "natural number"; (b) is an easy consequence of that definition; (c), (e), and (f) follow, as we have seen, from NPZ and P11; and (g) is just mathematical induction in the second, stronger form.

What of (d)?

The central argument of *Die Grundlagen der Arithmetik*, sketched in §§82–83 of that work, was intended to show that (d) follows from Hume's principle. In those two sections Frege asserts that (d) can be proved by deriving it from

(1) $\quad dP\#[x : xP^{*=}d] \wedge dPa \to aP\#[x : xP^* = a]$

and

(2) $\quad 0P\#[x : xP^* = x]$

with the aid of a version of the weaker-looking form of induction. But by the time he came to write out a proof of (d) in *Grundgesetze der Arithmetik*, Frege had changed his mind about the possibility of carrying out the one he had sketched in *Die Grundlagen*, for he asserts in *Grundgesetze* that (1) appears to be unprovable and instead derives (d) from

(1') $\quad d$ is finite $\wedge\, dP\#[x : xP^{*=}d] \wedge dPa \to aP\#[x : xP^{*=}a]$

and (2) with the aid of the stronger-looking form of induction (which of course can be derived from the weaker-looking form).[10]

In fact, "zero," "precedes" and "natural number" now having been defined, (d) follows from (a)–(c) and (e)–(g) in Log, as a slight modification of Frege's corrected argument shows.

It is remarkable that this should be so. For let e be an arbitrary individual constant and R an arbitrary predicate letter. In general, the analogue

$\quad \forall m(eR^{*=}m \to \exists n\, mRn)$

of (d) will not follow from the conjunction of the analogues of (a)–(c) and (e)–(g):

$\quad eR^{*=}e,$

$\quad \forall m \forall n(eR^{*=}m \wedge mRn \to eR^{*=}n),$

$\quad \forall m \forall n \forall n'(eR^{*=}m \wedge mRn \wedge mRn' \to n = n'),$

$\quad \forall m(eR^{*=}m \to \neg mRe),$

$\quad \forall m \forall m' \forall n(eR^{*=}m \wedge eR^{*=}m' \wedge mRn \wedge m'Rn \to m = m'),$

---

[10] For a fuller account, see Article 20 in this volume.

$$\forall X([Xe \wedge \forall m \forall n (eR^{*=}m \wedge Xm \wedge mRn \to Xn)] \to$$
$$\forall m[eR^{*=}m \to Xm]).$$

(The first, second, and last of these are logical truths, certainly.) These six sentences will all be true, but $\forall m(eR^{*=}m \to \exists n \, mRn)$ will be false, in the model in which e denotes the sole object in the domain and R is assigned the empty relation. Thus we cannot even prove $\exists n \, eRn$ or $\exists x \exists y \, x \neq y$ from their conjunction, and it might thus appear that we cannot prove anything significant in Log from the conjunction of (a)–(c) and (e)–(g). In particular, one might readily think that we should not be able to prove the seriality of *precedes* on the class of natural numbers, for how, one might wonder, do $\mathbf{0}$ and P differ from e and R? If a certain relation is one-one and nothing bears that relation to zero, how are we supposed to be able to prove without further assumptions that the relation has an infinite field?

What is surprising is that this is just not the case. We can prove $\exists n \, \mathbf{0} Pn$ and $\exists x \exists y \, x \neq y$ in Log from (c), (e) and (f), the relativizations to "natural number" of PM1, NPZ, and P1M and therefore prove from these that there are Dedekind infinitely many natural numbers. For if we define "natural number" as we have done, we can in fact prove the seriality of *precedes* on the class of natural numbers, and thus that the restriction of P to "natural number" defines a one-one function from that class onto one of its proper subclasses. It immediately follows that we can prove that there are Dedekind infinitely many natural numbers. All this, it is to be emphasized, can be done without further appeal to Hume's principle, or (the unrelativized forms of) NPZ, PM1, or P1M.

Here, then, is the proof; each of Lemmas 6 through 8 shows that in Log the proposition that follows the turnstile "⊢" can be proved from the non-logical hypotheses, if any, that precede it.

Lemma 6 (e), (f) ⊢ $\mathbf{0} P^{*=}a \to \neg aP^*a$.

Proof. $\mathbf{0} = \#[x : x \neq x]$. If $\mathbf{0}P^*\mathbf{0}$, then by Lemma 5, for some z, $zP\mathbf{0}$ and $\mathbf{0}P^{*=}z$. Thus z is finite and by (e) we have a contradiction. So $\neg \mathbf{0}P^*\mathbf{0}$.

Now assume d is finite, i.e., $\mathbf{0}P^{*=}d$, dPa and aP^*a. Then $\mathbf{0}P^{*=}a$. By Lemma 5, for some z, zPa and $aP^{*=}z$, whence zP^*z. Since $\mathbf{0}P^{*=}a$ and $aP^{*=}z$, z is finite. Since dPa, zPa, and d and z are finite, $z = d$ by (f), and therefore dP^*d. Thus d is finite $\wedge \neg dP^*d \wedge dPa \to \neg aP^*a$.

Lemma 6 now follows by (g). ∎

Lemma 7 (e) ⊢ $\mathbf{0}P\#[x : x$ is finite $\wedge xP^{*=}\mathbf{0}]$.

Proof. Since $\mathbf{0}$ is finite and $\mathbf{0}P^{*=}\mathbf{0}$, $\#[x : x$ is finite$\wedge xP^{*=}\mathbf{0}\wedge x \neq \mathbf{0}]P\#[x : x$ is finite$\wedge xP^{*=}\mathbf{0}]$, by the definition of P. Since $\mathbf{0} = \#[x : x \neq x]$, it suffices to show that $\#[x : x$ is finite $\wedge xP^{*=}\mathbf{0} \wedge x \neq \mathbf{0}] = \#[x : x \neq x]$, and hence by FE, to show that $\forall x(x$ is finite $\wedge xP^{*=}\mathbf{0} \wedge x \neq \mathbf{0} \leftrightarrow x \neq x)$. Now suppose x is finite, $xP^{*=}\mathbf{0}$, and $x \neq \mathbf{0}$. Then $xP^*\mathbf{0}$; by Lemma 5, for some z, $zP\mathbf{0}$ and $xP^{*=}z$; since x is finite, so is z, contra (e). ∎

Lemma 8   (c), (e), (f) $\vdash$ d is finite $\wedge$ $dPa \wedge$ $dP\#[x : x$ is finite and $xP^{*=}d] \rightarrow aP\#[x : x$ is finite and $xP^{*=}a]$.

Proof. Suppose d is finite, dPa and $dP\#[x : x$ is finite and $xP^{*=}d]$. Then a is finite, and by (c), $a = \#[x : x$ is finite $\wedge xP^{*=}d]$. Since a is finite and $aP^{*=}a$, it is immediate from the definition of P that $\#[x : x$ is finite $\wedge xP^{*=}a \wedge x \neq a]P\#[x : x$ is finite $\wedge xP^{*=}a]$. It thus suffices to show that $\#[x : x$ is finite $\wedge xP^{*=}a \wedge x \neq a] = \#[x : x$ is finite $\wedge xP^{*=}d]$, and thus by FE, to show that $\forall x(x$ is finite $\wedge xP^{*=}a \wedge x \neq a \leftrightarrow x$ is finite $\wedge xP^{*=}d)$.

Now suppose x is finite. Assume $xP^{*=}a$ and $x \neq a$. Then xP^*a, and so by Lemma 5, for some z, $xP^{*=}z$ and zPa. Since x is finite and $xP^{*=}z$, z is finite. Since zPa, dPa, and z and d are finite, $z = d$ by (f), and so $xP^{*=}d$. Conversely, assume $xP^{*=}d$. Since dPa, by facts about the ancestral, xP^*a, whence $xP^{*=}a$. But since a is finite, by Lemma 6, $\neg aP^*a$, whence $x \neq a$. ∎

From Lemmas 7 and 8 it follows by (g) in Log from (c), (e) and (f) that if m is a natural number, then m precedes the number of natural numbers that bear the weak ancestral of *precedes* to m: m is finite $\rightarrow$ $mP\#[x : x$ is finite and $xP^{*=}m]$. So from (c), (e) and (f), Log proves (d): m is finite $\rightarrow \exists n\, mPn$. Thus on Frege's definitions of "zero," "precedes" and "natural number," (a), (b) and (g), unsurprisingly, turn out to be logical truths; (d), surprisingly, turns out to be a consequence of the others.

The proof we have given that, on Frege's definitions, (d) follows from (c), (e) and (f) is a modification of Frege's proof from Hume's principle that every finite number precedes some number. In Frege's proof, Hume's principle is used only to obtain NPZ and P11. Once these are in hand, Hume's principle is used thereafter only to yield FE, which is an axiom of Log and which we have argued should count as a logical truth. The modification consists in relativizing a number of variables to the predicate "natural number," replacing almost every occurrence of a number term $\#[x : \varphi]$ with one of the term $\#[x : x$ is finite $\wedge \varphi]$ and appealing when necessary to the stronger-looking, rather than the weaker-looking, form of mathematical induction.

We conclude by showing that the Peano postulates, under Frege's definitions, are rather weak. We shall show that they do not imply even the *dis*junction of NPZ, PM1 and P1M. As before, we will construct a model $\langle N, f \rangle$: For any $A \subseteq N$, let fA

$$\begin{aligned}
&= \quad n+2 \text{ if } A \text{ is a finite set containing } n \text{ members} \\
&= \quad 2 \text{ if } A = N \\
&= \quad 0 \text{ if } A = N - \{0\} \text{ or } N - \{0,1\} \\
&= \quad 1 \text{ otherwise, in particular if } A = N - \{1\}.
\end{aligned}$$

So $f\emptyset = fN = 2$.

By Lemma 1, $\langle N, f \rangle \not\models$ NPZ: $fN = f\emptyset$. By Lemma 2, $\langle N, f \rangle \not\models$ PM1: let A, A', B and B' be $N - \{0,1\}$, $N - \{0\}$, $N - \{1\}$ and N, respectively. By Lemma 3, $\langle N, f \rangle \not\models$ P1M: let A, A', B, and B' be $N - \{0\}$, $N - \{1\}$, N and N, respectively.

Since $N \subseteq_1 A$ for no $A \subseteq N$, it is clear that in $\langle N, f \rangle$, $\mathbf{0}$ denotes 2, that if $j > 1$, then $\langle j, k \rangle$ satisfies mPn iff $k = j+1$, and that the objects satisfying "natural number" in $\langle N, f \rangle$ are the positive integers greater than 1. Thus the Peano postulates all hold in $\langle N, f \rangle$.

We have of course taken advantage of the unrestricted range of the variables in NPZ, PM1 and P1M in order to falsify these sentences while keeping the Peano postulates, with their (generally) restricted ranges, true.

18

Frege's Theorem and the Peano Postulates

Two thoughts about the concept of number are incompatible: that any zero or more things have a (cardinal) number, and that any zero or more things have a number (if and) only if they are the members of some one set. It is Russell's paradox that shows the thoughts incompatible: the sets that are not members of themselves cannot be the members of any one set. The thought that any (zero or more) things have a number is Frege's; the thought that things have a number only if they are the members of a set may be Cantor's and is in any case a commonplace of the usual contemporary presentations of the set theory that originated with Cantor and has become ZFC.

In recent years a number of authors have examined Frege's accounts of arithmetic with a view to extracting an interesting subtheory from Frege's formal system, whose inconsistency, as is well known, was demonstrated by Russell. These accounts are contained in Frege's formal treatise *Grundgesetze der Arithmetik* and his earlier exoteric book *Die Grundlagen der Arithmetik*. We may describe the two central results of the recent re-evaluation of his work in the following way: Let *Frege arithmetic* be the result of adjoining to full axiomatic second-order logic a suitable formalization of the statement that the *F*s and the *G*s have the same number if and only if the *F*s and the *G*s are equinumerous. Then (1) Frege himself succeeded in deriving arithmetic from Frege arithmetic and (2) Frege arithmetic is equiconsistent with full second-order arithmetic (and is thus consistent, with moral certainty). So Frege derived arithmetic from a single consistent and

Reprinted with the kind permission of the editors from *The Bulletin of Symbolic Logic* 1 (1995): 317–326. Copyright ©1995 Association for Symbolic Logic. All rights reserved. This reproduction is by special permission for this publication only.

The author is grateful to the Alexander von Humboldt Foundation for its support while work on this paper was begun.

obvious-seeming principle, if not from logic alone, as he had hoped he had. A number of the articles responsible for the re-evaluation of Frege's formal work are collected in William Demopoulos' anthology, *Frege's Philosophy of Mathematics*.[1]

Frege, we now see, thus provided a consistent theory of the natural numbers altogether different from those of Dedekind, Russell and Whitehead, Zermelo, von Neumann, and the several founders of combinatory logic and the lambda-calculus. Particularly noteworthy is the difference between Frege's account and that found in *Principia Mathematica*: for Frege, numbers are certain objects that lie at the bottom level of a type-theoretical hierarchy, while for Russell and Whitehead, they are situated two levels above the bottom of a similar hierarchy.

Here we investigate the Fregean conception of number; we are going to show that on Fregean definitions of *zero, (immediately) precedes,*[2] and *natural number,* but without use of the aforementioned principle, the Peano postulates contain a notable redundancy, whose discovery can, with only a small amount of exaggeration, be attributed to Frege, since the proof of redundancy is an adaptation of Frege's proof that the series of natural numbers is infinite. Thus the study of the details of Frege's work has yielded a result concerning the extremely familiar Peano axiomatization. We begin the investigation by recalling an old observation of Henkin's concerning a dependency among the Peano postulates.[3]

Suppose that the Dedekind–Peano postulates are formulated as usual in a second-order language containing a constant 0 (for zero) and a one-place function sign s (for successor):

(i) $\forall x 0 \neq sx$

(ii) $\forall x \forall y (sx = sy \rightarrow x = y)$

(iii)   Mathematical induction: $\forall F(F0 \wedge \forall x(Fx \rightarrow Fsx) \rightarrow \forall x\, Fx)$.

Henkin observed that (iii) *implies the disjunction of* (i) *and* (ii).

We may see this as follows: by (iii), the range of the first order variables is the class of denotations of $0, s0, ss0, \ldots$ Either these denotations are all distinct, in which case (i) and (ii) both hold, or they are not and for some least m, $s^m 0$ denotes the same object as $s^n 0$, for some $n > m$. If $m = 0$, (i) fails but (ii) holds; if $m > 0$, (ii) fails but (i) holds.

It is easy to construct models in which each of the seven conjunctions $\pm(i)\pm(ii)\pm(iii)$ other than $-(i)-(ii)+(iii)$ holds; so no other dependencies among (i), (ii), and (iii) await discovery.

Our main interest is in the Fregean conception of number and we shall

---

[1](Demopoulos, 1995).

[2]We use "precedes" to mean "immediately precedes" (rather than "is less than").

[3](Henkin, 1960).

accordingly follow Frege in not *assuming* that the relation *precedes* is the graph of a total function on the natural numbers, or even a functional relation. Accordingly, we formulate the Peano postulates in the language $\{\mathbf{0}, P, N\}$, with P a binary predicate letter for "precedes" and N a unary predicate letter for "is a natural number," as the universal closures of the following seven formulas:

(a) $\quad N\mathbf{0}$
(b) $\quad Nm \wedge mPn \rightarrow Nn$
(c) $\quad Nm \wedge mPn \wedge mPn' \rightarrow n = n'$
(d) $\quad Nm \rightarrow \exists n\, mPn$
(e) $\quad Nn \rightarrow \neg nP\mathbf{0}$
(f) $\quad Nm \wedge Nm' \wedge mPn \wedge m'Pn \rightarrow m = m'$
(g) $\quad F\mathbf{0} \wedge \forall m \forall n(Fm \wedge mPn \rightarrow Fn) \rightarrow Fn.$

The background logic of our investigation will be axiomatic second-order logic.

(g) is of course a formulation of mathematical induction. It is immediately implied by the stronger-looking

(h) $\quad F\mathbf{0} \wedge \forall m \forall n(Nm \wedge Fm \wedge mPn \rightarrow Fn) \rightarrow Fn.$

But (h) is equivalent to (g) in the presence of (a) and (b): Let $Gn \equiv Nn \wedge Fn$. Then (a), (b) and the antecedent of (h) yield $G\mathbf{0}$ and $\forall m \forall n(Gm \wedge mPn \rightarrow Gn)$; so by (g), Gn, whence Fn.

The conjunction of (a), (b) and (g) is equivalent to the celebrated Dedekind–Frege–Russell definition of N from $\mathbf{0}$ and P:

Def$N \quad \forall x(Nx \leftrightarrow \forall F(F\mathbf{0} \wedge \forall m \forall n(Fm \wedge mPn \rightarrow Fn) \rightarrow Fx)).$

We first examine dependencies among (c), (d), (e) and (f) under the assumption of Def$N$. It will help matters if we give (c), (d), (e), and (f) more suggestive designations. Thus we dub (c) **Functionality**, (d) **Totality**, (e) **Exiledom** (the sequence of natural numbers never returns to its point of origin), and (f) **Injectivity**.

We shall use: $\vdash^*$ to denote derivability from Def$N$ (in axiomatic second-order logic).

Formalizing Henkin's argument, we see that $\vdash^* \mathbf{F} \wedge \mathbf{T} \rightarrow \mathbf{E} \vee \mathbf{I}$.

In a similar vein, we may see that $\vdash^* \mathbf{F} \wedge \neg\mathbf{T} \rightarrow \mathbf{E} \wedge \mathbf{I}$, by formalizing the following argument: If $\mathbf{F}$ holds, every object satisfying Nx bears (the) P(-relation) to at most one object; if $\neg\mathbf{T}$ holds, some object satisfying Nx bears P to no object. By (a), (b) and (g), the objects satisfying Nx are just the denotation a_0 of $\mathbf{0}$, the unique object a_1, if it exists, to which a_0 bears P, the unique object a_2, if it exists, to which a_1 bears P, ... Now if $i < j$ and a_j exists, then a_i exists, but a_j does not bear P to a_i; otherwise we have a loop and so **T** holds; nor can every a_i exist, for then, since every a_i bears P to a_{i+1}, **T** again holds. So the a_is eventually give out without ever repeating. But then **E** and **I** both hold. So we have

$$\vdash^* \neg(\mathbf{F} \wedge \mathbf{T} \wedge \neg\mathbf{E} \wedge \neg\mathbf{I}) \qquad \text{(Henkin)},$$

as well as

$$\vdash^* \neg(\mathbf{F} \wedge \neg\mathbf{T} \wedge \mathbf{E} \wedge \neg\mathbf{I}),$$
$$\vdash^* \neg(\mathbf{F} \wedge \neg\mathbf{T} \wedge \neg\mathbf{E} \wedge \mathbf{I}), \qquad \text{and}$$
$$\vdash^* \neg(\mathbf{F} \wedge \neg\mathbf{T} \wedge \neg\mathbf{E} \wedge \neg\mathbf{I}).$$

We shall discuss the twelve other possibilities below. One of them, however, deserves mention here: It is not the case that $\vdash^* \mathbf{F} \wedge \mathbf{E} \wedge \mathbf{I} \to \mathbf{T}$, for (a), (b), (g), **F**, **E** and **I** are all true and **T** false in the model consisting of a single element which does *not* bear P to itself; of course, that single element will be the denotation of **0**, and it will therefore satisfy Nx.

The use of the constant **0**, rather than a monadic predicate letter Z (for "is a zero") in the formulation of **E** and Def$N$ is important. In particular, the assumption that there are not two zeros has significant consequences. For suppose we reformulate the Peano Postulates in the language $\{Z, P, N\}$ as the universal closures of:

(a′) $Zn \to Nn$

EZ $\exists n\, Zn$

UZ $Zm \wedge Zn \to m = n$

(b) $Nm \wedge mPn \to Nn$

F $Nm \wedge mPn \wedge mPn' \to n = n'$

T $Nm \to \exists n\, mPn$

E′ $Zm \wedge Nn \to \neg nPm$

I $Nm \wedge Nm' \wedge mPn \wedge m'Pn \to m = m'$

(g′) $\forall F(\forall x(Zx \to Fx) \wedge \forall d \forall a(Fd \wedge dPa \to Fa) \to \forall x(Nx \to Fx))$

and, correspondingly, redefine N:

Redef$N$ $\forall n(Nn \leftrightarrow \forall F(\forall x(Zx \to Fx) \wedge \forall d \forall a(Fd \wedge dPa \to Fa) \to Fn))$.

We can derive (a′) $\wedge$ (b) $\wedge$ (g′) from Redef$N$, and thus also derive (h′), i.e.,

$$\forall F(\forall x(Zx \to Fx) \wedge \forall d \forall a(Nd \wedge Fd \wedge dPa \to Fa) \to \forall x(Nx \to Fx)).$$

But not only can $\mathbf{F} \wedge \mathbf{E′} \wedge \mathbf{I} \to \mathbf{T}$ not be derived (in axiomatic second-order logic) from Redef$N$ $\wedge$ EZ $\wedge$ UZ, as the model with a single element a such that Za, $\neg aPa$, and Na shows, neither can any of

$$\mathbf{F} \wedge \mathbf{T} \to \mathbf{E′} \vee \mathbf{I},$$
$$\mathbf{F} \wedge \neg\mathbf{T} \to \mathbf{E′} \vee \mathbf{I},$$
$$\mathbf{F} \wedge \neg\mathbf{T} \to \mathbf{E′} \vee \neg\mathbf{I}, \qquad \text{and}$$
$$\mathbf{F} \wedge \neg\mathbf{T} \to \neg\mathbf{E′} \vee \mathbf{I}.$$

be derived from RedefN $\wedge$ EZ, as the following easy countermodels respectively show:

$$\langle \{a, b, c\}, \{a, b\}, \{aa, bc, cc\} \rangle,$$
$$\langle \{a, b, c, d\}, \{a, b, c\}, \{aa, bd, cd\} \rangle,$$
$$\langle \{a, b, c\}, \{a, b\}, \{aa, bc\} \rangle, \quad \text{and}$$
$$\langle \{a, b, c\}, \{a, b\}, \{ac, bc\} \rangle.$$

(a, b, c, d are any four objects. The members of the second set in each model are the objects satisfying Z there. The members of the third are the ordered pairs satisfying P. In each model all objects satisfy N.)

So, as will appear later, $\mathbf{F}, \mathbf{T}, \mathbf{E}, \mathbf{I}$ are totally independent if RedefN and EZ but nothing else is assumed.

We now add to the treatment of number given so far the distinctly Fregean ingredient. It will be of interest to stick with the language $\{Z, P, N\}$ for the time being.

Frege held that numbers are objects, but that what he called "a statement of number," i.e., a statement saying how many things there are of a certain kind,[4] contains an assertion about a concept. Concepts, according to Frege, are denotations of predicates, or values of monadic second-order variables; objects are denoted by terms that may appear on either side of the sign of identity. To express this aspect of Frege's account of number, we introduce a sign $\#$ for a totally defined function. $\#$ is read "the number of things that ..." and is a function-sign of "mixed type": when attached to a monadic second-order variable F, $\#$ produces a term $\#F$ of the same type as a first-order variable n. So $n = \#F$ is well-formed. We emphasize that $\#$ is to denote a total function and thus that $\forall F \exists! n\, n = \#F$ will be valid. We assume that the axioms and rules of second-order logic guarantee in some manner the derivability of $\forall F \exists! n\, n = \#F$.

We now define Z and P:

DefZ $\quad \forall n (Zn \leftrightarrow \exists F(n = \#F \wedge \forall x \neg Fx));$

DefP $\quad \forall m \forall n (mPn \leftrightarrow \exists F \exists G \exists y (Gy \wedge \forall x (Fx \leftrightarrow Gx \wedge x \neq y) \wedge m = \#F$
$\quad\quad \wedge\, n = \#G).$

DefP is a slight reformulation of Frege's definition of "precedes."

We shall use: $\vdash'$ to denote derivability from DefZ, DefP, and RedefN.

According to DefZ, a zero is the number belonging to some empty concept.

By DefZ alone, $\vdash'$ EZ, for $\exists F \forall x (Fx \leftrightarrow x \neq x)$ is one of the comprehension axioms of second-order logic, and therefore $\vdash' \exists F \forall x \neg Fx$, whence $\vdash' \exists n \exists F(n = \#F \wedge \forall x \neg Fx)$, i.e., $\vdash' \exists n\, Zn$. Also by DefZ alone,

$$\vdash' \text{ UZ} \leftrightarrow \forall F \forall G(\forall x \neg Fx \wedge \forall x \neg Gx \rightarrow \#F = \#G),$$

---

[4]One of Frege's examples is "Jupiter has four moons."

whose right side may be read: Empty concepts have the same number.

"Empty concepts ..." is an immediate consequence of *Functional Extensionality* (FE)

$$\forall x(Fx \leftrightarrow Gx) \rightarrow \#F = \#G,$$

which could well be considered a truth of logic, as it is valid under any extensional semantics for languages containing the function sign #. For if $\forall x(Fx \leftrightarrow Gx)$ is true, then the same class of objects will be assigned to the variables F and G, and then $\#F = \#G$ will be true.[5] The converse of FE is inconsistent, as the argument of Russell's paradox shows. But FE is itself an immediate consequence of *Hume's principle*

$$\#F = \#G \leftrightarrow F \approx G,$$

where $F \approx G$ is some standard second-order formula for equinumerosity. By now it is well known that Hume's principle is consistent. However, Hume's principle holds in no finite domain and cannot be considered a truth of logic.

Our principal result is that **T** can be derived from **F**, **E′** and **I** in ⊢′, i.e., that ⊢′ **F** ∧ **E′** ∧ **I** → **T**, which is perhaps surprising in view of the earlier observation that ⊬* **F** ∧ **E** ∧ **I** → **T**.

To prove this theorem we define the ancestral R^* of an arbitrary relation R:

$$xR^*y \equiv \forall F(\forall a(xRa \rightarrow Fa) \land \forall d\forall a(Fd \land dRa \rightarrow Fa) \rightarrow Fy).$$

We write: $xR^{*=}y$ to mean: $xR^*y \lor y = x$.

Among the consequences of the definitions of R^* and $R^{*=}$ are:

(1) $\forall x\forall y(xRy \rightarrow xR^*y)$,
(2) $\forall x\forall y(xR^*y \land yR^*z \rightarrow xR^*z)$,
(3) $\forall x\forall y(xR^*y \rightarrow \exists z(xR^{*=}z \land zRy))$,
(4) $\forall x\forall y(Fx \land xRy \rightarrow Fy) \rightarrow \forall x\forall y(Fx \land xR^{*=}y \rightarrow Fy)$,   and
(5) $\forall x\forall y(xR^{*=}y \leftrightarrow \forall F(Fx \land \forall d\forall a(Fd \land dRa \rightarrow Fa) \rightarrow Fy))$.

Frege's definition of "natural number" was $\forall n(Nn \leftrightarrow \mathbf{0}P^{*=}n)$, rather than Def$N$, but by (5), the definitions are equivalent.

We now prove that ⊢′ **F** ∧ **E′** ∧ **I** → **T**.

Lemma 1. ⊢′ **E′** ∧ **I** → $(Nx \rightarrow \neg xP^*x)$.

Proof. Suppose Zx and xP^*x. By (a′), Nx. By (3), for some z, $xP^{*=}z$ and zPx. By (b) and (4), Nz, contra **E′**. So $\forall x(Zx \rightarrow \neg xP^*x)$.

---

[5]FE is further discussed in Article 17 of this volume.

Suppose Nd, dPa and aP^*a. By (b), Na. By (3), for some z, $aP^{*=}z$ and zPa. By (1) and (2), zP^*z. By (b) and (4), Nz. By **I**, $z = d$. Thus dP^*d. So $\forall d \forall a (Nd \wedge \neg dP^*d \wedge dPa \rightarrow \neg aP^*a)$.

By (h′) (and the comprehension scheme of second-order logic), the lemma follows. (The proof does not cite Def$Z$ or Def$P$, by the way.) ∎

Lemma 2. $\vdash'$ $\mathbf{E}' \rightarrow (Za \rightarrow \exists G(aP\#G \wedge \forall x (Gx \leftrightarrow Nx \wedge xP^{*=}a)))$.

Proof. Suppose Za. Then by Def$Z$, for some F, $\forall x \neg Fx$ and $a = \#F$. By second-order logic, for some G, $\forall x (Gx \leftrightarrow Nx \wedge xP^{*=}a)$. To show $aP\#G$ it is enough, by Def$P$, to show that for some y, Gy and $\forall x(Fx \leftrightarrow Gx \wedge x \neq y)$. Let $y = a$. Since Za, Na; and trivially, $aP^{*=}a$. So Ga.

Suppose Gx and $x \neq a$. Then Nx and xP^*a. By (3), for some z, $xP^{*=}z$ and zPa. Since Nx and $xP^{*=}z$, Nz, by (b) and (4). So Nz, zPa, and Za, contra $\mathbf{E}'$. Since $\forall x \neg Fx$, $\forall x(Fx \leftrightarrow Gx \wedge x \neq y)$. ∎

Lemma 3. $\vdash'$ $\mathbf{F} \wedge \mathbf{E}' \wedge \mathbf{I} \rightarrow (Nd \wedge dPa \wedge dP\#F \wedge \forall x(Fx \leftrightarrow Nx \wedge xP^{*=}d) \rightarrow \exists G(aP\#G \wedge \forall x(Gx \leftrightarrow Nx \wedge xP^{*=}a)))$.

Proof. Suppose Nd, dPa, $dP\#F$ and $\forall x(Fx \leftrightarrow Nx \wedge xP^{*=}d)$. By second-order logic, for some G, $\forall x(Gx \leftrightarrow Nx \wedge xP^{*=}a)$. We must show $aP\#G$.

By $\mathbf{F}$, $a = \#F$. So it is enough to show $\#FP\#G$. Since Nd and dPa, Na; and trivially $aP^{*=}a$. So by Def$P$ it suffices to show $\forall x(Nx \wedge xP^{*=}a \wedge x \neq a \leftrightarrow Nx \wedge xP^{*=}d)$. Assume Nx.

Suppose $xP^{*=}a$ and $x \neq a$. Then xP^*a and by (3), for some z, $xP^{*=}z$ and zPa. Since Nx and $xP^{*=}z$, Nz by (b) and (4). By **I**, $z = d$, whence $xP^{*=}d$.

Conversely, suppose $xP^{*=}d$. Since dPa, xP^*a, whence $xP^{*=}a$. If $x = a$, then xP^*x, contra Lemma 1. So $x \neq a$. ∎

Theorem 1. $\vdash'$ $\mathbf{F} \wedge \mathbf{E}' \wedge \mathbf{I} \rightarrow \mathbf{T}$.

Proof. By comprehension, (h′), and Lemmas 2 and 3, $\vdash'$ $\mathbf{F} \wedge \mathbf{E}' \wedge \mathbf{I} \rightarrow (Nm \rightarrow \exists G(mP\#G \wedge \forall x(Gx \leftrightarrow Nx \wedge xP^{*=}m)))$. Thus $\vdash'$ $\mathbf{F} \wedge \mathbf{E}' \wedge \mathbf{I} \rightarrow (Nm \rightarrow \exists n\, mPn)$, i.e., $\vdash'$ $\mathbf{F} \wedge \mathbf{E}' \wedge \mathbf{I} \rightarrow \mathbf{T}$. ∎

Let us now write: $\vdash$ to mean derivability from "Empty concepts have the same number," Def0, Def$P$ and Def$N$, where Def0 is $\forall n(n = \mathbf{0} \leftrightarrow \exists F(n = \#F \wedge \forall x \neg Fx))$.

We then have

$\vdash \forall n(Zn \leftrightarrow n = \mathbf{0}) \rightarrow$ Def$Z \wedge$ Redef$N$.

Combining our results, we have

$\vdash \mathbf{F} \wedge \mathbf{T} \rightarrow \mathbf{E} \vee \mathbf{I}$,
$\vdash \mathbf{F} \wedge \neg \mathbf{T} \rightarrow \mathbf{E} \wedge \mathbf{I}$,   and
$\vdash \forall n(Zn \leftrightarrow n = \mathbf{0}) \rightarrow (\mathbf{F} \wedge \mathbf{E}' \wedge \mathbf{I} \rightarrow \mathbf{T})$.

Substituting $\cdot = \mathbf{0}$ for $Z \cdot$, we have

$$\vdash \mathbf{F} \wedge \mathbf{E} \wedge \mathbf{I} \to \mathbf{T}.$$

It follows (by propositional logic) that

$$\vdash \mathbf{F} \to \mathbf{T} \wedge (\mathbf{E} \vee \mathbf{I}).$$

We now show that this result is best possible in the sense that no stronger truth-functional compound of $\mathbf{F}$, $\mathbf{T}$, $\mathbf{E}$ and $\mathbf{I}$ can be derived.

Let

$$
\begin{aligned}
[1] &= \{\mathbf{F}, \mathbf{T}, \mathbf{E}, \mathbf{I}\}, \\
[2] &= \{\mathbf{F}, \mathbf{T}, \mathbf{E}, \neg\mathbf{I}\}, \\
[3] &= \{\mathbf{F}, \mathbf{T}, \neg\mathbf{E}, \mathbf{I}\}, \\
[4] &= \{\neg\mathbf{F}, \mathbf{T}, \mathbf{E}, \mathbf{I}\}, \\
[5] &= \{\neg\mathbf{F}, \mathbf{T}, \mathbf{E}, \neg\mathbf{I}\}, \\
[6] &= \{\neg\mathbf{F}, \mathbf{T}, \neg\mathbf{E}, \mathbf{I}\}, \\
[7] &= \{\neg\mathbf{F}, \mathbf{T}, \neg\mathbf{E}, \neg\mathbf{I}\}, \\
[8] &= \{\neg\mathbf{F}, \neg\mathbf{T}, \mathbf{E}, \mathbf{I}\}, \\
[9] &= \{\neg\mathbf{F}, \neg\mathbf{T}, \mathbf{E}, \neg\mathbf{I}\}, \\
[10] &= \{\neg\mathbf{F}, \neg\mathbf{T}, \neg\mathbf{E}, \mathbf{I}\}, \quad \text{and} \\
[11] &= \{\neg\mathbf{F}, \neg\mathbf{T}, \neg\mathbf{E}, \neg\mathbf{I}\}.
\end{aligned}
$$

Since $(\mathbf{F} \to \mathbf{T} \wedge (\mathbf{E} \vee \mathbf{I}))$ is truth-functionally equivalent to $\bigwedge[1] \vee \ldots \vee \bigwedge[11]$, in order to show that $\vdash A$ for no truth-functional compound A of $\mathbf{F}, \mathbf{T}, \mathbf{E}$ and $\mathbf{I}$ stronger than $(\mathbf{F} \to \mathbf{T} \wedge (\mathbf{E} \vee \mathbf{I}))$, it will suffice to provide a model $\langle D, f \rangle$ of each $[i]$, $1 \leq i \leq 11$. Henceforth we write: mPn to mean: P applies to m, n. In each of the eleven models, $f : \mathcal{P}D \to D$; and so "empty concepts" and indeed FE will hold; $\mathbf{0}$ will denote $f\emptyset$ and so Def$\mathbf{0}$ will hold; mPn iff there exist A, B and y in B such that $m = fA$, $n = fB$, $A \subseteq B$ and $A = B - \{y\}$, and so Def$P$ will hold; and N will apply to $f\emptyset$ and those elements of D to which $f\emptyset$ bears P^* and so Def$N$ will hold.

[1]: Let $D = N = \{\text{the naturals}\}$. Let $fA = n + 1$ if $|A| = n$, $fA = 0$ if $|A|$ is infinite. Then if $m > 0$, mPn iff $n = m + 1$ and $0Pn$ iff $n = 0$.

[2]: Let $D = \{0, 1\}$. Let $f\emptyset = 0$ and $f\{0\} = f\{1\} = f\{0, 1\} = 1$. Then $0P1$, $1P1$, not: $0P0$, and not: $1P0$.

[3]: Let $D = \{0\}$. Let $f\emptyset = f\{0\} = 0$. Then $0P0$.

[4]: The model for [4] is the most complicated one. Let $D = N, E = \{\text{the evens}\}, O = \{\text{the odds}\}$. Let $f\emptyset = 1$, $fE = 1$, $fA = 2n + 1$ if $|A| = n$, $fA = 2n$ if for some $B \subseteq O$, $0 < |B| = n$ and $A = E \cup B$, and $fA = 0$ otherwise. Then we have $1P3P5P7P$ etc. and $1P2P4P6P$

etc. Moreover, $0P0$ and $0P1$; and $0P2$, $0P4$, $0P6$, etc.; and P holds between no other pairs except those indicated.

F fails because $1P3$ and $1P2$. **T** clearly holds. But **E** and **I** hold too, because the objects to which N applies in this model are the positive integers, and the only pairs of positive integers among which P holds are those indicated in: $1P3P5P7P$ etc. and $1P2P4P6P$ etc.

[5]: Let $D = \{0, 1, 2, 3\}$. Let $f\emptyset = 0$, $f\{0\} = 1$, $f\{1\} = 2$ and $fA = 3$ otherwise. Then $0P1, 0P2, 1P3, 2P3, 3P3$ and nothing else holds.

[6]: Let $D = N$. Let $fA = n$ if $|A| = n$, $fA = 0$ if A is infinite. Then **0** denotes 0; $0P0$, $0P1$, $1P2$, $2P3$, ... and nothing else holds.

[7]: Let $D = \{0, 1\}$. Let $f\emptyset = 0$, $f\{0\} = 1$, $f\{1\} = 0$ and $f\{0, 1\} = 0$. Then $0P0, 0P1$, $1P0$, but not: $1P1$.

[8]: Let $D = N$. Let $f\emptyset = f(N - \{0\}) = 2$, $fN = 1$, $fA = n+2$ if $|A| = n$, and $fA = 0$ otherwise. Then $0P0, 0P2$, $0P1, 2P1, 2P3P4P$ ... and nothing else holds. Here **0** denotes 2 and the objects to which N applies are 2, 3, 4, etc. and also 1. Since $2P1$ and $2P3$, **F** fails; since for all a, not: $1Pa$, **T** fails. But **E** and **I** hold, since although $0P2$ and $0P1$, N does not apply to 0.

[9]: Let $D = \{0, 1, 2, 3\}$. Let $f\emptyset = 0$, $f\{0\} = 1$, $f\{1\} = 2$, $f\{0, 1\} = 1$, $f\{0, 2\} = 2$, $f\{0, 1, 2\} = 1$, $f\{1, 2, 3\} = 2$, $f\{1, 2\} = 2$, $f\{1, 3\} = 1$, $f\{0, 1, 2, 3\} = 3$, and $fA = 1$ otherwise. Then $0P1$, $0P2$, $1P1$, $2P2$, $1P2$, $2P1$, $1P3$, $2P3$ and nothing else holds.

[10]: Let $D = \{0, 1\}$. Let $f\emptyset = 0$, $f\{0\} = f\{1\} = 0$, and $f\{0, 1\} = 1$. Then $0P0$ and $0P1$, but not: $1P0$ and not: $1P1$.

[11]: Let $D = \{0, 1, 2\}$. Let $f\emptyset = 0$, $f\{0\} = 1$, $f\{1\} = 0$, $f\{2\} = 0$, $f\{0, 1\} = 0$, $f\{0, 2\} = 0$, $f\{1, 2\} = 1$ and $f\{0, 1, 2\} = 2$. Then $0P0, 0P1, 1P0, 0P2, 1P2$, and nothing else holds.

So if $\vdash A$, A a truth-functional compound of **F, T, E, I**, then $\vdash (\mathbf{F} \rightarrow \mathbf{T} \wedge (\mathbf{E} \vee \mathbf{I})) \rightarrow A$.

In any model for the language $\{\#\}$ of the form $\langle \{a\}, f \rangle$, it must be that aPa, since $f\emptyset = f\{a\} = a$. Thus the model described far above consisting of a single element not bearing P to itself, which shows that

$$\not\vdash^* \mathbf{F} \wedge \mathbf{E} \wedge \mathbf{I} \rightarrow \mathbf{T},$$

cannot be converted to a model for $\{\#, \mathbf{0}, P, N\}$ of $\{\mathbf{F}, \neg\mathbf{T}, \mathbf{E}, \mathbf{I}\}$ (and Def0, Def$P$, Def$N$ and "Empty concepts ...").

The central argument of Frege's *Foundations of Arithmetic* is a sketch of a proof that every natural number precedes some object (which will of course itself be a natural number). The argument sketched by Frege proceeds by showing that $0P\#[x : xP^{*=}0]$ and that if d is a natural number,[6] dPa and $dP\#[x : xP^{*=}d]$, then $aP\#[x : xP^{*=}a]$. Our proof that $\vdash' \mathbf{F} \wedge \mathbf{E'} \wedge \mathbf{I} \to \mathbf{T}$ is a modification of Frege's argument: instead of showing that if n is a natural number, $nP\#[x : xP^{*=}n]$, we in effect show from $\mathbf{F}, \mathbf{E'}$ and $\mathbf{I}$ that any natural number n is the number of some concept G such that $\forall x(Gx \leftrightarrow 0P^{*=}xP^{*=}n)$. Frege's proof appealed to Hume's principle, which, as Frege showed, is also sufficient for $\mathbf{F}, \mathbf{E}$, and $\mathbf{I}$; our derivation of $(\mathbf{F} \to \mathbf{T} \wedge (\mathbf{E} \vee \mathbf{I}))$ appeals only to "empty concepts have the same number" and the definitions from $\#$ of $\mathbf{0}$, P, and N.

---

[6]Frege realized that he needed this condition on d only in *Grundgesetze der Arithmetik* and not in *Die Grundlagen der Arithmetik*. For more details on his error, see Article 20 in this volume.

19

Is Hume's Principle Analytic?

The reduction, however, cuts both ways. It is not easy to see how Frege can avoid the seemingly frivolous argument that if his reduction is really successful, one who believes firmly in the synthetic character of arithmetic can conclude that Frege's logic is thus proved to be synthetic rather than that arithmetic is proved to be analytic.

— Hao Wang[1]

There are a number of issues on which Crispin Wright and I disagree, some of them substantive and some merely terminological. For example, we disagree over whether the term "analytic" can be suitably applied to HP and whether a derivation of arithmetic from HP would establish a doctrine appropriately called "logicism." I also have certain reservations, which I shall set out later, about his notions of *explanation* and *reconceptualization*. However, I think the areas of agreement about the interest of Frege's derivation of arithmetic are both wide-ranging and far more significant than those of disagreement. In particular I want to endorse Wright's closing suggestion that "the problems and possibilities of a Fregean foundation for mathematics remain [wide?] open" and the remark made earlier in his paper that

This article first appeared in Richard G. Heck, Jr., ed., *Logic, Language, and Thought*, Oxford: Oxford University Press, 1997. Reprinted by kind permission of Oxford University Press.

A version of this paper was presented to a 1994 American Philosophical Association symposium on the topic of logicism. Crispin Wright was the co-symposiast and Charles Parsons the commentator.

Michael Dummett much dislikes the designation "Hume's principle" because the remark in Hume's *Treatise* (I, III, I, para. 5), which Frege cited with approval and from which the name derives, presupposes the doctrine that a number is an item composed of units, a doctrine Frege is presumed to have refuted. Since this paper first appeared in a *Festschrift* for Michael, I used the designation "HP" instead. Cf. Chomsky and "LF".

[1](Wang, 1957), reprinted in (Wang, 1963), pp. 68–81. The quotation, together with other extremely interesting observations, appears on p. 80.

"the more extensive epistemological programme which Frege hoped to accomplish in *Grundgesetze* is still a going concern." I also want to emphasize that I consider Wright to have made a great scientific contribution in showing contemporary readers how the deduction of the Peano postulates from HP could be carried out and in formulating the conjecture, subsequently verified, that HP is consistent.[2]

The first issue I want to take up is whether a derivation of arithmetic from HP vindicates logicism.

My view is: *no logic, no logicism.*

It is clear what has to be established in order to show the truth of something we can call logicism with a clear conscience. Arithmetic has to be shown to be provable from an extension by definitions of a theory that is logically true. In technical parlance, arithmetic has to be interpreted in a logically true theory. It cannot be, trivially: Arithmetic implies that there are two distinct numbers; were the relativization of this statement to the definition of the predicate "number" provable by logic alone, logic would imply the existence of two distinct objects, which it fails to do (on any understanding of logic now available to us).

Wright states that if it has to be made out that HP is a truth of logic, then "the prospects are unimproved," the prospects, I take it, being those for establishing a species of logicism. I infer that he does not consider HP to be a truth of logic. Nor do I: the principle implies the existence of too many objects. So I do not conclude, as Wright does, that the proof of Frege's theorem by itself establishes logicism. It only shows the beautiful, deep, and surprising result that arithmetic is interpretable in Frege arithmetic, a theory whose sole non-logical axiom is HP.

Wright argues, though, that since HP is analytic, the proof yields "an upshot still worth describing as logicism, albeit rather different from the conventional understanding of the term." I might be prepared to agree that something describable as logicism in a different understanding of that term would have been established if HP had been shown to be analytic or akin to something properly called a definition. But I doubt that it can be.

Having to discuss whether HP is analytic is rather like having to consider whether hydrogen sulfide is dephlogisticated. One can certainly see reasons why one might be tempted to call H_2S dephlogisticated; but if I am right in thinking that to dephlogisticate is to combine with *oxygen*, there are conclusive reasons for not doing so.

The main reason why the notion of analyticity is all but useless in discussing propositions of mathematics like HP is that, although an analytic statement is supposed to be one that is true in virtue of the meanings of the

---

[2](Wright, 1983). The derivation is on pp. 154–168. The discussion of number-theoretic logicism III is on pp. 153–154.

terms contained in sentences expressing it (and syntactic features of those sentences), the phrase "true in virtue of meanings" leaves it indeterminate how much mathematics may be used to get from facts about meanings to the truth of the statement, or, more exactly, how much mathematics it is allowable to use in deriving the statement (or the statement that that statement is true) from reports of meanings. In brief, we are not told how strong the mathematics is that "in virtue of" permits. The stronger the mathematics permitted, the greater the number of analytic mathematical truths, of course. The point, in essence, is due to Gödel and is different from the objection raised by the question "Why mathematics rather than geology?"

In the interest of trying to get at what's really at issue between Wright and myself, however, I shall ignore the standard difficulties presented by "analytic," including the uncertainty what the interest or point of classifying a statement as analytic is and the worry that complex logical argumentation might itself *create* semantic content,[3] and suppose that I understand the concept sufficiently well, well enough at least to know what's meant by calling "all vixens are foxes" etc. analytic and by saying that there is a *semantic connection* between "vixen" and "fox."

At the outset, let me acknowledge that I have no knock-down argument that will persuade a diehard defender of the claim that HP is analytic to abandon the view. All I shall offer are what strike me as some rather, and perhaps sufficiently, weighty considerations against that position.

At first glance, HP might certainly *seem* analytic. In its statement "number" means "cardinal number," and, one would naturally wonder, isn't it a matter of the semantic connection between "cardinal number" and "one-one correspondence" that two concepts have the same cardinal number just when the things falling under one of them can be put in one-one correspondence with those falling under the other? Isn't the cardinal of x the same as that of y just when there's a one-one correspondence between x and y, and that because of what "cardinal number" means? So isn't the left hand side of HP close enough in meaning to the right hand side for it to count as analytic? Doesn't the left hand side have the same sense as the right?[4]

Let me begin to respond to this argument by recalling two features that analytic statements have been traditionally supposed to enjoy: first, they are true; secondly and roughly speaking, they lack content, i.e., they make no significant or substantive claims or commitments about the way the world is; in particular, they do not entail the existence either of particular objects or of more than one object. (It may be held that some analytic

---

[3] This possibility is suggested by a remark of Frege about condensation in §23 of his *Begriffsschrift*.

[4] Thanks here to Arthur Skidmore.

statement might entail the existence of at least one object, as will be the case if every logical truth counts as analytic.)

Some have been tempted by the idea of analytic statements that happen not to be true, e.g., "the present king of France is a royal." On the view in question, the semantic connection between "king" and "royal" suffices to ensure the analyticity of the entire statement, despite the failure of its subject to denote. But analytic statements are, and (since we are playing along) are analytically, analytic truths, and the view may be put aside. The example is worth noting, however, for, as I am going to suggest later, HP suffers from a defect similar to that of "the present king of France is a royal," which would not be analytic even if there were presently a (unique) king of France, since, of course, it would not be analytic that there is one.

The main significant worry for the defender of the analyticity of HP concerns the quite strong content that it appears to possess. HP has consequences having to do with certain features of the domain of objects over which its first-order variables range, in particular with the number of those objects there are. Much of the most interesting work in mathematical logic in the last twenty years or so has dealt with comparisons of strength of various logical and mathematical statements, examining which well-known theorems of mathematics can be derived from which logical principles (and vice versa!) in which background theories. We now know that Frege arithmetic is equi-interpretable with full second-order arithmetic, "analysis," and hence equi-consistent with it. Learning that HP is analytic would not help us in the slightest with the problem of assessing the strength of various theorems, fragments and subtheories of analysis, all of which would, I suppose, have to count as analytic. The first part of my worry about content is that HP, when embedded into axiomatic second-order logic, yields an incredibly powerful mathematical theory.

Wright will say: Hooray! Math is analytic after all. But we don't know what follows from its being so and we will have to study the sub-analytic to see what (logically) entails what just as hard as before. It is known that HP does not follow (a word I will not surrender) from the conjunction of two of its strong consequences: the (interesting) statements that nothing precedes zero and that *precedes* is a one-one relation. If HP is analytic, then it is strictly stronger (another non-negotiable term) than some of its strong consequences. It is also known that arithmetic follows from these two statements alone, and that arithmetic is strictly weaker than even their *dis*junction.[5] Faced with these results, how can we really want to call HP analytic?

Frege, for a lengthy stretch of his career, held that the existence of in-

---

[5]For proofs of all these results, see my "On the Proof of Frege's Theorem," reprinted as Article 17 in this volume.

finitely many objects could be seen to follow from a set of principles and definitions that could, by his lights, be counted as analytic. He abandoned the view in 1906, according to Dummett, when he realized that his attempted patch to Basic Law V would not work. It is doubtful that Russell could be considered a logicist in the full sense of the term while writing *Principia,* whose stated aim is to analyze the notions employed in mathematics, not to show arithmetic to be a branch of logic. Despite the Gödel incompleteness theorems and Russell's protestations that the axiom of infinity was no logical truth, it was a central tenet of logical positivism that the truths of mathematics were analytic. Positivism was dead by 1960 and the more traditional view, that analytic truths cannot entail the existence either of particular objects or of too many objects, has held sway since. Wright wishes to overthrow the tradition, but it should be asked how a statement that cannot hold if there are only finitely many objects can possibly be thought to be analytic, a matter of meanings or "conceptual containment."

On the symbolization that I prefer, HP reads:

$$\forall F \forall G (\#F = \#G \leftrightarrow F \approx G)$$

where "$F \approx G$" is an abbreviation for a second-order formula expressing that there is a one-one correspondence between the objects falling under the concept F and those falling under the concept G. We need not here write out the formula, but must remember that it contains some first-order quantifiers. We must also remember the grammatical category of "$\#$," "octothorpe": it is a function-sign, which when attached to a monadic second-order variable like "F," produces a term of the same type as individual variables that occur in "$F \approx G$." It is essential to the proof of Frege's theorem that octothorpe be so construed.

Thus octothorpe denotes a *total* function from concepts to objects. Logic, plus the convention that function signs like octothorpe denote total functions, will guarantee that $\forall F \exists! x \, \#F = x$ is true. It will not guarantee that HP is.

HP entails, as Wright has put it with exemplary force and Cartesian clarity, that there is a partition of concepts into equivalence classes, in which two concepts belong to the same class if and only if they are equinumerous. If there are only k objects, k a finite number, then, since there are $k + 1$ natural numbers $\leq k$, there will be $k + 1$ equivalence classes, *viz.* a class containing each concept under which zero objects fall, a class containing each concept under which exactly one object falls, ..., and a class containing each concept under which all k objects fall. (We need not here assume that concepts are individuated extensionally.) Thus, if there are only k objects, there is no function mapping concepts to objects that takes nonequinumerous concepts to different objects, for there won't be enough objects around to serve as values of the function, since $k + 1$ are needed. So if HP holds—even if only the left–right direction (the same direction as in the fatal Basic law V) holds—there must be infinitely many objects.

One person's tollens is another's ponens, and Wright happily regards the existence of infinitely many objects, and indeed, that of a Dedekind infinite concept, as analytic, since they are logical consequences of what he takes to be an analytic truth. He would also regard the existential quantification of HP (over the positions occupied by octothorpe) as analytic. But what guarantee have we that there is such a function from concepts to objects as HP and its existential quantification claim there to be?

I want to suggest that HP is to be likened to "the present king of France is a royal" in that we have no analytic guarantee that for every value of "F," there is an object that the open definite singular description "the number belonging to F" denotes. I shall also suggest that there may be some analytic truths in the vicinity of HP with which it is being confused. I hope that the suggestions will do justice both to the thought that there is a strong semantic connection between "the number of . . ." and "one-one correspondence" and to the traditional idea that analytic truths do not entail the existence of a lot of objects.

Our present difficulty is this: just how do we know, what kind of guarantee do we have, why should we believe, that there is a function that maps concepts to objects in the way that the denotation of octothorpe does if HP is true? If there is such a function then it is quite reasonable to think that whichever function octothorpe denotes, it maps nonequinumerous concepts to different objects and equinumerous ones to the same object, and this moreover because of the meaning of octothorpe, the number-of-sign or the phrase "the number of." But do we have any analytic guarantee that there is a function that works in the appropriate manner?

Which function octothorpe denotes and what the resolution is of the mystery how octothorpe gets to denote some one definite particular function that works as described are questions we would never dream of trying to answer. (Harold Hodes' article "Logicism and the ontological commitments of arithmetic"[6] contains much wisdom about these mysteries of mathematical reference.) Nevertheless, it would seem that if there is such a function, then whichever function octothorpe does denote, it also does the trick.[7]

Thus, I am moved to suggest, very tentatively and playing along, that the conditional whose consequent is HP and whose antecedent is its existential quantification might be regarded as analytic. The conditional will hold,

---

[6](Hodes, 1984).

[7]Hartry Field has made a similar suggestion in his review of Wright's *Frege's Conception of Numbers as Objects*, which is reprinted in (Field, 1989).

by falsity of antecedent, in all finite domains. By the axiom of choice, the antecedent will be true in all infinite domains, but then, we may suppose, nothing will prevent the consequent from being true.

I also find plausible the suggestion that the right-to-left half of HP, which states that if F and G are equinumerous, then their numbers are identical, is analytic. It is the left-to-right half, which states that if F and G are not equinumerous, then their numbers are distinct, that blows up the universe. (E.g., consider the concept *non-self-identical;* call its number zero. Now consider the concept *identical with zero;* call its number one. By the *left-to-right* half of HP: since the concepts are not equinumerous, zero is not one.)

The analogy with Basic Law V is obvious. Frege divided Basic Law V into Va, the left-to-right half, and its converse Vb. It was the left-to-right half that gave rise to Russell's paradox. Vb has considerable claim to being regarded as a logical truth: (a) it is valid under standard semantics, thanks to the axiom of extensionality; (b) if the Fs are the Gs, as the antecedent asserts, then whatever "extension" may mean, the extension of the Fs is the extension of the Gs; and (c) if the antecedent holds, then the concepts F and G bear a relation to each other that Frege called the analogue of identity. Thus under each of three familiar systems of formula-evaluation, Vb can never turn out false. In the case of both HP and Basic Law V, we have a principle whose left-to-right half requires that there be a function from concepts to objects respecting certain *non*-equivalences of those concepts. Unless enough objects exist, these non-equivalences cannot be respected. All that the right-to-left halves demand is that the equivalences be respected, as they can be trivially, by mapping all concepts to one and the same object. $\forall F \forall G(\forall x(Fx \leftrightarrow Gx) \rightarrow \#F = \#G)$, which has the same form as Basic Law Vb, can equally justifiably be claimed to be a logical truth, and the stronger $\forall F \forall G(F \approx G \rightarrow \#F = \#G)$ much more plausibly thought analytic, in virtue of the meaning of "$\#$," than its converse.

There is a further difficulty, or at any rate a further aspect of the same difficulty: If numbers belonging to concepts F and G are supposed to be identical if and only if F and G are equinumerous, then how do we know that, for every concept, there is such a thing as a number belonging to that concept? We should not be led astray by the concision, symmetry, and apparent familiarity and obviousness of

$$\#F = \#G \leftrightarrow F \approx G$$

into ignoring the fact that octothorpe is a function sign (for a function of higher type). Like constants and the usual sort of function sign, it may help in concealing significant existential commitments. (Perhaps because of that danger, Quine, concerned with ontology and logic's role in its study, almost entirely avoids constants and function signs in his textbook *Methods of Logic.*)

An analogy may help: if volumes are supposed to be translation- and rotation-invariant, finitely additive, and non-trivial, with singletons and balls of radius r having volumes 0 and $4\pi r^3/3$, respectively, then, as the "paradoxical" Banach–Tarski theorem shows, not every bounded set of points in 3-space has a volume. It would thus be illegitimate to introduce a sign for a totally defined function from bounded sets of points in 3-space to real numbers and assume that the function was translation-invariant, etc. And one had therefore better not say: it is analytic that volume is translation-invariant, etc., and it is analytic that there is always such a thing as the volume of any bounded set of points in 3-space, for the conjunction of the two statements claimed to be analytic is false.

Similarly, if numbers are supposed to be identical if and only if the concepts they are numbers of are equinumerous, what guarantee have we that every concept *has* a number?[8] Or, if we take ourselves to know that with every concept there is functionally associated some object, then how do we know that the associated object is a *number* belonging to F?

It will be useful here to formulate HP in a way that expressly brings out its existential commitments. Let *Numbers* be the statement: for every concept F, there is a unique object x such that for every concept G, x is a number belonging to G if and only if F is equinumerous with G. Is Numbers analytically true? I see no reason at all to believe that it is *analytic* that for every F, there is such a (unique) object x. To reply that it is, since Numbers follows from HP, and HP is analytic, would seem to beg a question that ought not to be begged.

Even more strongly, I don't see any reason to think that it's analytic that objects can be so assigned to concepts that any two concepts are assigned the same object if and only if they are equinumerous. It is not only the *existence* of a function of higher type making such an assignment of objects to concepts that seems synthetic to me: the weaker modal claim that objects *can* be so assigned strikes me as synthetic as well.

I repeat that one person's ponens is another's tollens and admit again that I don't have a *knock-down* argument against Wright's view.

I now want to raise some objections to Wright's notion of a reconceptualization and his use of the term "explanation."

Discussing Frege's (more-or-less) analogous case of directions and parallelism, Wright says, "we have the option ... of *re-conceptualizing,* as it were, the state of affairs which is described on the right. That state of affairs is initially given to us as the obtaining of a certain equivalence relation ... ;

---

[8]Profound thanks here to Peter Clark.

but we have the option, by stipulating that the abstraction is to hold, of so reconceiving such states of affairs that they come to constitute the identity of a new kind of thing ... of which, by this very stipulation, we introduce the concept."

Part of the problem with this suggestion is this: in HP, numbers belonging to concepts are themselves among the objects over which the first-order variables on the right hand side range. Talk of re-conceptualizing a state of affairs would be in order only if the objects supposedly introduced by stipulation were new, objects that had not been previously quantified over. Whether old objects can be chosen to be identical or not under the right conditions would not seem to be a matter that it could be up to us to decide. It is here that the analogy between directions and lines and numbers and concepts breaks down: no one supposes that directions are any sort of constituent of lines, but on the Fregean treatment of number, numbers quite definitely are objects that both fall under concepts and are associated with concepts, as their numbers. However, when the objects allegedly introduced by this sort of stipulation are already objects quantified over in the equivalence relation, unexpected, and sometimes unwelcome, results can occur when we attempt to identify certain of them. We can't, for example, stipulate that old objects be assigned to concepts in such a way that if some old object falls under one concept but not another, then the two concepts are to be assigned different objects.

Wright says, "The concept of direction is thus so introduced that that two lines are parallel *constitutes* the identity of their direction. It is in no sense a further substantial claim that directions exist and are identical under the described circumstances. But nor is it the case that, by stipulating that the principle is to hold, we thereby forfeit the right to a face-value construal of its left-hand side and thereby to the type of existential generalization which a face-value construal would license." All well and good for directions, maybe, but what if the objects introduced on the left are already among those discussed on the right? Could there not then be a danger that a "substantial further claim" about those very objects, *taken together,* would be entailed?

And of course there is such a danger: the *generalized* biconditional, or the biconditional with its free variables, taken as an *axiom,* might then entail that, e.g., there are many, many objects, too many for it to be capable of being regarded any longer as analytic.

One might think: but does that not automatically show that HP isn't analytic? How can an analytic truth be false in certain domains, indeed false in all the finite domains? There is of course a reply that is ready to hand, viz. that it's analytically false that the objects that exist constitute any one of those finite domains. The response strikes me as incredible, but again, I don't have a knock-down argument against the analyticity of HP, only a bunch of considerations.

(Heidegger would hardly have welcomed the response, "Because, analytically, there is always the number of things that there are; so there couldn't have been nothing rather than something.")

One final remark on reconceptualization. How can one call the left-hand side of HP a *reconceptualization* of the right if it can't always be made to hold whenever the right-hand side does? Of course if the variables range over a set, one can always pick some new objects to play the role of the numbers belonging to subsets of that set, but why is one so sure one can do this if there is no set of objects over which the variables range?

Wright's idea that the role of HP is that of an *explanation* also worries me.

In *Frege's Conception of Numbers as Objects,* Wright writes: "the fundamental truths of number theory would be revealed as consequences of an explanation : [note the colon] a statement whose role is to fix the character of a certain concept."[9] In the present paper, Wright calls HP "a principle whose role is to explain, if not exactly to define, the general notion of cardinal number."

Wright is impressed by the form of HP: a biconditional whose right limb is a formula defining an equivalence relation between concepts F and G and whose left limb is a formula stating when the cardinal numbers of F and G are the same. Since the sign for cardinal numbers does not occur in the right limb, can one not appropriately say that HP *explains* the concept of a cardinal number by saying what it is for two cardinal numbers, both referred to by expressions of the form "the number of ... " to be identical?[10]

Certainly. HP states a necessary and sufficient condition for an identity $\#F = \#G$ to hold. Moreover the formula defining the condition doesn't contain $\#$. So if one wants merely to sum up this state of affairs by saying that HP explains the concept of a cardinal number, I would not object.

However, it is hard to avoid the impression that more is meant, that Wright holds that to call a statement an *explanation of a concept* is to assign it an epistemological status importantly similar to the one it was once thought analytic judgments, including definitions, enjoy. It is to this further suggestion that I wish to demur. I can't help suspecting that Wright is using "explanation," at least in the phrase "explanation of a concept," as a term of art, as a member of the same family circle as "analytic," "definition" or "conceptual truth," that the only reason he does not call HP an "analytic definition" is that it is not of the form: Definiendum$(x) \equiv$ Definiens(x),

---

[9](Wright, 1983), p. 153.
[10]I am grateful to Wright and Richard Heck for helpful comments on the whole of this paper but am particularly grateful to them here.

and that he supposes it to be a super-hard truth like "all bachelors are unmarried" or "all equivalence relations are transitive."

The phrase "whose role" occurs in both quotations and may suggest that Wright thinks that HP has one and only one [pre-eminent] role, for "whose" seems in both places to mean "of which the" rather than "of which a." This thought seems to me to be incorrect. HP might be taken as an axiom, the sole (non-logical) axiom in some axiomatization of arithmetic. It might be a sentence we want to show to be needlessly strong for some purpose, e.g., deriving arithmetic. It might serve as something to be obtained from Basic Law V. It might be used as an example of a beautiful proposition. Etc. etc. But there's no such thing as *the* [unique] role of HP.

It is certainly true that *one* of the ways in which HP can be used is to fix the character of a certain concept. Here's how: lay Hume down. Then the concept *the number of ...* will have been fixed to be such that numbers belonging to concepts will be the same if and only if the objects falling under one of the concepts are in one-one correspondence with those falling under the other. But Hume is no different in this regard from any other statement that we might choose to take as an axiom. The axiom of choice fixes the concept of a set in a similar manner. Laid down, it determines that for any set of disjoint nonempty sets, there is a set with exactly one member in common with each of those sets. The principle of mathematical induction fixes the character of the natural numbers. The statement that bananas are yellow fixes the character of the concept of a banana. So nothing is said when it is said that one of the roles of HP is to fix the character of the concept of a cardinal number. And HP doesn't have a unique role.

Let me now defend myself about the "bad company" argument. What I think I was doing was illustrating that what is called (unfortunately, as Wright has stated) "contextual definition" is not, *in general,* a permissible way of introducing a concept. I didn't mean to be arguing that it never was and gave the example of the principle governing truth-values as another example of a legitimate contextual definition. Different examples had different purposes. I cited Hodes' splendid observation that the relation-number principle (the relation number belonging to R is identical with that belonging to S if and only if R and S are isomorphic relations) leads to the Burali–Forti paradox in order to point out that Basic Law V was not an isolated case and that HP might well be expected to be powerful if consistent (as it is). I gave the example of parities in order to show that one couldn't say that a contextual definition is OK if only it is consistent. (I had thought of nuisances, but I seemed to recall actually having heard of the "parity" of a set, and the notion is in any case a natural one.) The example of a principle true iff there are no more than two members was designed to show that one didn't need heavy involvement with set theory to find a contextual definition incompatible with HP. And did I ever say that it would be impossible to demarcate the good contextual definitions from the bad? I merely said that it would seem to be a problem we have no hope of solving at present. I have to reserve judgment on the question whether Wright has solved the problem, but I certainly hope he has.

Wright says I was wrong to say that there is no notion that V**, my revision of Frege's Basic Law V, is analytic of; what is true, he says, is "that there is no *prior*, no intuitively entrenched notion, no notion given independently, which V** is analytic of." I happily accept the correction.

I now want to make a somewhat conciliatory remark. I have been aspersing, at great length, the idea that HP is an analytic truth, all the while taking "analytic" to bear something like the sense it has in current philosophical discourse, namely, "truth in or by virtue of meanings." I think that is the sense in which Wright uses the term too. But there may be another notion of analyticity on which the analyticity of HP might well be more plausible.

It is the idea of Gödel's, as outlined in both his paper "Russell's mathematical logic" and his 1951 Gibbs lecture to the American Mathematical Society,[11] according to which a proposition is analytic if it is true "owing to the nature of the concepts occurring therein."[12] Concepts, he says in the Gibbs lecture, "form an objective reality of their own, which we cannot create or change, but only perceive and describe." By reflection, which of course includes philosophical or mathematical or other intellectual work, we can sometimes arrive at an understanding of the natures of certain concepts that is sufficient to enable us to see the truth of certain propositions in which they occur. With the passage of time, our understanding of those concepts may improve and the truth of ever more analytic propositions become evident to us. Perhaps, as Shoenfield has ironically suggested, the rejection of the "axiom" of constructibility is one example of improvement in our perception of the meaning (in Gödel's sense) of "set" or of the nature of sets.

The thought that understanding of abstract objects may be achieved through a sort of perception of them, which is crucial to Gödel's conception of the analytic, will certainly strike many contemporary philosophers as unacceptably mystical and at any rate highly implausible. (Perhaps, paradoxically, there is even a tinge of materialism in the suggestion that

---

[11]See also Article 7 in this volume.

[12]Gödel also describes propositions as analytic if they are true in virtue of the meanings of the terms expressing them, but it should be understood that his notion of meaning is much broader than that of "linguistic" meaning. For example, Gödel held that it is a matter of the *meanings* of "set" and "∈" that the axioms of set theory hold. The difference between the senses he attached to "meaning" and "concept" would not seem to be particularly significant.

our knowledge of abstract objects arises from "something like a perception" of them: could there not be ways in which we interact with abstracta that yield knowledge of them that are not at all like perception?) But if—*IF*—a Gödelian notion of analyticity could be made out, then HP might well be among the first candidates for this new sort of analytic truth. Perhaps by taking the thought in the right way, we can "see" that if nothing exists, then zero, at least, has to exist, for it is then the number of things there are, and therefore that something does exist after all, but then there have to exist at least two things, for ... This Fregean argument may strike one, as it does me, as a good example of the kind of reflection Gödel might have thought showed that the proposition that there are infinitely many natural numbers is analytic, on his understanding of "analytic," if not on that of most of those who use the word. Maybe in the end we could also thus "see" the truth of HP.

But even on such a Gödelian view of the analytic, at least two difficulties would confront the view that HP is analytic.

The first is that (it is *not* neurotic to think) we don't *know* that second-order arithmetic, which is equi-consistent with Frege Arithmetic, is consistent. Do we really know that some hotshot Russell of the 23rd Century won't do for us what Russell did for Frege? The usual argument by which we think we can convince ourselves that analysis is consistent—"Consider the power set of the set of natural numbers ..."—is flagrantly circular. Moreover, although we may think Gentzen's consistency proof for PA provides sufficient reason to think PA consistent, we have nothing like a similar proof for the whole of analysis, with full comprehension. We certainly don't have a constructive consistency proof for ZF. And it would seem to be a genuine possibility that the discovery of an inconsistency in ZF might be refined into that of one in analysis. Saying exactly which theories are known to be consistent is a difficult problem made even more difficult when one hears of respected mathematicians telling of their failed attempts to prove Q inconsistent, but ZF and analysis, and therefore also Frege arithmetic, are theories that are surely in the black area, not the grey. While we may regret that these theories may well be consistent and that it would probably be wise to bet on their consistency, we must not despair: we do not *know* that they are and need not yet give up the hope that someone will one day prove in one of them that 0=1. Uncertain as we are whether Frege arithmetic is consistent, how can we (dare to) call HP analytic?

One final worry, perhaps the most serious of all, although one that may at first appear to be dismissible as silly or trivial: as there is a number, zero, of things that are non-self-identical, so, on the account of number we have been considering, there must be a number of things that are self-identical, i.e., the number of all the things that there are. Wright has usefully dubbed this number, $\#[x : x = x]$, *anti-zero*. On the definition of $\leq$, according to which $m \leq n$ iff $\exists F \exists G(m = \#F \wedge n = \#G\wedge$ there is a one-one map of F into G), anti-zero would be a number greater than any other number.[13] Now the worry is this: *is* there such a number as anti-zero? According to Zermelo–Fraenkel set theory, there is no (cardinal) number that is the number of all the sets there are. The worry is that the theory of number we have been considering, Frege Arithmetic, is incompatible with Zermelo–Fraenkel set theory plus standard definitions, on the usual and natural readings of the non-logical expressions of both theories. To be sure, as Hodes once observed in conversation, if $\#\alpha$ is taken to denote the cardinal number of α when α is a set and some favorite object that is not a cardinal number when α is a proper class, then HP will be a theorem of von Neumann set theory. But on that definition of $\#$, $\#$ will not be translatable as "the cardinal number of." ZF and Frege arithmetic make incompatible assertions concerning what cardinal numbers there are. And of course, the response "Well, these are just formalisms; the question of their truth or falsity doesn't arise or makes no sense" is hardly available to one claiming that HP is analytic, i.e., an analytic *truth*. So one who seriously believes that it is has to be bothered by the incompatibility of the consequence of Frege arithmetic that there is such a number as anti-zero with the claim made by ZF + standard definitions (on the natural reading of its primitives) that there is no such number.

It is thus difficult to see how on any sense of the word "analytic," the key axiom of a theory that we don't know to be consistent and that contradicts our best-established theory of number (on the natural readings of its primitives) can be thought of as analytic.

---

[13]By the Schröder–Bernstein theorem, which can be proved in second-order logic, $\leq$ is antisymmetric: if $m \leq n \leq m$, then $m = n$.

20

Die Grundlagen der Arithmetik, §§82–83

(with Richard G. Heck, Jr.)

Reductions of arithmetic, whether to set theory or to a theory formulated in a higher-order logic, must prove the infinity of the sequence of natural numbers. In his *Was sind und was sollen die Zahlen?*, Dedekind attempted, in the notorious proof of Theorem 66 of that work, to demonstrate the existence of infinite systems by examining the contents of his own mind. The axioms of General Set Theory, a simple set theory to which arithmetic can be reduced, are those of Extensionality, Separation ("Aussonderung") and Adjunction: $\forall w \forall z \exists y \forall x (x \in y \leftrightarrow x \in z \lor x = w)$. It is Adjunction that guarantees that there are at least two, and indeed infinitely many, natural numbers. The authors of *Principia Mathematica*, after defining zero, the successor function and the natural numbers in a way that made it easy to show that the successor of any natural number exists and is unique, were obliged to assume an axiom of infinity on those occasions on which they needed the proposition that different natural numbers have different successors.

In §§70–83 of *Die Grundlagen der Arithmetik*, Frege outlines derivations of some familiar laws of the arithmetic of the natural numbers from principles he takes to be "primitive" truths of a general logical nature. In §§70–81, he explains how to define zero, the natural numbers, and the successor *relation*; in §78 he states that it is to be proved that this relation is one-one and adds that it does not follow that every natural number *has* a successor; thus by the end of §78, the existence, but not the uniqueness, of

First published in Matthias Schirn, ed., *Philosophy of Mathematics Today*. Oxford: Clarendon Press, 1997. Reprinted by kind permission of Oxford University Press.

We are grateful to Kathrin Koslicki and Jason Stanley. The first author also wishes to express his thanks to the Alexander von Humboldt Foundation.

the successor remains to be shown. Frege sketches, or attempts to sketch, such an existence proof in §§82–83, which would complete his proof that there are infinitely many natural numbers.

§§82–83 offer severe interpretive difficulties. Reluctantly and hesitantly, we have come to the conclusion that Frege was at least somewhat confused in these two sections and that he cannot be said to have outlined, or even to have intended, any correct proof there. We will discuss two (correct) proofs of the statement that every natural number has a successor which might be extracted from §§82–83. The first is quite similar to a proof of this proposition that Frege provides in *Grundgesetze der Arithmetik*, differing from it only in notation and other relatively minor respects. We will argue that fidelity to what Frege wrote in *Die Grundlagen* and in *Grundgesetze* requires us to reject the charitable suggestion that it was this (beautiful) proof that he had in mind when he wrote *Die Grundlagen*. The second proof we discuss conforms to the outline Frege gives in §§82–83 more closely than does the first. But if it had been the one he had in mind, the proof-sketch in these two sections would have contained a remarkably large gap that was never filled by any argument found in *Grundgesetze*. In any case, it is certain that Frege did not know of this proof.

We begin by discussing §§70–81.

In §70, Frege begins the definition of equinumerosity by explaining the notion of a relation, arguing that like (simple) concepts, relational concepts belong to the province of pure logic. In §71, he defines "the objects falling under F and G are correlated with each other by the relation φ." Using modern notation, but strictly following Frege's wording, we would write:

$$\forall a \neg (Fa \wedge \neg \exists b(a\varphi b \wedge Gb)) \wedge \forall a \neg (Ga \wedge \neg \exists b(Fb \wedge b\varphi a)).$$

To put the definition slightly more transparently, the objects falling under F and G are correlated by φ iff

$$\forall x(Fx \rightarrow \exists y(Gy \wedge x\varphi y)) \wedge \forall y(Gy \rightarrow \exists x(Fx \wedge x\varphi y)).$$

In §72 Frege defines what it is for the relation φ to be one-one ("beiderseits eindeutig," "single-valued in both directions"): it is for it, as we should say, to be a function, i.e., $\forall d \forall a \forall e(d\varphi a \wedge d\varphi e \rightarrow a = e)$, that is one-one, i.e., $\forall d \forall b \forall a(d\varphi a \wedge b\varphi a \rightarrow d = b)$. Frege then defines "equinumerous" ("gleichzahlig"): F is equinumerous with G iff there is a relation that correlates the objects falling under F one-one with those falling under G:

$$\exists \varphi [\forall x(Fx \rightarrow \exists y(Gy \wedge x\varphi y)) \wedge \forall y(Gy \rightarrow \exists x(Fx \wedge x\varphi y)) \wedge \\ \forall d \forall a \forall e(d\varphi a \wedge d\varphi e \rightarrow a = e) \wedge \forall d \forall b \forall a(d\varphi a \wedge b\varphi a \rightarrow d = b)].$$

We abbreviate this formula: $F \approx G$.

At the end of §72, Frege defines the number that belongs to F as the extension of the concept "equinumerous with the concept F." He also defines "n is a (cardinal) number": there is a concept such that n is the number that belongs to F. His next task, attempted in §73, is to prove a principle that Crispin Wright once called $N^=$ (for numerical equality),[1] Michael Dummett calls "the original equivalence,"[2] and we call "Hume's Principle": the number belonging to F is identical with that belonging to G iff F is equinumerous with G.

The trouble with the definition of number given in §72 and the proof of Hume's Principle given in §73 is that they implicitly appeal[3] to an inconsistent theory of extensions of second-level concepts. Russell of course demonstrated the inconsistency of Frege's theory, presented in *Grundgesetze der Arithmetik*, of extensions of first-level concepts; a routine jacking-up of Russell's argument shows that of the theory Frege tacitly appeals to in *Grundlagen*.[4] It is by now well known, however, that Frege Arithmetic, i.e., the result of adjoining a suitable formalization of Hume's Principle to axiomatic second-order logic, is consistent if second-order arithmetic is, and is strong enough to imply second-order arithmetic (as of course Frege can be seen as attempting to prove in *Grundlagen*). Indeed, Frege Arithmetic and second-order arithmetic are equi-interpretable; in Appendix 2 we show how to interpret Frege Arithmetic in second-order arithmetic.

Writing: $\#F$ to mean: the number belonging to the concept F, we may symbolize Hume's Principle: $\#F = \#G \leftrightarrow F \approx G$.

The development of arithmetic sketched in §§74–81 makes use only of Frege Arithmetic and can thus be reconstructed in a consistent theory (or one we believe to be so!). Nothing will be lost and much gained if we henceforth suppose that Frege's background theory is Frege Arithmetic.

In §74, Frege defines 0 as the number belonging to "not identical with itself": $0 = \#[x : x \neq x]$. ($[x : \ldots x \ldots]$ is the concept *being an object x such that $\ldots x \ldots$*.) Frege notes that it can be shown on logical grounds alone that nothing falls under $[x : x \neq x]$. In §75, he states that $\forall x \neg Fx \rightarrow (\forall x \neg Gx \leftrightarrow F \approx G)$ has to be proved, from which $\forall x \neg Fx \leftrightarrow 0 = \#[x : Fx]$

---

[1](Wright, 1983).

[2](Dummett, 1991).

[3]The appeal is made when Frege writes "In other words:" at the end of the second paragraph of §73.

[4]Let (V) be $\forall \mathcal{F} \forall \mathcal{G}('\mathcal{F} = '\mathcal{G} \leftrightarrow \forall X(\mathcal{F}X \leftrightarrow \mathcal{G}X))$. Then (V) is inconsistent (in third-order logic). For let $\mathcal{F}$ be $[X : \forall \mathcal{H}(\forall x(Xx \leftrightarrow x = '\mathcal{H}) \rightarrow \neg \mathcal{H}X)]$ and let X be $[x : x = '\mathcal{F}]$. Suppose $\mathcal{F}X$. Then $\forall \mathcal{H}(\forall x(Xx \leftrightarrow x = '\mathcal{H}) \rightarrow \neg \mathcal{H}X)$. So $\forall x(Xx \leftrightarrow x = '\mathcal{F}) \rightarrow \neg \mathcal{F}X$, whence $\neg \mathcal{F}X$ by the definition of X. Thus $\neg \mathcal{F}X$. So for some $\mathcal{H}$, $\forall x(Xx \leftrightarrow x = '\mathcal{H})$ and $\mathcal{H}X$, and then $X'\mathcal{F} \leftrightarrow '\mathcal{F} = '\mathcal{H}$. By the definition of X again, $'\mathcal{F} = '\mathcal{F} \leftrightarrow '\mathcal{F} = '\mathcal{H}$, $'\mathcal{F} = '\mathcal{H}$, and by (V), $\forall X(\mathcal{F}X \leftrightarrow \mathcal{H}X)$, contra $\neg \mathcal{F}X$ and $\mathcal{H}X$. (We use "$'$" to mean "the extension of" and "$[: \ldots]$" to denote concepts (of whatever level).)

follows. These have easy proofs; Frege outlines that of the former in detail.
§76 contains the definition of "n follows immediately after m in the 'natürliche Zahlenreihe' '":

$$\exists F \exists x (Fx \wedge \#F = n \wedge \#[y : Fy \wedge y \neq x] = m).$$

It is advisable, we think, to regard the relation defined in this section as going from m to n, despite the order of "n" and "m" in both the definiens and definiendum of "n immediately follows m in the natural series of numbers." Thus we shall symbolize this relation: mPn ("P" for "(immediately) precedes").

Call a concept *Dedekind infinite* if it is equinumerous with a proper sub-concept of itself; equivalently, if it has a subconcept equinumerous with the concept *being a natural number*. With the aid of the equivalence of these definitions of Dedekind infinity, it is not difficult to see that nPn if and only if n is the number belonging to a Dedekind infinite concept. Thus the number of finite numbers, which Frege designates: ∞, but which we shall as usual denote: $\aleph_0$, follows itself in the "natürliche Zahlenreihe," in symbols, $\aleph_0 P \aleph_0$. Since $\aleph_0$ is not a finite, i.e. natural, number, we shall translate "in der natürlichen Zahlenreihe" as "in the natural sequence of numbers."[5]

§77 contains the definition of 1, as $\#[x : x = 0]$, and a proof that $0P1$. In §78, Frege lists a number of propositions to be proved:

1. $0Pa \rightarrow a = 1$;

2. $1 = \#F \rightarrow \exists x \, Fx$;

3. $1 = \#F \rightarrow (Fx \wedge Fy \rightarrow x = y)$;

4. $\exists x \, Fx \wedge \forall x \forall y (Fx \wedge Fy \rightarrow x = y) \rightarrow 1 = \#F$;

5. P is one-one ("beiderseits eindeutig"), i.e., $mPn \wedge m'Pn' \rightarrow (m = m' \leftrightarrow n = n')$.[6]

---

[5] Timothy Smiley observed that "in the natural series of numbers" is to be preferred as a translation of "in der natürlichen Zahlenreihe" to Austin's "in the series of natural numbers" (Smiley, 1988). We have substituted "sequence" for "series" throughout.

[6] Frege does not indicate what proof of 78.5 he might have intended. Here is an obvious one that he might have had in mind.

Suppose mPn and $m'Pn'$. Then for some F, F', x, x', Fx, $F'x'$, $\#F = n$, $\#F' = n'$, $\#[y : Fy \wedge y \neq x] = m$, and $\#[y' : F'y' \wedge y' \neq x'] = m'$.

Assume $m = m'$. Then by Hume's Principle, there is a one-one correspondence φ between the objects y such that Fy and $y \neq x$ and the objects y' such that $F'y'$ and $y' \neq x'$. We may assume that if $y\varphi y'$, then Fy, $y \neq x$, $F'y'$, and $y' \neq x'$. Let $y\psi y'$ iff $(y\varphi y' \vee [y = x \wedge y' = x'])$. Then ψ is a one-one correspondence between the objects falling under F and those falling under F', and so by Hume's Principle, $n = n'$.

Assume $n = n'$. By Hume's Principle, let ψ be a one-one correspondence between the

Frege observes that so far it has not yet been stated that every number immediately follows or is followed by another. He then states:

6. Every number except 0 immediately follows a number in the natural sequence of numbers.

It is clear from §44 of *Grundgesetze*[7] that Frege did not take (6) to imply that 0 does not immediately follow a number, that $\neg zP0$. This proposition is proved separately in *Grundgesetze*, as Proposition 108, and will be used later on here, at a key point in the argument.

§79 contains the definition of the strong ancestral of φ, "x precedes y in the φ-sequence" or "y follows x in the φ-sequence":

$$\forall F(\forall a(x\varphi a \to Fa) \to \forall d\forall a(Fd \to d\varphi a \to Fa) \to Fy),$$

which was Definition (76) of the *Begriffsschrift*. Frege will use this definition in §81 to define "member of the natural sequence of numbers ending with n." We shall use the standard abbreviation: $x\varphi^* y$ for the strong ancestral. To prove that if $x\varphi^* y$, then $\ldots y \ldots$, it suffices, by the comprehension schema $\exists F\forall a(Fa \leftrightarrow \ldots a \ldots)$ of second-order logic, to show that $\forall a(x\varphi a \to \ldots a \ldots)$ and $\forall d\forall a(\ldots d \ldots \to d\varphi a \to \ldots a \ldots)$. We call this method of proof *Induction 1*. (Induction 2 and Induction 3 will be defined below.)

Here and below, we associate iterated conditionals to the right. Thus, e.g., "$A \to B \to C$" abbreviates "$(A \to (B \to C))$." This convention provides an easy way to reproduce in a linear symbolism one major notational device of both *Begriffsschrift* and *Grundgesetze*.

Frege mentions in §80 that it can be deduced from the definition of "follows" that if b follows a in the φ-sequence and c follows b, then c follows a; the transitivity of the strong ancestral is Proposition (98) of the *Begriffsschrift*. The proof Frege gives there can be formalized in second-order logic only with the aid of the comprehension schema (or something to the same effect); however, there is an easier proof that makes use only of the ordinary quantifier rules, applied to the universal second-order quantifier in the definition of φ^*. For the proof in §§82–83, Frege will also need Proposition (95) of *Begriffsschrift*: if $x\varphi y$, then $x\varphi^* y$, which easily follows from the definition of φ^*.

At the very end of §80 Frege states that only by means of the definition of following in a sequence is it possible to reduce the method of inference

objects falling under F and those falling under F'. We may assume that if $y\psi y'$, then Fy and $F'y'$. Let $y\varphi y'$ iff $(Fy \wedge y \neq x \wedge F'y' \wedge y' \neq x' \wedge [y\psi y' \vee (y\psi x' \wedge x\psi y')])$. Then φ is a one-one correspondence between the objects y such that Fy and $y \neq x$ and the objects y' such that $F'y'$ and $y' \neq x'$, and so by Hume's Principle, $m = m'$.

[7] All references to sections of *Grundgesetze der Arithmetik* are to sections of Volume I of that work. The numbering of sections starts over again in Volume II.

("Schlussweise," which Austin mistranslates as "argument") from n to $n+1$, to the general laws of logic. Of course, the method of inference from n to $n+1$ is what we call mathematical induction; Frege's remark may be taken to be a claim that mathematical induction can be proved with the aid of the definition of the ancestral of P.

In §81 Frege defines the weak ancestral: "y is a member of the φ-sequence beginning with x" and "x is a member of the φ-sequence ending with y" are to mean: $x\varphi^*y \lor y = x$. We shall use the abbreviation: $x\varphi^{*=}y$. He states at the beginning of the section that if φ is P, then he will use the term "natural sequence of numbers" instead of "φ-sequence." We thus have five terms: "y follows x in the natural sequence of numbers," "x precedes y ...," "y immediately follows x ...," "x is a member of the natural sequence of numbers ending with y," and "y is a member ... beginning with x." We shall abbreviate these as: xP^*y, xP^*y, xPy, $xP^{*=}y$, and $xP^{*=}y$, respectively.

Induction 2 is the following method of proof, in which weak ancestrals occur as hypotheses: To prove that if $x\varphi^{*=}y$, then $\ldots y \ldots$, it suffices to prove

(i) $\ldots x \ldots$

and

(ii) $\forall d \forall a(\ldots d \ldots \rightarrow d\varphi a \rightarrow \ldots a \ldots)$.

Induction 2 quickly follows from Induction 1: if (i) and (ii) hold, then so does $\forall a(x\varphi a \rightarrow \ldots a \ldots)$; thus if $x\varphi^*y$, then $\ldots y \ldots$, by Induction 1. But if $x = y$, then by (i), $\ldots y \ldots$ again. Frege proves Induction 2 as Proposition 144 of *Grundgesetze*.

A basic fact about the weak ancestral, to which we shall repeatedly appeal, is that $x\varphi^*a$ and thus $x\varphi^{*=}a$, provided that $x\varphi^{*=}d$ and $d\varphi a$, for then either $x\varphi^*d\varphi a$, $x\varphi^*d\varphi^*a$, and $x\varphi^*a$, or $x = d\varphi a$, $x\varphi a$, and $x\varphi^*a$, by (95) and (98) of the *Begriffsschrift*. That $x\varphi^*a$ if $x\varphi^{*=}d$ and $d\varphi a$ is Proposition 134 of *Grundgesetze der Arithmetik*; that $x\varphi^{*=}a$ if $x\varphi^*a$ is Proposition 136.

Frege has not yet defined finite, or natural, number. He will do so only at the end of §83, where "n is a finite number" is defined as "n is a member of the natural sequence of numbers beginning with 0," i.e., as: $0P^{*=}n$. By Induction 2, to prove that $\ldots n \ldots$ if n is finite, it suffices to prove $\ldots 0 \ldots$ and $\forall d \forall a(\ldots d \ldots \rightarrow dPa \rightarrow \ldots a \ldots)$.

In the formalism in which we are supposing Frege to be working the existence and uniqueness of 0, defined in §74 as $\#[x : x \neq x]$, are given by the comprehension scheme for second-order logic and the standard convention of logic that function signs denote *total* functions. Thus $\#$ denotes a total function from second- to first-order entities and the existence of $\#[x : x = x]$, that of $\#[x : x \neq x]$, and that of $\#[x : x = \#[x : x \neq x]]$

will count as truths of logic. The propositions that 0 is a natural number and that any successor of a natural number is a natural number follow immediately from the definition of "natural number"; 78.5 says that P is functional and one-one. So apart from the easily demonstrated statement that nothing precedes zero, by the end of §81 Frege can be taken to have established the Dedekind–Peano axioms for the natural numbers, except for the statement that every natural number *has* a successor.

Using the notation we have introduced, we may condense §§82–83 as follows:

> §82. It is now to be shown that—subject to a condition still to be specified—(0) $nP\#[x : xP^{*=}n]$. And in thus proving that there exists a Number k such that nPk, we shall have proved at the same time that there is no last member of the natural sequence. ...
>
> ...It is to be proved that (1) $dPa \to dP\#[x : xP^{*=}d] \to aP\#[x : xP^{*=}a]$.
>
> It is then to be proved, secondly, that (2) $0P\#[x : xP^{*=}0]$. And finally it is to be deduced that $(0')\ 0P^{*=}n \to nP\#[x : xP^{*=}n]$. The method of inference ("Schlussweise") here is an application of the definition of the expression "y follows x in the natural sequence of numbers," taking $[y : yP\#[x : xP^{*=}y]]$ for our concept F.[8]
>
> §83. In order to prove (1), we must show that (3) $a = \#[x : xP^{*=}a \wedge x \neq a]$. And for this again it is necessary to prove that (4*) $[x : xP^{*=}a \wedge x \neq a]$ has the same extension as $[x : xP^{*=}d]$. For this we need the proposition that (5′) $\forall a(0P^{*=}a \to \neg aP^{*}a)$. And this must once again be proved by means of our definition of following in a sequence, on the lines indicated above.
>
> We are obliged hereby to attach to the proposition that $nP\#[x : xP^{*=}n]$, the condition that $0P^{*=}n$. For this there is a convenient abbreviation ...: n is a finite number. We can thus formulate (5′) as follows: no finite number follows itself in the natural sequence of numbers.

(We have added some reference numbers; (1) is Frege's own. Primes ′ indicate the presence of a finiteness condition in the antecedent; the asterisk

---

[8]This sentence seems to throw Austin. But we take its last half to mean: when one takes for the concept F what is common to the statements about d and about a, about 0 and about n, and thus that the concept in question is $[y : yP\#[x : xP^{*=}y]]$. (Austin's translation makes it sound as if some binary relation holding between d and a and also between 0 and n were meant. However good his German and English may have been, Austin was no logician. It is time for a reliable English translation of *Grundlagen*.)

* in (4*) indicates (what at least appears to be) a reference to extensions.)

It might appear that Frege proposes in these two sections to prove, not (0), but (0′), as follows: First prove (5′) by an appeal to the definition of P^*. Then derive (4*) from (5′) and (3) from (4*). From (3) derive (1). Prove (2). Then, finally, infer (0′) from (2) and (1), by a similar appeal to the definition of P^*.

However, it will turn out that this precise strategy cannot succeed. It cannot be (4*) and (3) that Frege wishes to derive—(3), e.g., is false if $a = \aleph_0$, as we shall see—but rather certain conditionals (4′) and (3′), whose consequents are (4*) (or rather an equivalent of it) and (3).

We do not, of course, know how Frege might have tried to fill in the details of this proof-sketch at the time of composition of *Die Grundlagen*. In particular, we do not exactly know how he would have proved (5′). (We can be reasonably certain that his proof of (2), however, would have been at least roughly like the proof we shall give below.) But, since he later proved a version of the following lemma as Proposition 141 of *Grundgesetze*, it seems plausible to us to speculate that he might have intended to appeal to it or something rather like it in his proof of (5′). The lemma is a logicized version of the arithmetical truth: if $i < k$, then for some j, $j + 1 = k$ and $i \leq j$.

Lemma $xP^*y \rightarrow \exists z(zPy \wedge xP^{*=}z)$.

Proof. Let $Fa \equiv \exists z(zPa \wedge xP^{*=}z)$. Then $xPa \rightarrow Fa$, for if xPa, then certainly Fa: take $z = x$. And $Fd \wedge dPa \rightarrow Fa$: Suppose Fd and dPa. Then for some z, zPd and $xP^{*=}z$. By the basic fact about the weak ancestral, $xP^{*=}d$. But since dPa, Fa. The lemma follows by Induction 1. ∎

With the aid of the lemma, we can now use Induction 2 to prove (5′).

Proof. $0 = \#[x : x \neq x]$. By Hume's Principle and the definition of P, $\forall z \neg zP0$, and therefore by the lemma, $\neg 0P^*0$.

Now suppose dPa and aP^*a. Then by the lemma, for some z, zPa and $aP^{*=}z$, i.e., either aP^*z or $a = z$, and therefore either $zPaP^*z$ or $zPa = z$. In either case zP^*z. Since dPa and zPa, $z = d$ by 78.5 and so dP^*d. Thus $\neg dP^*d \wedge dPa \rightarrow \neg aP^*a$.

(5′) now follows by Induction 2. ∎

(5′) merits a digression. The part of *Die Grundlagen der Arithmetik* entitled "Our definition completed and its worth proved" begins with §70 and ends with §83; the concluding sentence of §83 reads, "We can thus formulate the last proposition above as follows: No finite number follows itself in the natural sequence of numbers." Apart from its position in the book and the fact that Frege mentions it in both the table of contents and the recapitulation of the book's argument at the end of *Die Grundlagen*, there are a number of reasons for thinking that Frege regarded this proposition as especially significant.

First, there is, according to Frege, an interesting connection with *counting*. When we count, he points out in §108 of *Grundgesetze*, we correlate the objects falling under a concept $\Phi(\xi)$ with the number words in their normal order from "one" up to a certain one, "N"; N is then the number of objects falling under $\Phi(\xi)$. Since correlating relations between concepts are not in general unique, "the question arises whether one might arrive at a different number word 'M' with a different choice of this relation. By our stipulations, M would then be the same number as N, but at the same time one of the two number words would follow after the other, e.g., 'N' after 'M'. Then N would follow in the series of numbers after M, i.e., after itself. Our Proposition [(5′)] excludes this for finite numbers." We find this argument of considerable interest, but will not enter into a discussion of its correctness here.

Secondly, one of Frege's major philosophical aims, as is well known, was to show that reason, under the aspect of logic, could yield conclusions for which many philosophers of his day might have supposed some sort of Kantian intuition to be necessary. The proof of (5′) is a paradigm illustration of how the role of intuition in delivering knowledge can be played by logic instead.

One might think that the truth of (5′) could be seen by the following mixture of reason and intuition: (5′) says that there is no (non-null) loop of P-steps leading from a back to a whenever a is a finite number. So if a is finite but not zero and there is a loop from a to a, then within the loop, there is some number x that (immediately) precedes a, and therefore there is a loop from x (through a, back) to x. But since a is finite, there is a finite sequence of P-steps from zero to some number d preceding a; since *precedes* is one-one, $d = x$, and therefore there is a loop from d to d. Thus a loop "rolls back" from a to d, and then all the way back to zero. But there is no loop from zero to zero; otherwise some number would precede zero, and that is impossible.

Of course, Frege's proof of Proposition 145 avoids any appeals to intuition like those found in the foregoing argument.

Finally, in Proposition 263 of *Grundgesetze*, Frege shows that any structure satisfying a certain set of four conditions is isomorphic to that of the natural numbers. We find it quite plausible to think that Frege realized that the statements that the natural numbers satisfy these conditions constitute an axiomatization of arithmetic and regarded them as *the* basic laws of arithmetic, the *Grundgesetze* of the title of his book.[9] Since one of these conditions is the one that (5′) shows to be satisfied, there is considerable reason to think that Frege regarded (5′) as one of the basic laws of arithmetic. End of digression.

(4*) at least appears to mention extensions of (first-level) concepts and may well do so. But (4*) is unlike the definition of cardinal number and proof of Hume's Principle in that any mention of extensions it contains is readily eliminable without loss: Frege could have written to exactly the same point, "a member of the natural sequence of numbers ending with a, but not identical with a is a member of the natural sequence of numbers ending with d, and vice versa."

It is evident that Frege cannot in fact be proposing to derive (4*) or the equivalent

$$(4) \qquad \forall x([xP^{*=}a \wedge x \neq a] \leftrightarrow xP^{*=}d)$$

from (5′) since both of these contain free occurrences of "d". Since the supposition of §82 that dPa is clearly still in force, it might be thought that Frege wishes to derive

$$(4^{\dagger}) \qquad dPa \rightarrow \forall x([xP^{*=}a \wedge x \neq a] \leftrightarrow xP^{*=}d)$$

from (5′).

However, if $d = a = \aleph_0$, then, as we have observed, dPa; and then, since $aP^{*=}a$, $(4^{\dagger})$ has a true antecedent and false consequent. Thus it cannot be $(4^{\dagger})$ that Frege is proposing to derive from (5′).

We may note, though, that $\forall x([xP^{*=}a \wedge x \neq a] \leftrightarrow xP^{*=}d)$ can be derived from dPa and $\neg aP^{*}a$. So we may take it that Frege is proposing to derive (4′) $0P^{*=}a \rightarrow dPa \rightarrow \forall x([xP^{*=}a \wedge x \neq a] \leftrightarrow xP^{*=}d)$ from (5′).

$$(4') \quad 0P^{*=}a \rightarrow dPa \rightarrow \forall x([xP^{*=}a \wedge x \neq a] \leftrightarrow xP^{*=}d).$$

Proof. Suppose $0P^{*=}a$ and dPa. Assume $xP^{*=}a \wedge x \neq a$. Then $xP^{*}a$. By the lemma, for some c, cPa and $xP^{*=}c$. By 78.5, $c = d$. Thus $xP^{*=}d$. Conversely, assume $xP^{*=}d$. Since dPa, $xP^{*}a$ by the basic fact about the weak ancestral, and so $xP^{*=}a$. If $\neg aP^{*}a$, then also $x \neq a$. But since $0P^{*=}a$, it follows from (5′) that indeed $\neg aP^{*}a$. Thus (4′) is proved. ∎

Nor could Frege be proposing to derive (3) $a = \#[x : xP^{*=}a \wedge x \neq a]$ from any proposition he takes himself to have demonstrated. For (3) is false if "a" has $\aleph_0$ as value. In fact, $\#[x : xP^{*=}\aleph_0 \wedge x \neq \aleph_0] = 0$. For if $xP\aleph_0$, then since $\aleph_0 P\aleph_0$, $x = \aleph_0$, by 78.5. Let S be the converse of

---

[9]For elaboration of this suggestion, see the second author's "The development of arithmetic in Frege's *Grundgesetze der Arithmetik*" (Heck, 1993).

P. Then if $\aleph_0 Sx$, $x = \aleph_0$. Thus if $\aleph_0 S^* x$, $x = \aleph_0$. (Let $Fa \equiv a = \aleph_0$ in the definition of S^*.) But the ancestral is the converse of the ancestral of the converse. So if $xP^*\aleph_0$, then $x = \aleph_0$. Thus $xP^{*=}\aleph_0$ iff $x = \aleph_0$, and therefore for *no* x, $xP^{*=}\aleph_0 \wedge x \neq \aleph_0$. By a proposition given in §75, $\#[x : xP^{*=}\aleph_0 \wedge x \neq \aleph_0] = 0$.

However, it is important to observe that at this point it is not only the conjunct dPa of the antecedent of (1) that is assumed to be in force; the other conjunct $dP\#[x : xP^{*=}d]$ is also assumed to hold. (It is easy to be oblivious to this further assumption since "$a = \#[x : xP^{*=}a \wedge x \neq a]$" does not mention d. But it is supposed at this point that a is such that dPa, and it is likewise also supposed that d is such that $dP\#[x : xP^{*=}d]$.) Since $a = \#[x : xP^{*=}a \wedge x \neq a]$ follows from these two conjuncts and the consequent of ($4'$), we may take it that Frege wishes to prove ($3'$):

($3'$) $0P^{*=}a \rightarrow dPa \rightarrow dP\#[x : xP^{*=}d] \rightarrow a = \#[x : xP^{*=}a \wedge x \neq a]$.

Proof. Suppose dPa and $dP\#[x : xP^{*=}d]$. Then by 78.5 (the other way), $a = \#[x : xP^{*=}d]$. Suppose further that $0P^{*=}a$. Then by ($4'$), $\forall x([xP^{*=}a \wedge x \neq a] \leftrightarrow xP^{*=}d)$. By Hume's Principle, $\#[x : xP^{*=}a \wedge x \neq a] = \#[x : xP^{*=}d]$. Thus $a = \#[x : xP^{*=}a \wedge x \neq a]$. ■

We come now to the difficult question how Frege proposes to derive (1) from ($3'$). Frege tells us that to prove (1), we must show (3). But (3) is not unconditionally true. However, ($3'$), whose consequent is (3) and whose antecedent contains a conjunct stating that the value of "a" satisfies the condition of finiteness, can be proved. Thus it might seem reasonable to think that Frege may be proposing, as in the case of (4) and (3), not to derive (1) from (3), but some conditional instead whose antecedent expresses a finiteness condition and whose consequent is (1). Moreover, since dPa is one of the clauses of the antecedent, if we take $0P^{*=}d$ as another conjunct of the antecedent, we need not also take $0P^{*=}a$. So we let ($1'$) be $0P^{*=}d \rightarrow dPa \rightarrow dP\#[x : xP^{*=}d] \rightarrow aP\#[x : xP^{*=}a]$. ($1'$) follows readily from ($3'$):

($1'$) $0P^{*=}d \rightarrow dPa \rightarrow dP\#[x : xP^{*=}d] \rightarrow aP\#[x : xP^{*=}a]$.

Proof. Suppose that $0P^{*=}d$, dPa and $dP\#[x : xP^{*=}d]$. By the basic fact about the weak ancestral, $0P^{*=}a$. By ($3'$), $a = \#[x : xP^{*=}a \wedge x \neq a]$. Since $a = a$, $aP^{*=}a$. By the definition of P, $\#[x : xP^{*=}a \wedge x \neq a]P\#[x : xP^{*=}a]$. Thus $aP\#[x : xP^{*=}a]$. ■

It may be useful to recapitulate here our (somewhat intricate) derivation of ($1'$) from ($5'$) and the other propositions to which Frege appeals:

Suppose $0P^{*=}d$, dPa and $dP\#[x : xP^{*=}d]$. By the basic fact about the weak ancestral, $0P^{*=}a$, and thus by ($5'$), $\neg aP^*a$. If $xP^{*=}a$ and $x \neq a$, then

xP^*a, and so by the lemma, for some z, $xP^{*=}z$ and zPa. By one half of 78.5, $z = d$, and so $xP^{*=}d$; conversely if $xP^{*=}d$, then by the basic fact, xP^*a, whence $x \neq a$ (since $\neg aP^*a$) and $xP^{*=}a$. Thus $\forall x(xP^{*=}a \wedge x \neq a \leftrightarrow xP^{*=}d)$, which is (4), and so by Hume's Principle, $\#[x : xP^{*=}a \wedge x \neq a] = \#[x : xP^{*=}d]$, and therefore $dP\#[x : xP^{*=}a \wedge x \neq a]$. By the other half of 78.5, $a = \#[x : xP^{*=}a \wedge x \neq a]$, which is (3). Since $aP^{*=}a$ (trivially), by the definition of P, $\#[x : xP^{*=}a \wedge x \neq a]P\#[x : xP^{*=}a]$, and therefore $aP\#[x : xP^{*=}a]$.

(2) is proved much more easily.

(2) $0P\#[x : xP^{*=}0]$.

Proof. $0 = \#[x : x \neq x]$. By Hume's Principle and the definition of P, $\forall z \neg zP0$. By the lemma, $\forall x \neg xP^*0$, and so $\forall x \neg(xP^{*=}0 \wedge x \neq 0)$. By a result of §75 mentioned above, $\#[x : xP^{*=}0 \wedge x \neq 0] = 0$. But $0P^{*=}0$, whence $\#[x : xP^{*=}0 \wedge x \neq 0]P\#[x : xP^{*=}0]$, and therefore $0P\#[x : xP^{*=}0]$. ∎

(0′) must now be derived from (1′) and (2). It is not possible to appeal to Induction 2 because of the presence of "$0P^{*=}d$" in the antecedent of (1′). But, it might be supposed, Frege can appeal here to Induction 3, which he explicitly demonstrated in *Grundgesetze* as Proposition 152: To prove that if $x\varphi^{*=}y$, then $\dots y \dots$, it suffices to prove

(i) $\dots x \dots$

and

(ii) $\forall d \forall a(x\varphi^{*=}d \rightarrow \dots d \dots \rightarrow d\varphi a \rightarrow \dots a \dots)$.

Note the formula "$x\varphi^{*=}d$," whose presence weakens (ii) and thereby strengthens the method. The derivation of Induction 3 from Induction 2 is significantly more interesting than that of Induction 2 from Induction 1. It appeals to the comprehension scheme of second-order logic and uses a technique sometimes called "loading the inductive hypothesis." (At the beginning of §116 of *Grundgesetze*, Frege writes, "To prove proposition (γ) of §114, we replace the function mark '$F(\xi)$' with '$\neg(aP^{*=}\xi \rightarrow \neg F(\xi))$.' ")

Proof of Induction 3. Suppose (i) and (ii). Let $Ga \equiv \dots a \dots \wedge x\varphi^{*=}a$ (second-order comprehension). Now, $x\varphi^{*=}x$ trivially; thus by (i), Gx. We now show $\forall d \forall a(Gd \rightarrow d\varphi a \rightarrow Ga)$: Suppose $d\varphi a$ and Gd, i.e., $\dots d \dots$ and $x\varphi^{*=}d$. By (ii), $\dots a \dots$. By the basic fact about the weak ancestral, $x\varphi^{*=}a$. Thus $\forall d \forall a(Gd \rightarrow d\varphi a \rightarrow Ga)$. By Induction 2, Gy, whence $\dots y \dots$ ∎

(0′) is now immediate from (2) and (1′) by Induction 3: let $x = 0$, $\varphi = P$, $y = n$, and $\dots y \dots$ iff $yP\#[x : xP^{*=}y]$.

We believe that no one will seriously dispute that this proof of $(0')$, which features a derivation of $(1')$ from $(5')$ and an appeal to Induction 3, is Fregean in spirit, ingenious, and of a structure that fits the proof-sketch found in §§82–83 rather well. But there are a number of strong reasons for doubting that Frege had *it* in mind while writing these two sections. Accordingly, we shall refer to it as the *conjectural* proof.[10]

First of all, Frege twice *says* that (1) is to be proved, once in §82 and again in §83. He says, moreover, "The method of inference here is an application of the definition of the expression 'y follows x in the natural sequence of numbers', taking $[y : yP\#[x : xP^{*=}y]]$ for our concept F." It would thus seem natural to take Frege as arguing by appeal to Induction 1 or Induction 2 (with P as φ). Frege mentions the condition that n be finite, but does not also mention, as he might easily have done, the need to assume that d (or a) is finite as well. Thus it would seem overly charitable to assume that the argument he really intended proceeds via Induction 3.

Secondly, notice that Frege says in §83 that $(5')$, which Frege proved in *Grundgesetze* by appeal to Induction 2, "must likewise ('ebenfalls') be proved by means of our definition of following in a series, as indicated above." It seems plain that Frege does not intend to use Induction 3 to prove $(5')$; "ebenfalls" suggests that the induction used to prove $(0')$ would be like the one used for $(5')$.

The most telling objections to the suggestion that Frege was intending to sketch the conjectural proof in *Die Grundlagen*, however, arise from a close reading of Section H (eta) of Part II of *Grundgesetze*. We quote and comment upon part of Section H.[11]

H. **Proof of the Proposition**

$$0P^{*=}b \rightarrow bP\#[x : xP^{*=}b]$$

§114. **Analysis**

We wish to prove the proposition that the Number that belongs to the concept

member of the number-series ending with b

follows after b in the number-series if b is a finite number. Herewith, the conclusion that the number-series is infinite follows at once; i.e., it follows at once that there is, for each finite number, one immediately following after it.

---

[10] Of course, what is conjectural is whether the proof is Frege's, not whether it is a (correct) proof.

[11] The present translation is based upon one due to the second author and Jason Stanley. We have changed Frege's notation to our own and added some material in brackets.

We first attempt to carry out the proof with the aid of
Proposition (144) [viz., $aq^{*=}b \rightarrow \forall \mathbf{d}(F\mathbf{d} \rightarrow \forall \mathbf{a}(\mathbf{d}q\mathbf{a} \rightarrow F\mathbf{a}) \rightarrow$
$(F\mathbf{a} \rightarrow F\mathbf{b})$], replacing the function-mark "$F(\xi)$" with "$\xi P \# [x :$
$xP^{*=}\xi]$." For this we need the proposition "$dP\#[x : xP^{*=}d] \rightarrow$
$dPa \rightarrow aP\#[x : xP^{*=}a]$".

That is to say, one's "first" idea might be to prove $(0')$ by applying Induction
2 to the concept $[y : yP\#[x : xP^{*=}y]]$, which would, among other things,
require a proof of (1). (A footnote, to which we shall return, is attached to
this last sentence.)

Substituting ... in (102) [viz., $\#[x : Fx \wedge x \neq b] = c \rightarrow$
$Fb \rightarrow cP\#F$], ... we thus obtain "$\#[x : xP^{*=}a \wedge x \neq a] =$
$a \rightarrow aP^{*=}a \rightarrow aP\#[x : xP^{*=}a]$," from which we can remove
the subcomponent "$aP^{*=}a$" by means of (140) [viz., $aq^{*=}a$].
The question arises whether the subcomponent "$\#[x : xP^{*=}a \wedge$
$x \neq a] = a$" can be established as a consequence of "dPa" and
"$dP\#[x : xP^{*=}d]$".

Put differently, the problem reduces to that of proving

$(3^{\dagger})$ $dPa \rightarrow dP\#[x : xP^{*=}d] \rightarrow a = \#[x : xP^{*=}a \wedge x \neq a]$,

which is $(3')$ without the finiteness condition $0P^{*=}a$, and which, together
with the relevant instance of Frege's Proposition 102, implies (1).

$\cdot$ By the functionality of progression in the number-series ...,
we have "$dP\#[x : xP^{*=}d] \rightarrow dPa \rightarrow \#[x : xP^{*=}d] = a$" ... We
thus attempt to determine whether "$\#[x : xP^{*=}a \wedge x \neq a] =$
$\#[x : xP^{*=}d]$" can be shown to be a consequence of "dPa". ...
For this it is necessary to establish "$[bP^{*=}a \wedge b \neq a] \leftrightarrow bP^{*=}d$"
as a consequence of "dPa" ...

That is, $(3^{\dagger})$ will follow from

$(4^{\dagger})$ $dPa \rightarrow \forall x([xP^{*=}a \wedge x \neq a] \leftrightarrow xP^{*=}d)$,

an easy consequence of Hume's Principle, and the one-one-ness of P.

For this it is necessary to establish "$bP^{*=}d \rightarrow [bP^{*=}a \wedge b \neq$
$a]$" and "$[bP^{*=}a \wedge b \neq a] \rightarrow bP^{*=}d$" as consequences of "$dPa$".
But it turns out that another condition must be added if "$b \neq a$"
is to be shown to be a consequence of "$bP^{*=}d$" and "dPa." By
(134) we have "$bP^{*=}d \rightarrow dPa \rightarrow bP^*a$". If b coincided with

a, then the main component would transform into "aP^*a". By (145) ⟦our (5′)⟧, this is impossible if a is a finite number. Thus the subcomponent "$0P^{*=}a$" is also added.

Admittedly, the desired application of (144) thereby becomes impossible; but with (137) ⟦viz., $aq^{*=}e \to eqm \to aq^{*=}m$⟧ we can replace this subcomponent with "$0P^{*=}d$" and derive from (144) the Proposition ⟦152⟧, "$(aq^{*=}b \to \forall\mathbf{d}(F\mathbf{d} \to aq^{*=}\mathbf{d}$ $\to \forall\mathbf{a}(\mathbf{d}q\mathbf{a} \to F\mathbf{a})) \to (Fa \to Fb))$" ..., which takes us to our goal.

That is, to establish the first half of (4†), we need to know that $\neg aP^*a$; this will follow from (5′) and the additional assumption that a is finite. However, this new assumption must then be carried along throughout the proof, transforming (4†) into (4′), (3†) into (3′), and (1) into "$0P^{*=}a \to dP\#[x : xP^{*=}d] \to dPa \to aP\#[x : xP^{*=}a]$," from which (1′) easily follows. The attempt to prove (0′) via Induction 2 then fails, since we simply have not proved (1), though we can still complete the proof by making use of Induction 3 instead.

It is, we think, difficult to read these paragraphs without supposing that they reveal Frege's *second thoughts* about his idea in *Die Grundlagen* of applying Induction 2 to prove $(0P^{*=}n \to nP\#[x : xP^{*=}n])$ by substituting $[y : yP\#[x : xP^{*=}y]]$ for F. The attempt won't work, he says, because we need the hypothesis that a is finite in order to derive $\neg aP^*a$, which is needed for $(bP^{*=}d \to dPa \to b \neq a)$, which is in turn necessary for the rest of the proof. Read side by side with §§82–83 of *Die Grundlagen*, Frege's discussion in these paragraphs strikes one as penetrating and direct criticism of his earlier work. Moreover, the criticism suggests a way in which the conjectural proof can be regarded as Frege's after all: it is the proof obtained on amending the proof-sketch of §§82–83 in the way suggested in this section of *Grundgesetze*.

It is striking that the formal proof Frege actually gives in *Grundgesetze*, though closely related to the conjectural proof, is not quite the same proof. The formal proof,[12] given in §§115, 117, and 119, does proceed by deriving (0′) by means of Induction 3 (Frege's Proposition 152), from (1′) ($= \varepsilon 150$[13]) and (2) ($=154$). And the proof of (1′) does begin with a derivation of (4′) ($= \varepsilon 149$), from (5′) ($= 145$). But (1′) is not derived from (4′) via (3′); the argument is slightly different.

---

[12]For a fuller account, see (Heck, 1993). For the benefit of interested readers, Appendix 3 of the present paper contains translations into our notation of relevant parts of Frege's proofs.

[13]By Proposition xn we mean the proposition labeled with Greek letter x which occurs *during*, as opposed to *after*, the proof of proposition number n.

This part of the *Grundgesetze* proof, translated into English plus our notation, runs as follows: By the basic fact about the weak ancestral it suffices to show that if $0P^{*=}a$, dPa, and $dP\#[x : xP^{*=}d]$, then $aP\#[x : xP^{*=}a]$. By (4′) and (an easy consequence of) Hume's Principle, we have that $\#[x : xP^{*=}a \wedge x \neq a] = \#[x : xP^{*=}d]$ (cf. 149). But substituting into Proposition (102) quoted above, we have $\#[x : xP^{*=}a \wedge x \neq a] = \#[x : xP^{*=}d] \rightarrow aP^{*=}a \rightarrow \#[x : xP^{*=}d]P\#[x : xP^{*=}a]$. Hence, by (140), $\#[x : xP^{*=}d]P\#[x : xP^{*=}a]$ (cf. β150). Since dPa and $dP\#[x : xP^{*=}d]$, $a = \#[x : xP^{*=}d]$ (cf. γ150), whence $aP\#[x : xP^{*=}a]$ (cf. δ150) and we are done.

Comparing this argument with the relevant portion of the conjectural proof, one sees immediately how little they differ from each other; one might therefore overlook (or ignore) the fact that (3′) does not actually appear in the proof given in *Grundgesetze*. But the omission of (3′) is significant, since the "proof" discussed in §114 explicitly highlights (3†) as what must be proved if (1) is to be derived from (4†). The typical point of a section of *Grundgesetze* headed "Analysis" is to describe a formal proof found in "Construction" sections that follow it. Thus on reading §114, one would naturally expect the following proof to include, not just proofs of the results of adding a finiteness condition to (4†) and to (1), but also, as part of the derivation of the latter from the former, a proof of a proposition similarly related to (3†). As we said, however, the derivation of (1′) from (4′) in §115 does not go via (3′). That (3†) is so much as mentioned in §114 is therefore bound to seem mysterious unless one reads it as we have suggested, i.e., as criticism of Frege's *own* "first attempt" to prove (0′) in §§82–83 of *Die Grundlagen*, for (3†) or (3′) is indeed an intermediate step in *that* proof.

This observation concerning how the *Grundgesetze* proof differs from the conjectural proof also suggests a plausible explanation of the origin of the mistake of which we have accused Frege. Consider the two lists of propositions:

(1′) $0P^{*=}d \rightarrow dPa \rightarrow dP\#[x : xP^{*=}d] \rightarrow aP\#[x : xP^{*=}a]$
(3′) $0P^{*=}a \rightarrow dPa \rightarrow dP\#[x : xP^{*=}d] \rightarrow a = \#[x : xP^{*=}a \wedge x \neq a]$
(4′) $0P^{*=}a \rightarrow dPa \rightarrow \forall x([xP^{*=}a \wedge x \neq a] \leftrightarrow xP^{*=}d)$
(5′) $0P^{*=}a \rightarrow \neg aP^{*}a$

(1) $dPa \rightarrow dP\#[x : xP^{*=}d] \rightarrow aP\#[x : xP^{*=}a]$
(3†) $dPa \rightarrow dP\#[x : xP^{*=}d] \rightarrow a = \#[x : xP^{*=}a \wedge x \neq a]$
(4†) $dPa \rightarrow \forall x([xP^{*=}a \wedge x \neq a] \leftrightarrow xP^{*=}d)$
(5†) $\neg aP^{*}a$

As we have seen, $(4')$ follows from $(5')$, $(3')$ from $(4')$, and $(1')$ from $(3')$. But notice also that $(4^\dagger)$ follows from $(5^\dagger)$, $(3^\dagger)$ from $(4^\dagger)$, and (1) from $(3^\dagger)$, as obvious modifications of our proofs show. Frege, able to prove $(5')$ and desirous of proving (1), may well have lost sight of the need for a finiteness condition somewhere in the middle of his argument—perhaps he had not yet fully written out the argument in *Begriffsschrift*—and mistakenly concluded that he could deduce (1) from $(5')$. If forced to guess, we would suppose that it was between $(4')$ and (1), i.e., at $(3^\dagger)$ or $(3')$, that the finiteness condition vanished, for it is there that the *Grundgesetze* proof differs from the conjectural proof.

The first sentence of the second paragraph of §83 calls for some discussion. Frege writes there that we are obliged "hereby" ("hierdurch") to attach to the proposition that $nP\#[x : xP^{*=}n]$ the condition that $0P^{*=}n$. One might be forgiven for thinking that, in so stating, Frege is indicating that this condition is required by the presence of the finiteness condition in $(5')$, since it is with an indication of how $(5')$ is to be proven that the previous paragraph ends. But this thought cannot be right. Frege says in §82 that, once (1) and (2) are proved, "it is to be deduced that $0P^{*=}n \rightarrow nP\#[x : xP^{*=}n]$" by means of Induction 2. Thus what subjects n in (0) to a finiteness condition is not the presence of such a condition in $(5')$, but the kind of proof of $(0')$ being given in the first place. "Hierdurch" refers to the use in the proof of $(0')$ of the "definition of following in a series, on the lines indicated above," that is, as was discussed in §82.

There is one final piece of textual evidence to which we should like to draw attention. As we said earlier, a footnote is attached to Proposition (1) when it is first mentioned in §114: "This proposition is, as it seems, unprovable, but it is not here being asserted as true, since it stands in quotation marks." The natural explanation for this remark of Frege's is that he once *did* believe (1) to be provable, namely when he wrote *Die Grundlagen*, and any defender of the view that Frege was outlining the conjectural proof in §§82–83 will have the occurrence of this remark to explain away.

Apart from the light it may throw on the question whether Frege made a reparable error, the footnote is astonishing. Note that Frege says, not that (1) seems to be false, but that it "seems to be *unprovable*" [itals. ours]. There is, moreover, reason to suppose Frege believed (1) to be not false, but *true*. For one thing, had Frege believed it to be false, he presumably would have said so. Furthermore, Frege's difficulty was probably not that he did not know how to prove (1), but rather that he did not know how to prove it *in his formal system*. There is a very simple proof of (1) which depends only upon $(1')$, Dedekind's claim that every infinite number is (the number of a concept that is) Dedekind infinite,[14] and the observation, made earlier, that "d is Dedekind infinite" is equivalent to "dPd". We may take $(1')$ to be one half of a dilemma, the other half of which is:

$$\neg 0P^{*=}d \to dPa \to dP\#[x : xP^{*=}d] \to aP\#[x : xP^{*=}a].$$

This proposition may be proved as follows: suppose the subcomponents. Since d is not finite, it is Dedekind infinite. So dPd, and since dPa, $d = a$ and (1) follows immediately.

This proof is one Frege might well have known. It is not at all difficult and once $(1')$ has been proved, a proof of (1) by dilemma immediately suggests itself. Moreover, Frege was familiar with Dedekind's claim and, at least while he was working on Part II of *Grundgesetze*, believed it to be true.[15] As for the observation, not only is it easily proved, it is natural, in Frege's system, just to use "dPd" as a *definition* of "d is Dedekind infinite" (cf. *Grundgesetze*, Proposition 426). We conclude that Frege believed (1) to be a true but unprovable formula of Frege Arithmetic.

Frege's belief that (1) is unprovable in Frege Arithmetic was mistaken, however. A proof of (1) can be given that makes use of techniques that are different from any found in §§82–83 of *Die Grundlagen* or in relevant sections of *Grundgesetze*, but with which Frege was familiar. What we shall prove is that the hypothesis $0P^{*=}d$ of $(1')$, that d is finite, is dispensable. More precisely, we shall prove that if $dP\#[x : xP^{*=}d]$, then $\#[x : xP^{*=}d]$ is finite, from which it follows that d is finite, since by Proposition (143) of *Grundgesetze* (viz., $dPb \to aP^*b \to aP^{*=}d$), any predecessor of a finite number is finite.

Theorem (FA)[16] *Suppose $dP\#[x : xP^{*=}d]$. Then $\#[x : xP^{*=}d]$ is finite.*

Proof. In FA, define $h : [x : 0P^{*=}x] \to [x : xP^{*=}d]$ by:

$$h(0) = d; \quad h(n+1) = \begin{cases} y & \text{if} \quad yPh(n) \\ h(n) & \text{if} \quad \neg\exists y\, yPh(n). \end{cases}$$

The definition is OK since P is one-one.

Since in general $yR^*z \leftrightarrow z(R^\cup)^*y$,[17] $\forall x(xP^{*=}d \leftrightarrow d(P^\cup)^{*=}x)$, and so h is onto. Therefore $[x : xP^{*=}d]$ is countable, i.e., either finite or countably

---

[14]Of course in set theory without the axiom of choice, i.e., ZF as opposed to ZFC, this claim cannot be proved.

[15]See Frege's review, published in 1892, of Cantor's "Zur Lehre vom Transfiniten." Of course, if Frege did know of this proof and believed (1) to be unprovable, then he must have believed Dedekind's result too to be unprovable, which he (rightly) did. For further discussion, see the second author's "The Finite and the Infinite in Frege's *Grundgesetze der Arithmetik*" (Heck, 1997).

[16]This result is due to the second author; the present proof, to the first.

[17]$R^\cup$ is the converse of R.

infinite. If the latter, then $\#[x : xP^{*=}d] = \aleph_0$ and by the supposition of the theorem, $dP\aleph_0$. But as we saw just after the proof of $(4')$, $xP^{*=}\aleph_0 \leftrightarrow x = \aleph_0$. Since $dP\aleph_0$, $dP^{*=}\aleph_0$, $d = \aleph_0$, and $\#[x : xP^{*=}d] = 1$, contra $\#[x : xP^{*=}d] = \aleph_0$. Therefore $\#[x : xP^{*=}d]$ is finite. ∎

Thus Frege could have proved (1) after all and thus appealed to Induction 2 to prove $(0')$. Of course the technology borrowed from second-order arithmetic used in the proof just given, particularly the inductive definition of h, is considerably more elaborate than that needed to derive Induction 3 from Induction 2. The conjectural proof is unquestionably to be preferred to this new one on almost any conceivable grounds.

So. Frege erred in §§82–83 of *Die Grundlagen*, where an oversight marred the proof he outlined of the existence of the successor. Mistakes of that sort are hardly unusual, though, there are four or five ways the proof can be patched up, and Frege's way of repairing it cannot be improved on. But even if one ought not to make too much of Frege's mistake, there is lots to be made of his belief that (1) was true but unprovable in his system. One question that must have struck Frege is: If there are truths about numbers unprovable in the system, what becomes of the claim that the truths of arithmetic rest solely upon definitions and general logical laws? Another that may have occurred to him is: Can the notion of a truth of logic be explained otherwise than via the notion of provability?

Appendix 1. Counterparts in *Grundgesetze* of some propositions of *Die Grundlagen*

Proposition of this paper	Proposition of *Grundgesetze*
Hume's Principle	32, 49
$\forall x \neg Fx \leftrightarrow 0 = \#[x : Fx]$	94, 97
78.1	114
78.2	113
78.3	117
78.4	122
78.5	90
78.6	107
$\neg z P0$	108
Induction 1	123
The basic fact about the weak ancestral	134, 136
The Lemma	141
Induction 2	144
$(5')$	145
$(4')$	α149
$(1')$	150
Induction 3	152
(2)	154
$(0')$	155

Appendix 2. Interpreting Frege arithmetic in second-order arithmetic

The language (of second-order arithmetic) contains variables $x, y, z, \ldots$ over natural numbers; variables $\alpha, \beta, \gamma, \ldots$ over sets of numbers; and variables $\rho, \sigma, \ldots$ over binary relations of numbers. (We do not need variables over n-place relations for $n > 2$.) Its non-logical symbols are $0, s, +, \times, <$. Terms t are built up out of $0, s, +, \times$ as usual; the atomic formulas are $t = t'$, $t < t'$, αt, $\rho t t'$; formulas are then built up as usual.

The axioms of second-order arithmetic are induction: $(\alpha 0 \wedge \forall x (\alpha x \rightarrow \alpha s x) \rightarrow \alpha x$ [a single formula]; the recursion axioms for successor, plus and times and the definition of less-than: $0 \neq sx$, $sx = sy \rightarrow x = y$, $x + 0 = x$, $x + sy = s(x + y)$, $x \times 0 = 0$, $x \times sy = (x \times y) + x$, $x < y \leftrightarrow \exists z \, sz + x = y$; and the comprehension axioms (which are axioms of standard second-order logic): $\exists \alpha \forall x (\alpha x \leftrightarrow A)$, $\exists \rho \forall x \forall y (\rho x y \leftrightarrow B)$, A a formula in which α is not free and B a formula in which ρ is not free.

Since we have $+$ and $\times$, we could have dispensed with binary relation variables; and since we have binary relation variables, we could have dispensed with $+$ and $\times$ and set variables: $J, = \lambda xy \imath z(x^2 + 2xy + y^2 + 3x + y = 2z)$, is an onto pairing function. Thus if we have $+$ and $\times$, we have J, and so we can replace $\rho tt'$ by $\alpha J(t, t')$. And we can define $x + y = z$ and $x \times y = z$ from 0 and s using binary relation variables: $x + y = z$ iff $\forall \rho(\rho 0x \wedge \forall u \forall v(\rho uv \rightarrow \rho susv) \rightarrow \rho yz)$; $x \times y = z$ iff $\forall \rho(\rho 00 \wedge \forall u \forall v(\rho uv \rightarrow \rho su(v + x)) \rightarrow \rho yz)$. And, uninterestingly, we can replace αx by ρxx.

The least number principle can be proved from the induction axioms as usual.

Introduce the notation: $\{x : A\}$ as usual: $\{x : A\}t$ abbreviates $\exists \alpha(\forall y[\alpha y \leftrightarrow \exists x(x = y \wedge A)] \wedge \alpha t)$, or, equivalently, $\forall \alpha(\forall y[\alpha y \leftrightarrow \exists x(x = y \wedge A)] \rightarrow \alpha t)$. $[\alpha, y$ new$]$.[18]

Definition $\alpha \approx \beta$ *via* $\rho \equiv \forall x(\alpha x \leftrightarrow \exists y(\beta y \wedge \rho xy)) \wedge \forall y(\beta y \leftrightarrow \exists x(\alpha x \wedge \rho xy)) \wedge \forall x \forall y \forall a \forall b(\rho xy \wedge \rho ab \rightarrow (x = a \leftrightarrow y = b))$

Definition $\alpha \approx \beta \equiv \exists \rho(\alpha \approx \beta$ *via* $\rho)$.

Definition $N = \{i : i = i\}$.

1. $\vdash \approx$ is an equivalence relation.

2. $\vdash \{i : i < m\} \approx \{i : i < n\} \rightarrow m = n$.

3. $\vdash \neg(N \approx \{i : i < n\})$.

4. $\vdash N \approx \alpha \vee \exists n(\{i : i < n\} \approx \alpha)$.

1 will be used without citation below; it is evident how to prove it. Sketches of proofs of 2, 3, 4 in analysis are given below.

5. $\vdash \exists! x[\exists n(\{i : i < n\} \approx \alpha \wedge x = n + 1) \vee (N \approx \alpha \wedge x = 0)]$.

Proof. By 2, 3, 4. ∎

Definition $\#e = \imath x[\exists n(\{i : i < n\} \approx \alpha \wedge x = n + 1) \vee (N \approx \alpha \wedge x = 0)]$.

Theorem $\vdash \#\alpha = \#\beta \leftrightarrow \alpha \approx \beta$.

Proof. Case 1. $\#\alpha = 0 = \#\beta$. Then $\alpha \approx N \approx \beta$. Case 2. $\#\alpha = 0$, $\#\beta = n + 1$. Then $\alpha \approx N$, $\beta \approx \{i : i < n\}$, and $\neg(\alpha \approx \beta)$, by 3. Case 3. $\#\alpha = m + 1$, $\#\beta = 0$. Similar to case 2. Case 4. $\#\alpha = n + 1 = \#\beta$. Then $\alpha \approx \{i : i < n\} \approx \beta$. Case 5. $\#\alpha = m + 1$, $\#\beta = n + 1$, $m \neq n$. Then $\alpha \approx \{i : i < m\}$, $\beta \approx \{i : i < n\}$, and by 2, $\neg(\alpha \approx \beta)$. ∎

---

[18]Thanks here to Roy Dyckhoff.

Proof of 2.   Suppose not. By the least number principle, let n be least such that for some least $m < n$, $\{i : i < m\} \approx \{i : i < n\}$, via some ρ. $m, n \neq 0$, hence for some j, k, $m = sj$, $n = sk$. So $j < k < n$. Let $b = \rho(j)$; $a = \rho^{-1}(k)$; let $\sigma = ((\rho - \{\langle j, b \rangle, \langle a, k \rangle\}) \cup \{\langle a, b \rangle\}) - \{\langle j, k \rangle\}$. Then $\{i : i < j\} \approx \{i : i < k\}$ via σ, contra leastness of n. ∎

Proof of 3.   Suppose $N \approx \{i : i < n\}$, n least. Let $N \approx \{i : i < n\}$ via ρ. Clearly $n = m + 1$, some m. $m < n$. Let i be such that $\rho(i) = m$. Let $\sigma(j) = \rho(j)$ if $j < i$ and let $\sigma(j) = \rho(j + 1)$ if $j \geq i$. Then $N \approx \{i : i < m\}$ via σ, contra leastness of n. ∎

Proof of 4.   Suppose that for no n, $\{i : i < n\} \approx \alpha$. We prove the existence of a one-one function ρ mapping N onto α.

Say that σ is good to n if σ is a function; domain$(\sigma) = \{i : i < n\}$; for all $i < n$, $\sigma(i) \in \alpha$; for all $j < i < n$, $\sigma(j) < \sigma(i)$; and for all $i < n$, if $k \in \alpha$ and $k < \sigma(i)$, then for some $j < i$, $k = \sigma(j)$. [Officially: σ is good to n iff $\forall i \forall p \forall q (\sigma i p \wedge \sigma i q \to p = q) \wedge \forall i (\exists p\, \sigma i p \leftrightarrow i < n) \wedge \forall i \forall p (\sigma i p \to \alpha p) \wedge \forall j \forall i \forall p \forall q (j < i \wedge \sigma j q \wedge \sigma i p \to q < p) \wedge \forall i \forall k \forall p (\alpha k \wedge \sigma i p \wedge k < p \to \exists j (j < i \wedge \sigma j k))$.]

$\emptyset$ is good to 0. If σ is good to n, then $\sigma \cup \{\langle n, \text{least member of } \alpha$ not in the range of $\sigma \rangle\}$ is good to sn (the existence of the least member of α not in the range of σ follows from our supposition). By induction, for every n, some σ is good to n. If σ is good to n, σ' is good to n', and $n > n'$, then by induction for every $i < n$, $\sigma(i) = \sigma'(i)$. By comprehension, let $\rho = \{\langle n, k \rangle : \exists \sigma (\sigma$ is good to sn and $\sigma(n) = k)\}$. It is sufficiently clear that ρ is a function; domain$(\rho) = N$; for all i, $\rho(i) \in \alpha$; if $j < i$, $\rho(j) < \rho(i)$, and if $k \in \alpha$, $k < \rho(i)$, then for some $j < i$, $k = \rho(j)$. We must show that ρ is onto α. Since $\rho(j) < \rho(i)$ whenever $j < i$, for every k, $k \leq \rho(k)$, otherwise for some least k, $\rho(k) < k$, and then $\rho(\rho(k)) < \rho(k)$, contra leastness of k. Thus if $k \in \alpha$, then $k \leq \rho(k)$, and so for some $j \leq k$, $k = \rho(j)$. Thus ρ is onto α. ∎

Appendix 3. Translations into present-day notations of §§115, 117, 119, and part of §113 of *Grundgesetze*

Below we use boldface instead of Fraktur and notation introduced above in place of Frege's own, we omit the signs indicating which rule of inference is applied and reference numbers to axioms of Frege's system, and we utilize certain easy equivalences (e.g., $p \wedge q$ for $\neg (p \to \neg q)$).

§113 (part)

$$aq^*b \to \exists \mathbf{e}(\mathbf{e}qb \wedge aq^{*=}\mathbf{e}) \tag{141}$$
$$\forall \mathbf{e}(\mathbf{e}qb \to \neg aq^{*=}\mathbf{e}) \to \neg aq^*b \tag{142}$$

$$\neg aP^{*=}d \to d = c \to \neg aP^{*=}c$$

(88):
$$\neg aP^{*=}d \to dPb \to cPb \to \neg aP^{*=}c$$
$$\neg aP^{*=}d \to dPb \to \forall \mathbf{e}(\mathbf{e}Pb \to \neg aP^{*=}\mathbf{e})$$

(142):
$$\neg aP^{*=}d \to dPb \to \neg aP^*b$$
$$aP^*b \to dPb \to aP^{*=}d \tag{143}$$

§115

130
$$bq^{*=}a \to \neg bq^*a \to a = b$$
$$\neg bq^*a \to bq^{*=}a \to b = a \tag{146}$$
$$bq^{*=}a \wedge b \neq a \to bq^*a \tag{147}$$

147
$$bP^{*=}a \wedge b \neq a \to bP^*a$$

(143):
$$dPa \to (bP^{*=}a \wedge b \neq a) \to bP^{*=}d$$
$$(bP^{*=}d \to (bP^{*=}a \wedge b \neq a)) \to dPa \to ((bP^{*=}a \wedge b \neq a)$$
$$\leftrightarrow bP^{*=}d) \tag{β}$$

134
$$bP^{*=}d \to dPa \to bP^*a$$
$$\neg aP^*a \to bP^{*=}d \to dPa \to b \neq a$$

(145):
$$0P^{*=}a \to bP^{*=}d \to dPa \to b \neq a$$
$$bP^{*=}a \to 0P^{*=}a \to bP^{*=}d \to dPa \to (bP^{*=}a \wedge b \neq a)$$

(137):
$$dPa \to 0P^{*=}a \to bP^{*=}d \to (bP^{*=}a \wedge b \neq a)$$

(β):
$$dPa \to 0P^{*=}a \to ((bP^{*=}a \wedge b \neq a) \leftrightarrow bP^{*=}d)$$

(77):
$$dPa \to 0P^{*=}a \to ([x : xP^{*=}a \wedge x \neq a]b \leftrightarrow bP^{*=}d) \tag{148}$$
$$dPa \to 0P^{*=}a \to \forall \mathbf{a}([x : xP^{*=}a \wedge x \neq a]\mathbf{a} \leftrightarrow \mathbf{a}P^{*=}d)$$

(96):
$$dPa \to 0P^{*=}a \to \#[x : xP^{*=}a \wedge x \neq a] = \#[x : xP^{*=}d] \tag{149}$$

(102):
$$dPa \to 0P^{*=}a \to aP^{*=}a \to \#[x : xP^{*=}d]P\#[x : xP^{*=}a]$$

(140):
$$dPa \to 0P^{*=}a \to \#[x : xP^{*=}d]P\#[x : xP^{*=}a]$$
$$\#[x : xP^{*=}d] = a \to dPa \to 0P^{*=}a \to aP\#[x : xP^{*=}a]$$

(70):
$$dP\#[x : xP^{*=}d] \to dPa \to 0P^{*=}a \to aP\#[x : xP^{*=}a]$$

(137):
$$dP\#[x : xP^{*=}d] \to 0P^{*=}d \to dPa \to aP\#[x : xP^{*=}a]$$
$$\forall \mathbf{d}(dP\#[x : xP^{*=}\mathbf{d}] \to 0P^{*=}\mathbf{d} \to \forall \mathbf{a}(dPa$$
$$\to \mathbf{a}P\#[x : xP^{*=}\mathbf{a}])) \tag{150}$$

§117

$$\forall \mathbf{d}(F\mathbf{d} \to aq^{*=}\mathbf{d} \to \forall \mathbf{a}(dq\mathbf{a} \to F\mathbf{a})) \to (Fd \to aq^{*=}d$$
$$\to \forall \mathbf{a}(dq\mathbf{a} \to F\mathbf{a}))$$
$$\forall \mathbf{d}(F\mathbf{d} \to aq^{*=}\mathbf{d} \to \forall \mathbf{a}(dq\mathbf{a} \to F\mathbf{a})) \to (Fd \to aq^{*=}d$$
$$\to dqb \to Fb)$$
$$\forall \mathbf{d}(F\mathbf{d} \to aq^{*=}\mathbf{d} \to \forall \mathbf{a}(dq\mathbf{a} \to F\mathbf{a})) \to (\neg Fb \to aq^{*=}d$$
$$\to dqb \to \neg Fd)$$
$$\forall \mathbf{d}(F\mathbf{d} \to aq^{*=}\mathbf{d} \to \forall \mathbf{a}(dq\mathbf{a} \to F\mathbf{a})) \to ((aq^{*=}b \to \neg Fb)$$
$$\to aq^{*=}b \to aq^{*=}d \to dqb \to \neg Fd)$$

(137):
$$\forall \mathbf{d}(F\mathbf{d} \to aq^{*=}\mathbf{d} \to \forall \mathbf{a}(dq\mathbf{a} \to F\mathbf{a})) \to ((aq^{*=}b \to \neg Fb)$$

$$\to dqb \to (aq^{*=}d \to \neg Fd))$$
$$\forall\mathbf{d}(F\mathbf{d} \to aq^{*=}\mathbf{d} \to \forall\mathbf{a}(dq\mathbf{a} \to F\mathbf{a})) \to ((aq^{*=}d \land Fd)$$
$$\to dqb \to (aq^{*=}b \land Fb))$$
$$\forall\mathbf{d}(F\mathbf{d} \to aq^{*=}\mathbf{d} \to \forall\mathbf{a}(dq\mathbf{a} \to F\mathbf{a})) \to \forall\mathbf{d}((aq^{*=}\mathbf{d} \land F\mathbf{d})$$
$$\to \forall\mathbf{a}(dq\mathbf{a} \to (aq^{*=}\mathbf{a} \land F\mathbf{a})) \tag{151}$$

140 $\quad aq^{*=}a$

$\qquad Fa \to (aq^{*=}a \land Fa)$

(144): $\quad aq^{*=}b \to \forall\mathbf{d}((aq^{*=}\mathbf{d} \land F\mathbf{d}) \to \forall\mathbf{a}(dq\mathbf{a} \to (aq^{*=}\mathbf{a} \land F\mathbf{a})))$
$\qquad\quad \to (Fa \to (aq^{*=}b \land Fb))$

(151): $\quad aq^{*=}b \to \forall\mathbf{d}(F\mathbf{d} \to aq^{*=}\mathbf{d} \to \forall\mathbf{a}(dq\mathbf{a} \to F\mathbf{a}))$
$\qquad\quad \to (Fa \to (aq^{*=}b \land Fb))$

$\qquad aq^{*=}b \to \forall\mathbf{d}(F\mathbf{d} \to aq^{*=}\mathbf{d} \to \forall\mathbf{a}(dq\mathbf{a} \to F\mathbf{a}))$
$$\to (Fa \to Fb) \tag{152}$$

150 $\quad \forall\mathbf{d}(\mathbf{d}P\#[x : xP^{*=}\mathbf{d}] \to 0P^{*=}\mathbf{d} \to \forall\mathbf{a}(\mathbf{d}P\mathbf{a}$
$$\to \mathbf{a}P\#[x : xP^{*=}\mathbf{a}])) $$

(152): $\quad 0P^{*=}b \to 0P\#[x : xP^{*=}0] \to bP\#[x : xP^{*=}b] \tag{153}$

§119

126 $\quad \neg aP^*0$

(130): $\quad aP^{*=}0 \to 0 = a$

$\qquad aP^{*=}0 \to a = 0$

(58): $\quad \neg[x : xP^{*=}0 \land x \neq 0]a$

$\qquad \forall\mathbf{a}\neg[x : xP^{*=}0 \land x \neq 0]\mathbf{a}$

(97): $\quad \#[x : xP^{*=}0 \land x \neq 0] = 0$

(102): $\quad 0P^{*=}0 \to 0P\#[x : xP^{*=}0]$

(140): $\quad 0P\#[x : xP^{*=}0] \tag{154}$

(153): $\quad 0P^{*=}b \to bP\#[x : xP^{*=}b] \tag{155}$

Propositions cited but not proved above:

(58) $\qquad F(fa) \to F(\neg[x : \neg fx]a)$

(70) $\qquad ePd \to ePa \to d = a$

(77) $\qquad F(fa) \to F([x : fx]a)$

(88) $\qquad dPe \to aPe \to d = a$

(96) $\qquad \forall\mathbf{a}(u\mathbf{a} \leftrightarrow v\mathbf{a}) \to \#u = \#v$

(97) $\qquad \forall\mathbf{a}\neg u\mathbf{a} \to \#u = 0$

(102) $\qquad \#[x : ux \land x \neq c] = m \to uc \to mP\#u$

(130) $\qquad F(aq^{*=}c) \to F(aq^*c \lor c = a)$

(137) $\qquad aq^{*=}e \to eqm \to aq^{*=}m$

(140) $\qquad aq^{*=}a$

(144) $\qquad aq^{*=}b \to \forall\mathbf{d}(F\mathbf{d} \to \forall\mathbf{a}(dq\mathbf{a} \to F\mathbf{a})) \to (Fa \to Fb)$

(145) $\qquad 0P^{*=}b \to \neg bP^*b$

21

Constructing Cantorian Counterexamples

Cantor's theorem states that there is no one-to-one correspondence between any set A and the power set $\mathcal{P}A$ of A, i.e., that there is no one-one function mapping any set A onto $\mathcal{P}A$. Cantor's theorem is an immediate consequence of either of two propositions, one, "$\neg g : A \rightarrow_{\text{onto}} \mathcal{P}A$," to the effect that no function maps any set A *onto* $\mathcal{P}A$; the other, "$\neg f : \mathcal{P}A \rightarrow_{1-1} A$," stating that no *one-one* function maps $\mathcal{P}A$ into A. Let us call the former "Not Onto" and the latter "Not 1–1."

Not 1–1 follows directly from Not Onto: Suppose $f : \mathcal{P}A \to A$. For x in A, let $g(x) =$ the unique B such that $f(B) = x$, if there is such a B; and let $g(x) = A$ otherwise. Then $g : A \to \mathcal{P}A$. By Not Onto, some subset B of A is not in the range of g. But then for some $C \neq B$, $f(C) = f(B)$ [otherwise $B = g(f(B))$] and f is thus not one-one.

Not Onto can, of course, be proved by the exceedingly familiar diagonal argument: Suppose $g : A \to \mathcal{P}A$. Let $B = \{x \in A : x \notin g(x)\}$. $B \subseteq A$. But $B = g(x)$ for no x in A, and so g is not onto A. Note that this proof of Not Onto provides an explicit definition, viz.: $\{x \in A : x \notin g(x)\}$, from g, of a subset of A that is not in the range of g.

The derivation we gave of Not 1–1 from Not Onto does not, however, similarly provide an explicit definition from f of a pair B, C, of sets such that $B \neq C$ and $f(B) = f(C)$. If we define g from f as above and take $B = \{x \in A : x \notin g(x)\}$, we may conclude that for some $C \neq B$, $f(C) = f(B)$, but the proof gives us no hint as to the identity of any such C.

There is a familiar direct proof of Not 1–1, again a diagonal argument: Suppose $f : \mathcal{P}A \to A$. Let $D = \{x \in A : \exists E(f(E) = x \land x \notin E)\}$. $D \subseteq A$, so for some $x \in A$, $f(D) = x$. If $x \notin D$, $x \in D$. So $x \in D$ and thus for

This article and the accompanying note by Vann McGee were first published in *The Journal of Philosophical Logic* 26 (1997): 237–239. Both are here reprinted with the kind permission of Kluwer Academic Publishers.

some E, $f(E) = x$ and $x \notin E$, whence $D \neq E$. Since $f(D) = x = f(E)$, f is not one-one. (In a variant argument one considers $\{x \in A : \forall E(f(E) = x \to x \notin E)\}$.)

But again, this direct proof, although it does give an explicit definition of D and show that for *some* E, $D \neq E$ and $f(D) = f(E)$, does not explicitly define any such E. (And the variant argument does no better.)

Thus although we can, given $g : A \to \mathcal{P}A$, explicitly define a counterexample to the statement that g is onto $\mathcal{P}A$, it may well appear that we cannot, analogously given $f : \mathcal{P}A \to A$, explicitly define a counterexample to the statement that f is one-one. Not Onto might seem to be "constructively" provable in a way that Not 1–1 is not.

Not so.

Take $f : \mathcal{P}A \to A$. For any relation r, let $r_x = \{y : yrx \wedge y \neq x\}$. Let us call a relation r *good* iff r is a (reflexive) well-ordering of a subset of A and for every x in the field $\mathcal{F}(r)$ of r, $f(r_x) = x$.

Let R be the union of all good r. If r and r' are good, then one of r and r' is an initial section of the other; therefore R is itself good.

Let $C = \mathcal{F}(R)$. $C \subseteq A$. Let $x = f(C)$, and let $B = R_x$. C, x and B are all explicitly defined from f.

If $x \notin C$, then $R \cup \{(y, x) : y \in C \text{ or } y = x\}$ is good and therefore $x \in C$. Thus $x \in C$.

Since $x \notin \{y : yRx \wedge y \neq x\} = B$, $B \neq C$. Since R is good, $x = f(R_x) = f(B)$. But $x = f(C)$. Thus f is not one-one. So there is a proof that defines a counterexample after all.

We note that since $R_x \subseteq \mathcal{F}(R)$, we have proved a nonobvious strengthening of Non 1–1: If $f : \mathcal{P}A \to A$, then for some B, C, $B \neq C$, $B \subseteq C$ and $f(B) = f(C)$.

Editorial note (by Vann McGee):

This paper was written in the spring of 1996, shortly before the author's death. In another version of the paper, Professor Boolos gave a somewhat less direct proof that is interesting because it helps illuminate the connection between Not 1–1 and the set-theoretic paradoxes. (That there is such a connection is clear; Basic Law V, the source of all Frege's woe, was the denial of an instance of the second-order version of Not 1–1.) Here is a sketch of the alternative proof:

Given $f : \mathcal{P}A \to A$, define a function H from the universe of hereditarily well-founded sets into A by setting $H(x)$ equal to $f(\{H(y) : y \in x\})$. An induction shows that, if f is one-one, H is one-one, so that H embeds the whole universe of hereditarily well-founded pure sets into the set A. But this, as we shall see, is impossible.

One way to see that H cannot be one-one is to observe that, if it were, then applying replacement to H^{-1} would give us a set consisting of all hereditarily well-founded sets, impossible on account of Mirimanoff's paradox.

For present purposes, however, a more useful demonstration is to note that, because $\{H(\alpha) : \alpha$ a (von Neumann) ordinal$\}$ is a set (since it is included in A), it follows by replacement that, if H is one-one, then the image under H^{-1} of $\{H(\alpha) : \alpha$ an ordinal$\}$ is a set. But this gives us a set of all ordinals, impossible on account of Burali–Forti's paradox.

Thus the restriction of H to the ordinals is not one-one, so that there exists an ordinal γ such that, for some $\beta < \gamma$, $H(\beta) = H(\gamma)$. Finding the least such γ and setting $B = \{H(\alpha) : \alpha < \beta\}$ and $C = \{H(\alpha) : \alpha < \gamma\}$ gives us our counterexample (the same one as before) to Not 1–1; $B \neq C$, but $f(B) = f(C)$.

III Various Logical Studies and Lighter Papers

Introduction to Part III

Mathematical Induction

According to the principle of *mathematical induction*, (0) and (1) below imply (2) below, where §*n* indicates the immediate successor of *n*:

(0) $P(0)$ holds.
(1) For every natural number *n*, if $P(n)$ holds then $P(\S n)$ holds.
(2) For every natural number *n*, $P(n)$ holds.

The principle is central in mathematics. To refer back to systems mentioned in the introductory notes to Parts I and II, mathematical induction figures as a central axiom in first-order and second-order Peano arithmetic PA^1 and PA^2; deriving mathematical induction was a central goal in the attempts of Frege and of Russell to provide a logical foundation for mathematics, as it also is a central goal when, following Zermelo, mathematics is developed in a set-theoretic framework like Z or ZFC. And the status of the principle is the central question for two articles in the present part. Article 24 asks how best to justify the principle. Article 22 considers paradoxes that have led some to doubt whether it is justified at all.

Article 24 notes that the principle of mathematical induction can be derived from the *well-ordering* principle, according to which, if there is any natural number *n* such that $P(n)$ fails, then there is a least such. Also, it can be derived from Frege's or Russell's logical definitions of natural number, since these have built in the inductive property as part of what it is to be a natural number. (The idea in these definitions is to define *zero* and *immediate successor* first, and then define a *natural number* as anything that is zero or stands to zero in the ancestral of the immediate succession relation.) Also the principle can be derived from set-theoretic definitions of natural number, because the axioms of set theory include a *well-foundedness* principle for elementhood ∈ that subsumes the well-ordering principle for order < on natural numbers. (The idea in set-theoretic definitions is that a number is identified with the set of its predecessors, so that $0 = \emptyset, 1 = \{0\} = \{\emptyset\}$, and so on; thus the order relation < on natural numbers just *is* elementhood ∈ as it applies to natural numbers.) But in all these cases mathematical induction is being derived from assumptions that themselves are of a more or less transparently inductive character. The question raised by the article is whether it can be derived from assumptions for which this is not so. The answer suggested by the article is that Scott's work, mentioned already in Article 6 of Part I, provides such a derivation.

Article 22 considers the ancient *sorites* paradox, which arises when the principle of mathematical induction is applied to a vague condition $P(n)$. More specifically, there is a vague division of natural numbers into "large" and "small," with "small" meaning something like "so small that n grains of sand don't make a heap" or "so small that n hairs on a man's head won't save him from being bald." The paradox consists simply in the fact that the following, taken together, are inconsistent with mathematical induction:

(0′)   0 is small.
(1′)   For every natural number n, if n is small then $\S n$ is small.
(2′)   Not every natural number n is small.

Or rather, the paradox consists in that fact plus facts about the status of each of (0′), (1′), and (2′) taken separately. (0′) is obviously true. (1′) is widely held to be at least not obviously false. (2′) again is obviously true, since for instance the number that has been called *googolplex* is not small.

Clearly, one can derive from (0′) and (1′) an explicit contradiction to the assertion that googolplex is not small. One can do so in googolplex steps, by first deriving that $1 = \S 0$ is small, then deriving that $2 = \S 1 = \S\S 0$ is small, and so on. What the article shows is that given adequate notational resources, one can arrive at an explicit contradiction very much more speedily than that. And what the article concludes is that because (1′) leads so speedily to an inconsistency with obvious truths, it ought to be regarded as obviously false, contrary to widely held opinion.

By "adequate notational resources" are meant notations for the addition, multiplication, and exponentiation functions, along with the usual recursion equations that define addition as repeated succession, multiplication as repeated addition, and exponentiation as repeated multiplication:

$$
\begin{aligned}
x + 0 &= x \\
x + \S y &= \S(x + y) \\
x \cdot 0 &= 0 \\
x \cdot \S y &= x + x \cdot y \\
x^0 &= 1 \\
x^{\S y} &= x \cdot x^y
\end{aligned}
$$

It is the fact that exponentiation is fast-growing, that it can give large values for small arguments, that underlies the results of the article. The list could in fact be extended to include still faster-growing functions, by defining super-exponentiation as repeated exponentiation, super-duper-exponentiation as repeated super-exponentiation, and so on. If we write $\langle 0 \rangle$ for addition, $\langle 1 \rangle$ for multiplication, $\langle 2 \rangle$ for exponentiation, $\langle 3 \rangle$ for super-exponentiation, we can define the still faster-growing *Ackermann function* α by $\alpha(n) = n\langle n \rangle n$. So we have:

$$
\begin{array}{rcccl}
\alpha(0) & = & 0 + 0 & = & 0 \\
\alpha(1) & = & 1 \cdot 1 & = & = 1 \\
\alpha(2) & = & 2^2 & = & 4 \\
\alpha(3) & = & 3^{3^3} & = & 7,625,597,484,987 \\
& \vdots
\end{array}
$$

and then the fast growth really begins. A number of matters pertaining to such fast-growing functions that are in the background in Article 22 come to the foreground in later articles.[1]

Lengths of Proofs

First some notation. Given a formal system S and a formula φ, let $\lambda(S, \varphi)$ be the length of the *shortest* proof of φ in S, if φ is provable in S, and ∞ if φ is not provable in S. Now consider two formal systems \$ and $\mathcal{L}$, where the latter is a restriction or sub-system of the former—or to put the matter positively, the former is an extension or super-system of the latter. That is virtually the relationship between ZFC and Z, or Z and PA^2, or PA^2 and PA^1. It may happen that though there is a proof of φ in both systems, $\lambda(\mathcal{L}, \varphi)$ is dramatically greater than $\lambda(\$, \varphi)$ so that there is a dramatic slow-down in the proof of φ when one passes from \$ to $\mathcal{L}$—or to put the matter positively, $\lambda(\$, \varphi)$ is dramatically less than $\lambda(\mathcal{L}, \varphi)$, so that there is a dramatic speed-up in the proof of φ when one passes from $\mathcal{L}$ to \$.

Of course, the extreme case would be when $\lambda(\$, \varphi)$ is finite while $\lambda(\mathcal{L}, \varphi)$ is infinite; in other words, φ is provable in \$ but not in $\mathcal{L}$. Specifically, that happens with the sequence ZFC, Z, PA^2, PA^1: This is a sequence of systems of strictly decreasing strength. Generally, the methods used by mathematical logicians to prove slow-down/speed-up results are refinements of those used to prove unprovability/provability results. And the prototype for the latter are the methods used by Gödel to prove his celebrated incompleteness theorems, of which more later. It is hardly an exaggeration to describe the whole subject of lengths of proofs as a spin-off from Gödel's theorems. All the remaining articles in this part are directly or indirectly linked to Gödel's theorems. Two are so linked by being about lengths of proofs.

---

[1] To mention some other work of the author not unrelated to the material in this section, (Boolos, 1974a) and (Boolos, 1975b) are more technical, mathematical papers on formal arithmetic.

Articles 23 and 25 have the following format in common. In the background there is a result, known in a general way to specialists, about lengths of proofs. The first aim of the article is to present an example, accessible to non-specialists, where the short proof in one system is small enough that it can actually be written down in a few pages, while the long proof in another system is literally astronomically long. The second aim is to raise, if not to settle, the question of the philosophical and/or pedagogical bearing of such examples. In Article 23, the example is based on the exponential function, in Article 25, on the Ackermann function.

Article 23 concerns Gerhard Gentzen's *cut-elimination* theorem. This connects two formulations of first-order logic, an ostensibly stronger one with a certain rule "cut" and an ostensibly weaker one without "cut." The theorem says that any derivation in the ostensibly stronger system can be transformed into a derivation in the ostensibly weaker system, which is therefore not really weaker after all. Derivations in the system without cut are in various ways much more "direct" than derivations in the system with cut, and that is what makes the cut-elimination theorem a useful lemma in proving many substantial theorems in mathematical logic. But derivations in the system without cut can also be very much longer than derivations in the system with cut, and the first aim of the article is to provide an accessible example. As explained in the article, the relationship between the systems without and with cut is paralleled by the relationship between the tree method, used in some introductory logic texts, in its versions without and with a certain rule of "splitting."

Article 25 concerns the relationship between first-order and second-order logic. Anything that is expressible in first-order terms and is a theorem of pure second-order logic is a theorem of pure first-order logic, since if it is theorem in pure second-order logic, then it is valid, and anything expressible in first-order terms that is valid is a theorem of first-order logic. Nonetheless, the derivation in first-order logic can be very much longer than the derivation in second-order logic, and the first aim of the article is to provide an accessible example.

A pedagogical question is raised by these articles. The usual practice in introductory courses is to teach only first-order and not second-order logic (and where the tree method is used, only the version without splitting). Is this an appropriate choice, given that practically speaking second-order methods can lead to speedier results than can first-order methods (let alone the tree method without splitting)? A philosophical question is also raised. Many have taken our grasp of the meaning of the logical particles in natural language to be mentally represented by a grasp of the rules for something like a system of first-order logic (and some have taken it to be representable more specifically by a grasp of the rules for something like a system without cut). Is this a plausible representation, given that we can grasp the validity of arguments like those in the examples, where a first-order derivation (to say nothing of one without cut) would be astronomically long?[2]

The Incompleteness Theorems and Semantic Paradoxes

While nominalists and finitists may have hailed Russell's and other set-theoretic paradoxes as exposing the evils of "abstract entities" and "completed infinities," Russell himself thought that what was responsible for them was a kind of vicious circularity that can be found at work even in cases where there is no abstraction or infinity involved, notably in the so-called *semantic* paradoxes. Among these the one with the most direct analogy to Russell's paradox is the *heterological* or *Grelling* paradox. An adjective is *autological* if it is true of itself, like "short" or "polysyllabic" or "English," and *heterological* if it is untrue of itself, like "long" or "monosyllabic" or "French." The paradox results when we ask whether *heterological* is heterological. The answer seems to be that it is if and only if it isn't. This is recognizably the analogue for adjectives of the Russell paradox for sets.

But it is also is recognizably an analogue for adjectives of the ancient *liar* or *Epimenides paradox* for sentences. This paradox arises when we ask whether the sentence, "This very sentence is false," is true. The answer seems to be that it is true if and only if it is false. The liar paradox occurs in various disguises, for instance the *preface paradox*. Even if an author is committed to the truth of every sentence in the body of a book, it seems rational to add in the preface an acknowledgment of human fallibility on the order of, "Some sentence in this book is false." But if every *other* sentence in the book happens to be true, this one will have the same status as "This very sentence is false." The same general family as the Grelling and Epimenides paradoxes has other members, including those called the *Berry paradox*, the *Richard paradox,* and so on.

There is a relationship between the liar paradox and Gödel's original proof of his first incompleteness theorem. The hard work in Gödel's proof consists in showing how in a language like that of PA^1, which was designed for talking about numbers and their arithmetic properties, we can in effect

---

[2]Again to mention not unrelated work, another remediable limitation of the tree method as usually presented in textbooks is treated in (Boolos, 1984f). In a larger sense, concern for pedagogy is reflected in the successive editions of his co-authored textbook, *Computability and Logic,* as well as in reviews of popularizations, textbooks, and courseware, (Boolos, 1986), (Boolos, 1990c), and (Boolos, 1990d). Everyone hopes that experimental psychology will eventually produce results applicable to issues in the teaching of logic, as of other subjects; but how far this is from being the case at present is suggested by a note in *Cognition,* (Boolos, 1984d).

talk instead about formulas and their syntactic properties. Gödel assigns to every formula α of the language of PA^1 a code number, and thus since numbers have numerals, a code numeral $\ulcorner\alpha\urcorner$. Gödel also constructs a formula $\pi(x)$ expressing that x is the code number of a formula that is provable in PA^1. For instance, $\neg\pi(\ulcorner\bot\urcorner)$ expresses that the constant falsehood $\bot$ is not provable in PA^1, which is a way of asserting the consistency of PA^1.

The resources available in a language like that of PA^1 for thus talking about itself are not quite as rich as the resources available in a language like English for talking about itself, and this is just as well, in view of how the resources available in English can lead to paradoxes like that of the liar. One cannot in the language of PA^1 produce a formula that says of itself that it is not *true*. But Gödel shows that one can do something *analogous*, producing a formula $\gamma = \gamma(PA^1)$ that says of itself that it is not *provable* in PA^1. It says this in the precise sense that $\gamma \leftrightarrow \neg\pi(\ulcorner\gamma\urcorner)$ is provable in PA^1. What's more, Gödel shows that, if PA^1 is consistent, then γ *isn't* provable in PA^1, so that what γ says of itself is quite true, and γ is an unprovable truth. This is his first incompleteness theorem, slurring over technicalities.

Though $\gamma = \gamma(PA^1)$ is unprovable in PA^1, it is provable in PA^2 or Z or ZFC, and that is how one shows that these theories are strictly stronger than PA^1. But the incompleteness theorem applies to these stronger systems, too, and they have Gödel sentences $\gamma(PA^2), \gamma(Z), \gamma(ZFC)$ of their own. In fact the incompleteness theorem applies to any stronger theory at least as strong as PA^1, provided it is a theory of the usual type, where there is a "mechanical" rule for determining whether a formula counts as an axiom of the theory or not. In the case of PA^1, we can see, or so we think, that the rules only count as axioms things that are true, and so, since truth implies consistency, we conclude that PA^1 is consistent and $\gamma(PA^1)$ true.

Article 26 presents a new proof of the first incompleteness theorem roughly satisfying the proportion:

new proof : original proof :: Berry paradox : Epimenides paradox

The afterword underscores that what is new about the new proof is its being based on a new principle or new paradox, not its being any shorter. (The article is indeed shorter than Gödel's classical paper, but that is because the hard work of setting up the code numbering and apparatus related thereto, which is the same in both proofs, is omitted.)

Article 27 originally appeared as one of several invited commentaries on a position paper by Roger Penrose on the implications of Gödel's theorem, related to his best-seller, *The Emperor's New Mind* (Penrose, 1989). Penrose is only the latest of many who have attempted to draw from the theorem implications for the philosophy of mind, and in particular, the implication that no "mechanical" procedure could match our abilities to "see" the truth of mathematical assertions. The thought is that if a mechanical procedure generates *only* mathematical assertions that we are able to see to be true, then it cannot generate *all* such mathematical assertions. For consider the theory T whose axioms are the mathematical assertions generated by the mechanical procedure in question. The thought is that we can see that $\gamma(T)$ is true, much as we can see that $\gamma(PA^1)$ is true, whereas T itself cannot prove that $\gamma(T)$ is true, by Gödel's theorem.

Logicians are virtually unanimous in their judgment that there is a fallacy in this line of thought, but it is at least a point in favor of Penrose that they are not unanimous in their diagnoses as to just *where* the fallacy lies. A first line, taken by many, would say the mistake lies in overlooking the possibility that it might in actual fact *be* the case that the procedure generates only mathematical assertions we can see to be true, without our commanding a clear enough view of what the procedure generates to enable us to *see* that this is the case. A second line, taken by others, would say that even if we do see that the procedure generates only mathematical assertions we think we see are true, it might be rational to acknowledge human fallibility by refraining from concluding that the procedure generates only mathematical assertions that are in actual fact true. (This latter line recalls the paradox of the preface.) Gödel's thoughts about this sort of question are indicated in Article 7 of Part I (and in the lecture to which that article is an introduction). The author's thoughts about it are indicated in the article under discussion. *Very* roughly, Gödel's view is more like the first line of response just indicated, and the author's more like the second line, as might have been expected from his expressions of partial skepticism about set theory in some of the articles in Part I.

Article 28 concerns the most systematic device available to a natural language like English for speaking about itself: quotation. An elegant version of the liar paradox due to W. V. Quine illustrates the use of this device:

'yields a falsehood when appended to its own quotation'
yields a falsehood when appended to its own quotation

Logicians from Frege to Quine have complained that the conventional rules for use of quotation marks are somewhat illogical and subject to ambiguity, and have proposed reformed rules. The starting point of the article is the observation by one of the author's students that the reforms proposed so far do not suffice to preclude all ambiguity. Michael Ernst's paradox is that the following:

'b' appended to 'a'

ambiguously denotes either of the following two expressions:

ab
b' appended to 'a

After explaining just what is desired—an unambiguous rule of quotation for a language whose expressions are linear sequences of symbols from a finite alphabet—the article presents a solution.

Article 29 concerns a spin-off from Raymond Smullyan's popularization of Gödel theory by way of logical puzzles. Smullyan has us imagine an island inhabited by just two types of people, *knights,* who always speak truly, and *knaves* who always speak falsely. Some of the knights have official certificates of knighthood, while others don't. Suppose now an islander says, "I am uncertified." What can we conclude? A knave wouldn't make this statement, because coming from a knave it would be true, while a certified knight wouldn't make this statement, because coming from a certified knight it would be false. So the speaker must be an uncertified knight. The analogy with the Gödel sentence, which says it is unprovable, and is an unprovable truth, should be plain. But the puzzles about knights and knaves have a life of their own, and Smullyan complicates some of them by supposing there also exist normals, who may speak truly or falsely at will, or that the inhabitants may take yes–no questions in English but answer them in their own tongue, of which all that is known is that either "bal" means "yes" and "da" means "no," or the reverse. The article takes up the ultimate complication along such lines, and presents a solution.

Article 30 concerns Gödel's *second* incompleteness theorem for PA^1. The resources available in a language like that of PA^1 for talking about itself are rich enough that the result of the first incompleteness theorem, in the hypothetical form of the assertion that if PA^1 is consistent, then γ isn't provable, or $\neg\pi(\ulcorner\bot\urcorner) \to \neg\pi(\ulcorner\gamma\urcorner)$, or equivalently $\neg\pi(\ulcorner\bot\urcorner) \to \gamma$, is provable in PA^1. The first incompleteness theorem says that if PA^1 is indeed consistent, then the consequent of this conditional isn't provable in PA^1; it follows that the antecedent can't be provable in PA^1, which is to say that the consistency of PA^1 isn't provable in PA^1. That is the second incompleteness theorem, slurring over technicalities. But there is no need to say more by way of explanation of the second incompleteness theorem here, since explaining it is precisely the aim of the article.

It may just be mentioned that the theorem dashed the hopes of David Hilbert, who had hoped to silence finitist criticism of classical mathematics by producing a proof using only finitist mathematics (something weaker than PA^1) to prove the consistency of classical mathematics (something stronger than PA^1). It means that the most one can hope for in general is to produce is a finitist proof that *if* one system is consistent, *then* so is another: a *relative* consistency proof.

The article consists of two parts: first, a monosyllabic explanation of the statement of the theorem as promised by the article's title; second, a polysyllabic explanation of the proof thereof, added at the request of the editor of *Mind,* where the article first appeared. The *cognoscenti,* familiar with other expositions of the incompleteness theorems, will note what is the distinguishing feature of the the account in this article. What is known as *modal* logic adds to classical logic one-place connectives $\Box$ and $\diamond$ intended to represent necessity and possibility in some sense. In discussing Gödel's theorems about what is or isn't a theorem, his proofs about what is or isn't provable, it seems a natural idea to use the symbol $\Box$ for provability and the symbol $\diamond$ (definable as $\neg\Box\neg$) for consistency (definable as undisprovability). The distinctive feature of the account in this article is the use of modal symbolism to simplify the exposition.

The serious, sustained application of the methods of modal logic to the study of the proof theory in general and the incompleteness phenomenon in particular is known as *provability logic.* A brief and inevitably incomplete description of this field, including the author's work in it, is contained in an afterword at the end of this volume.[3]

---

[3] Again to mention not unrelated work, (Boolos, 1988a) on alphabetical order relates to that topic somewhat as Article 28 relates to quotation. Article (Boolos, 1980e) is a review of (Smullyan, 1978) emphasizing the intrinsic interest of its puzzles even apart from their use in popularizing Gödel's theorems.

22

Zooming Down the Slippery Slope

The principle of mathematical induction asserts that a predicate applies to every natural number if it applies both to zero and to the successor of every natural number to which it applies. We use "$'$" and "$\#$" to denote, respectively, the successor function and zero—our reason for using "$\#$" rather than "0" will become apparent later on—and use "x" as a variable ranging over the natural numbers. We call the statement expressed by the result of inserting a predicate into the induction schema:

$$\ldots \# \ldots \wedge \forall x(\ldots x \ldots \rightarrow \ldots x' \ldots) \rightarrow \forall x \ldots x \ldots$$

induction with respect to that predicate.

Let us call a natural number *small* if it is below the vicinity of one billion ($= 1000000000$). Because of the presence of the imprecise term "vicinity" in its definition, "is small" is a vague predicate. Despite the vagueness of "is small," it is nevertheless clear that 17 and 0 are small and that 2^{30}, $= 1073741824$, and 1000000000 are not. As I understand "vicinity," one million is not in the vicinity of one billion; thus I suppose that one million is small.

Are vague predicates of numbers, like "is small," counterexamples to the principle of mathematical induction? Call the statement that for every natural number n, if n is small, $n + 1$ is small, the *induction premiss*. Since zero is small but one billion is not, "is small" is a counterexample to the principle of mathematical induction if and only if the induction premiss is true. But is the premiss true? That is, is the successor of a small number always a small number?

Reprinted with kind permission of the editors from *Noûs* 25 (1991): 695–706.

I am grateful to many people for helpful conversation about the content of this paper, which dates from 1981, but especially so to my colleagues Ned Block and the late James Thomson, and to Paul Benacerraf, Philip Kitcher, Harry Lewis, and Rohit Parikh.

I think that on reflection most would agree that one cannot regard the induction premiss as true and would give as a ground for thinking it false some such reason as: the premiss has a false consequence, e.g. the statement that if zero is small, so is one billion. I do not here want to take up the interesting questions why one might be tempted to regard the induction premiss as true or what true proposition or propositions there might be which we are confusing with the induction premiss; instead I want to look at the character of the reasons we have for thinking it false.

An important aspect of the induction premiss is that its falsity is not apparent. The argument:

Zero is small.
If zero is small, then one billion is small.
Therefore, one billion is small.

presents no paradox; it is an uninteresting (valid) argument with an obviously false second premiss.

Unlike the second premiss of this argument, the induction premiss does not seem to be *obviously* false. It *is* false—it has to be false, since in conjunction with a truth it implies a falsehood. The trouble is that it looks true; for after all, isn't a number merely one less than a number in the vicinity of a billion itself in the vicinity of a billion? A large part of what makes the paradoxes of vagueness paradoxes is that although we know that the troublesome premisses have false consequences, those premisses look true anyway, and the conclusion that they are false is one we feel *constrained* to accept. Can we dispel the sense of constraint? Is there a way to change the way the induction premiss strikes us, perhaps by showing its falsity in a way that makes it look *evidently* false?

Using "Sx" to abbreviate "x is small," we may symbolize the induction premiss: $\forall x(Sx \rightarrow Sx')$. And letting $[i]$ be the result of writing down i consecutive occurrences of "$'$", we may assert that $S\#$ is true (since zero is small) and $S\#[1000000000]$ is false (since one billion is not small).

Why do we think that $\forall x(Sx \rightarrow Sx')$ is false? We may reason: if it is true, then so are all members of the set $\{(S\#[i] \rightarrow S\#[i+1]) : i < 1000000000\}$, which contains one billion U(niversal) I(nstantiation) instances of $\forall x(Sx \rightarrow Sx')$; therefore so is the "spanning" conditional $(S\# \rightarrow S\#[1000000000])$; but this is absurd, since its antecedent is true and its consequent is false.

Let us look at the middle step in this reasoning according to which if for all $i < 1000000000$, $(S\#[i] \rightarrow S\#[i+1])$ is true, then so is $(S\# \rightarrow S\#[1000000000])$. Why do we find this step convincing?

One possible answer is that we can *prove* that if all those instances are true, then so is the spanning conditional. But it would appear that in order for a proof to have any chance of actually showing this, it must also show that for any n (and not just $n = 1000000000$), if all members of $\{(S\#[i] \to S\#[i+1]) : i < n\}$ are true, i.e. if for all $i < n$, $i + 1$ is small if i is, then $(S\# \to S\#[n])$ is true, i.e. n is small if 0 is.

And this stronger statement *can* be proved, by the following inductive proof: if $n = 0$, it is trivial that if for all $i < n$, $i + 1$ is small if i is, then n is small if 0 is. Moreover, the statement holds for $n + 1$ if it holds for n, for then if for all $i < n + 1$, $i + 1$ is small if i is, for all $i < n$, $i + 1$ is small if i is and thus n is small if 0 is; and furthermore, $n + 1$ is small if n is (since $n < n + 1$), and therefore $n + 1$ is small if 0 is. By induction, for every n, if for all $i < n$, $i + 1$ is small if i is, then n is small if 0 is.

The trouble with this line of argument against the induction premiss is that the appeal to induction made in the argument enjoys no dialectical advantage over an "appeal" to the negation of the induction premiss. The proof shows the truth of $\forall x(\forall y < x(Sy \to Sy') \to (S\# \to Sx))$ by an appeal to induction with respect to the predicate $(\forall y < x(Sy \to Sy') \to (S\# \to Sx))$, a formula compounded from the vague S. The statement we are trying to prove is itself obtained by inserting S into a slightly less familiar version of the induction schema:

$$\forall x(\forall y < x(\ldots y \ldots \to \ldots y' \ldots) \to (\ldots \# \ldots \to \ldots x \ldots)).$$

It is hard to see why we would be any more justified in rejecting the induction premiss by appeal to induction with respect to the compound predicate than we would by appeal to the result of inserting the simpler predicate "is small" into the slightly less familiar version of the induction schema. And then of course it is equally hard to see why we wouldn't be just as justified in rejecting the induction premiss more simply by appealing to induction with respect to "is small," or still *more* simply by citing the negation of the induction premiss.

Another possible answer is that we know that the spanning conditional is a *logical consequence* of that billion-membered set and that it follows that the spanning conditional is true if all its members are true. We would thus seemed to be equipped with a perfectly adequate reason for regarding the induction premiss as false, viz. that logical consequences of truths are truths.

A sentence is a logical consequence of a set of sentences if it is true in all interpretations in which all members of the set are true. And we can of course prove in set theory together with the usual familiar definitions that the spanning conditional is true in every interpretation in which all instances of the induction premiss are true. In order to reach our desired conclusion we need only show that there is an interpretation in which the domain is the set N of natural numbers, "$\#$" and "$'$" denote zero and successor, and S has as its extension the set of all small numbers.

It is a theorem of set theory (plus definitions) that for any set M of natural numbers there is an interpretation whose domain is N, in which "#" and " ′ " denote zero and successor and S has M as its extension. We thus need only see that there is a set containing all and only the small natural numbers.

Of course if there is such a set M, then the notions of *consequence* and *interpretation* need not have been invoked in the first place, for then $N - M$ also exists and is nonempty, since 1000000000 is a member. Like any nonempty set of natural numbers, it has a least member, which since zero is small, is not zero and is therefore a successor of which the predecessor, a small number whose successor is not small, is a counterexample to the induction premiss.

Well, *is* there such a set? Isn't $\{n \in N : n \text{ is small}\}$ such a set? Yes, certainly, if it exists. But how do we know that it exists?

It might be answered that if "<" and "10^9" are eliminated in the standard way, then $\forall x(Sx \to x < 10^9) \to \exists y \forall x(x \in y \leftrightarrow Sx)$ is a logical consequence of the axioms of set theory, ZFC. This conditional says that if everything small is less than a billion (as is the case), then there is a set containing the small numbers.

The hitch here is that in order to establish the truth of the conditional (assuming the truth of the axioms of set theory) without appeal either to the contention that there is a set of small numbers or to the claim that there is an interpretation in which S has as its extension the set of all small numbers, it would appear that we have to derive this conditional from the axioms of set theory. But unless there is some way of doing this other than by formalizing the argument that first comes to mind—if everything small is less than a billion, then "Sx" is coextensive with one of the 2^n, where $n = 10^9$, disjunctions of different formulae $x = \mathbf{i}$, $\mathbf{i}$ a set-theoretic term for some number less than a billion ... —we are far better off attempting to use UI and modus ponens a billion times to show the falsity of the induction premiss.

Our difficulty is that since there is no obviously correct way to define "is small" in the language of set theory, there is no obviously correct extension of set theory by a definition, in which we can appeal to an Aussonderungsaxiom (an axiom of separation) to show the existence of $\{n \in N : n \text{ is small}\}$.

In any case it seems that we regard the induction premiss as false either because we accept set-theoretical axioms governing the vague predicate "is small" or because we simply accept certain instances (or familiar variants) obtained by inserting "is small" or its compounds therein.[1]

---

[1] An alternative suggestion, made to me by Philip Kitcher, is that we regard the induction premiss as false because we take ourselves to have a procedure which when applied to any number generates an argument that shows the number to be small if zero

Is there, however, a reason of a different character for rejecting the premiss, one which rests neither on set-theoretical principles involving "is small" nor on a tacit or explicit appeal to induction with respect to predicates containing "is small"? It would be nice to have a reason that might actually be used to confute a skeptic who claimed, scampishly, to believe in the induction premiss as well as in all of mathematics and logic.

I think there is. I've asked a number of people the following question: Take a perfectly standard system of natural deduction, say the one given in (Mates, 1972).[2] Offhand and trying as best as you can to ignore the fact that I am asking you this question, how many lines would you say the shortest deduction in Mates' system of $S\#[1000000]$ from $S\#$ and $\forall x(Sx \to Sx')$ contains? I've usually gotten answers on the order of a million; people have occasionally guessed two million, perhaps forgetting that T (truth-functional inference) is a rule of the usual standard systems, like Mates', and that a million conditionals, so arranged that the consequent of each but the last is the antecedent of the next, together with the antecedent of the first, imply the consequent of the last by one single application of T. Thus after a million or so instances of UI, one grand application of T delivers the conclusion.

Actually, the correct answer is less than 70, as we'll see in a moment. The fact that the answer most often given is far above the correct one is explained by a failure to notice a certain "explosiveness" of sentences with the form of the induction premiss which, when it is seen in action, makes the premiss appear to be the falsehood we are convinced it must

and successors of small numbers are small. The argument yielded by the procedure may be diagrammed:

$$\forall x(Sx \to Sx')$$
$$S\#$$
$$(S\# \to S\#')$$
$$S\#'$$
$$\vdots$$
$$S\#[n].$$

But what entitles us to suppose that we do have a procedure that works for *any* n? Mathematical induction, applied to a precise predicate, can be used to show that there is an argument (i.e. a sequence of sentences) of the form described. But why do we think that all subconclusions of the argument are *true* if its premisses are? Here again, we make appeal to induction with respect to a predicate compounded from the vague "is small," an appeal that can be justified by an extension of the Aussonderungsschema, or, perhaps, we just take it for granted that we do have such an argument-generating procedure.

[2]In what follows I have tried to adhere to Mates' syntactical conventions; but I have used "$\forall x$," "$\exists x$," "$\neg$," and "$\wedge$" instead of "(x)," "$(\exists x)$," "$-$," and "$\&$," and also used "$\#$," "$'$," "0," "1," and "$+$" as operation symbols. Mates calles UI US, by the way.

be. *Just as learning of a hitherto unknown false consequence may make us change our opinion concerning the truth-value of a statement or coming to understand the proof of a surprising theorem may render the theorem less surprising, seeing that an evident falsehood, known to be a consequence of a statement, is a less remote consequence than we had suspected may change our minds about the obviousness of the falsity of the statement.* The speed with which false consequences can be derived from the induction premiss is a remarkable feature of it, appreciation of which may enable us to regard the premiss as patently false and to dispel any sense we may have that we are *compelled* to consider it false.

Here's the short (Mates-style) derivation of $S\#[1000000]$ from $S\#$ and $\forall x(Sx \to Sx')$:

$\{1\}$	(1)	$\forall x(Sx \to Sx')$	P
$\{1\}$	(2)	$(Sa \to Sa')$	1 UI
$\{1\}$	(3)	$(Sa' \to Sa'')$	1 UI
$\{1\}$	(4)	$(Sa \to Sa'')$	2, 3 T
$\{1\}$	(5)	$\forall x(Sx \to Sx'')$	4 UG
$\{1\}$	(6)	$(Sa'' \to Sa[4])$	5 UI
$\{1\}$	(7)	$(Sa \to Sa[4])$	4, 6 T
$\{1\}$	(8)	$\forall x(Sx \to Sx[4])$	7 UG
$\{1\}$	(9)	$(Sa[4] \to Sa[8])$	8 UI
$\{1\}$	(10)	$(Sa \to Sa[8])$	7, 9 T
$\{1\}$	(11)	$\forall x(Sx \to Sx[8])$	10 UG

$$\vdots$$

$\{1\}$	(57)	$(Sa[262144] \to Sa[524288])$	56 UI
$\{1\}$	(58)	$(Sa \to Sa[524288])$	55, 57 T
$\{1\}$	(59)	$\forall x(Sx \to Sx[524288])$	58 UG
$\{1\}$	(60)	$(S\# \to S\#[524288])$	59 UI
$\{1\}$	(61)	$(S\#[524288] \to S\#[786432])$	56 UI
$\{1\}$	(62)	$(S\#[786432] \to S\#[917504])$	53 UI
$\{1\}$	(63)	$(S\#[917504] \to S\#[983040])$	50 UI
$\{1\}$	(64)	$(S\#[983040] \to S\#[999424])$	44 UI
$\{1\}$	(65)	$(S\#[999424] \to S\#[999936])$	29 UI
$\{1\}$	(66)	$(S\#[999936] \to S\#[1000000])$	20 UI
$\{2\}$	(67)	$S\#$	P
$\{1, 2\}$	(68)	$S\#[1000000]$	60–67 T

There is a similar derivation of $S\#[1073741824]$ from $S\#$ and $\forall x(Sx \to Sx')$ that takes 95 lines in Mates' system. In general, if $n < 2^k$, there is a derivation of $S\#[n]$ from $S\#$ and $\forall x(Sx \to Sx')$ in Mates' system that contains $\leq 4k + 1$ lines. And the compression is not much altered if T is so restricted that no application may contain more than two premisses; the bound $4k + 1$ need not then be raised beyond $5k$.

We are thus not in the predicament of *having* to make an appeal to induction with respect to a vague predicate or to a new set-theoretical principle in order to justify rejecting the induction premiss. Paralleling the derivation in Mates' system, we can construct a shortish (but admittedly tiresome) argument beginning: "Suppose that for every natural number n, if n is small, $n + 1$ is small. Let n be an arbitrary number. If n is small, $n + 1$ is small, and if $n + 1$ is small, $n + 2$ is small. Thus if n is small, $n + 2$ is small. But n was arbitrary. So for every natural number n, if n is small, $n + 2$ is small. Thus if $n + 2$ is small, $n + 4$ is small ..." and concluding " ... So for every natural number n, if n is small, $n + 1073741824$ is small. This is absurd, as 0 is small and 1073741824 is not."

A pair of related thoughts may occur to one in connection with this argument and its formalization in Mates' system: the derivation cannot actually be written out, for its last line alone contains over a billion characters; one can of course describe that 95-line derivation with complete precision, but in doing so one will *employ* decimal (or perhaps binary) notation. Might there be some illicit use of induction somewhere in the offing, perhaps in the use of decimal notation to describe the derivation? Also, what about the use of arithmetical principles in the informal argument? We do make use of facts from arithmetic when we argue: "So if n is small, then $n + 8192$ is small; and if $n + 8192$ is small, $(n + 8192) + 8192$ is small. So (since $(n + 8192) + 8192 = n + 16384$) if n is small, $n + 16384$ is small. But n was arbitrary ..." What about that appeal to the fact that $(n + 8192) + 8192 = n + 16384$?

In reply, we can and should admit that the informal argument does make tacit appeal to a small number (c. 30) of arithmetical facts, of the sort indicated. But we can ignore skeptical doubts (if that's what they are) about the truth of these facts. Our skeptic was supposed to accept all of mathematics, and we need not worry about convincing a *nut*, someone prepared to deny plain arithmetical facts, that the induction premiss is false. It remains the case that there is a shortish argument, which appeals to certain arithmetical truths (which can be established without appeal to principles governing vague predicates), that shows that if $n + 1$ is small whenever n is, then $n + 2^{30}$ is small whenever n is. *Some conclusions are not as remote as they appear.*

But an interesting question about feasibility has arisen. (The classical discussion of feasibility in logic is (Parikh, 1971).) Although we cannot actually write down the derivation of $S\#[2^{30}]$ from $S\#$ and $\forall x(Sx \rightarrow Sx')$, could we not—perhaps by using binary or some other notation—express the proposition that 2^{30} is small by some other sentence of which we could actually produce a derivation from $S\#$ and $\forall x(Sx \rightarrow Sx')$ in Mates' system? In fact, binary notation is almost perfectly suited to our purposes when viewed from a reverse Polish perspective: take the digits "0" and "1" to be *postposed unary function signs*, denoting the functions $n \mapsto 2n$ and $n \mapsto 2n+1$ respectively. Then the string that results when a "$\#$" is attached immediately to the left of an ordinary binary numeral is a term that denotes the number represented in binary by that numeral. For example, "1101" denotes 13 in binary and "$\#1101$" denotes $(((0 \times 2)+1) \times 2+1) \times 2) \times 2+1$, which $= 13$. (The same style of parsing also works for the decimal system, of course. Take "0" to denote the function $n \mapsto 10n + 0$, "1," $n \mapsto 10n + 1$, ..., and "9," $n \mapsto 10n + 9$. Then, e.g. "$\#1492$" denotes $((((0 \times 10) + 1) \times 10 + 4) \times 10 + 9) \times 10 + 2 = 1492$, "$\#30$" denotes $(0 \times 10 + 3) \times 10 + 0 = 30$, and "$\#0$" denotes $0 \times 10 + 0 = 0$.)

Our question is then: can we not somehow derive

$$S\#100000000000000000000000000000000$$

from $S\#$ and $\forall x(Sx \rightarrow Sx')$ in Mates' system via a derivation that a sufficiently motivated human being could actually write down?

In order for a "yes" answer even to be conceivable, we will obviously need to add to $S\#$ and $\forall x(Sx \rightarrow Sx')$ some "bridge" principles or axioms involving $\#$ and $'$ on the one hand and 0 and 1 on the other. To be interesting, our bridge principles should be as weak as possible: we don't want to take as an axiom the conditional with antecedent $(S\# \wedge \forall x(Sx \rightarrow Sx'))$ and consequent the sentence above. We don't want instances of induction. We want, in fact, only principles that could (reasonably) be called "meaning postulates" for our version of binary notation and which have as little mathematical content as seems compatible with their possibly implying the conclusion: we don't want to assume the commutativity or associativity of addition, for example.

There is an obvious, natural set of non-cheating principles to take:

(1) $\forall x\, x0 = xx+$
(2) $\forall x\, x1 = xx+'$
(3) $\forall x\, x\#+ = x$
(4) $\forall x\forall y\, xy'+ = xy+'$

(We have postposed the addition sign $+$ in order to insure unique readability. Had we not done so, $+\#'\#0$ would have been well-formed; but the denotation of this term is ambiguous as between 1 and 2.) We thus define the functions denoted by "0" and "1" in our "logicized" version of binary from the addition and successor functions in the obvious way; addition is then linked with successor and zero via the familiar recursion axioms.

So our new question is: can we actually produce a derivation of the above sentence expressing the smallness of 2^{30} from (1)–(4), (5) $S\#$, and (6) $\forall x(Sx \rightarrow Sx')$?

Believing that the notational compression effected by binary would probably have to be compensated for by an increase in the size of proofs utilizing this notation, I was surprised to find out that there *is* a humanly producible derivation, one containing, according to my count, fewer than 3100 characters (not counting the characters in the premiss-set designations, the line numbers, or the annotations). The derivation can be found by refining the one sketched below for $n = 2^{30}$. The thirty or so arithmetical principles to which we appealed in our informal argument from the induction premiss to the spanning conditional are, similarly, all feasibly derivable from "axioms." Indeed, there is a single feasible derivation that contains all the principles as subconclusions.

From our point of view, 2^{30} is a rather special number, as its binary numeral has a form that is extremely simple to describe: one 1 followed by thirty 0s. Because of this simple form it might be suspected that the proof that 2^{30} is small (from our bridge principles, the premiss that zero is small, and the induction premiss) is significantly smaller than any that could be provided for almost any other number in the vicinity of one billion. Let us conclude by looking at the question to what extent this is so. If **n** is the "canonical" term in our version of binary for some arbitrary number n in the vicinity of a billion is there always a derivation of $S\mathbf{n}$ from (1)–(6) that we could actually produce, if we had to?

The answer, it turns out, is yes. In fact, if n is a natural number whose binary numeral is $b_1 \ldots b_k$ (each b_i is either 0 or 1, with $b_1 = 1$ if $n \geq 1$), so that $2^{k-1} \leq n < 2^k$ if $n \geq 1$, **n** *is our canonical term* $\#b_1 \ldots b_k$ for n, and $f(j) = (j + 64)(j + 28)$, then there is a derivation of $S\mathbf{n}$ from (1)–(6) containing fewer than $f(k)$ characters.

Here is a sketch of the derivation. After the premisses (1)–(6) come proofs of $\forall x(Sx \rightarrow Sx\#+)$ and $\forall x\forall y\, xy\# + + = xy + \#+$. Letting $C(z)$ be the formula

$$\exists w\big(w = z \wedge (Sw \wedge (\forall x(Sx \rightarrow Sxw+) \wedge \forall x\forall y(xyw + + = xy + w+)))\big),$$

we quickly infer $C(\#)$. (We use $C(z)$ instead of its simpler and shorter equivalent

$$Sz \wedge (\forall x(Sx \rightarrow Sxz+) \wedge \forall x\forall y(xyz + + = xy + zy+))$$

because $C(z)$ contains only one free occurrence of the variable z instead of four, making for shorter instances.) We then establish $\forall z(C(z) \rightarrow C(z0))$ and $\forall z(C(z) \rightarrow C(z1))$. (This is the crucial step of the proof. We cannot, apparently, prove either $\forall z(Sz \to Szz+)$ or

$$\forall z((Sz \wedge \forall x(Sx \to Sxz+)) \to Szz+ \wedge \forall x(Sx \to Sxzz++)).$$

But since

$$\forall z(\forall x \forall y\, xyz++ = xy+z+ \to \forall x \forall y\, xyzz+++ = xy+zz++)$$

is a *logical* truth, we can prove $\forall z(C(z) \to C(zz+))$. This much of the derivation takes somewhat under 1700 characters. Next come k lines, growing to $86 + 2k$ characters long, all obtained by UI from $\forall z(C(z) \to C(z0))$ and $\forall z(C(z) \to C(z1))$, of which the $i^{\text{th}}(1 \leq i \leq k)$ is

$$(C(\#b_1 \ldots b_{i-1}) \to C(\#b_1 \ldots b_i)).$$

$C(\#)$ is the antecedent of the first of these conditionals; $C(\mathbf{n})$, the consequent of the kth. Thus these k conditionals, together with $C(\#)$, truthfunctionally imply $C(\mathbf{n})$. We conclude the derivation by inferring $C(\mathbf{n})$ by T and then quickly deriving $S\mathbf{n}$ from $C(\mathbf{n})$.

2^{40} is greater than one trillion, and $f(40) = 7072$; thus it would be perfectly feasible if rather boring for someone to write down a derivation from (1)–(6) of $S\mathbf{n}$, where n is any number less than a trillion. However, the suspicion that in general $S\mathbf{n}$ has a significantly smaller derivation if n is a power of 2 may well be correct: as we have seen, if n is an arbitrary number $< 2^k$, then there is a derivation containing no more than $c_1 k^2$ characters, for some constant c_1. If k^2 cannot be reduced to $k \cdot \log(k)$, then the suspicion is correct, for when $n = 2^k$, by regarding $\forall z(C(z) \to C(z0))$ in the same light as $\forall x(Sx \to Sx')$ and employing the same compression technique used in the 68-line derivation of $S\#[1000000]$, we can obtain a derivation of $S\mathbf{n}$ from (1)–(6) containing no more than $c_2 k \cdot \log(k)$ characters, for some constant c_2. Further, the analogy between $\forall x(Sx \to Sx')$ and $\forall z(C(z) \to C(z0))$ enables us to see that for some constant c_3, if n is a power 2^k of 2, where k is itself a power of 2, then there is a derivation of $S\mathbf{n}$ from (1)–(6) containing no more than $c_3 k$ characters.

It may be worth while to point out that on any reasonable understanding of *carrying out*, our "exponentially compressed" derivations cannot be carried out in the usual formulations of the tree (tableau) method, as formulated in e.g. (Jeffrey, 1981), (Smullyan, 1968), or (Hodges, 1977). The method is of course *complete*, as are the standard methods of natural deduction, but to call a (sound) method complete is merely to say that it can generate some demonstration of each valid statement; it is not to say that it can generate (a replica of) every demonstration that can be generated by any other sound method.

In order for our compressed derivations to be replicated in a tree-style method, we must adjoin to a standard formulation of the method some such rule as the version of modus ponens that (Jeffrey, 1981) calls XM, for "excluded middle": split any open branch in two and add any sentence to one of the two new branches and its negation to the other. Pictorially:

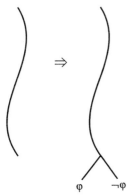

$$\varphi \qquad \neg\varphi$$

In the absence of XM or something like it, a derivation of $S\#[1000000]$ from $S\#$ and $\forall x(Sx \rightarrow Sx')$, i.e. a closed tree with $S\#$, $\forall x(Sx \rightarrow Sx')$, and $\neg S\#[1000000]$ at the top, would have to contain c. two million lines.[3] But with the aid of XM, we may immediately let $\varphi = \forall x(Sx \rightarrow Sx'')$, use $\forall x(Sx \rightarrow Sx')$ to close the subtree beginning with $\neg\forall x(Sx \rightarrow Sx'')$, then reapply XM with $\varphi = \forall x(Sx \rightarrow Sx'''')$, use $\forall x(Sx \rightarrow Sx'')$ to close the subtree beginning with $\neg\forall x(Sx \rightarrow Sx'''')$, etc. We thus see once again that adjunction of modus ponens or XM permits the development and subsequent employment of information capable of sharply lessening the inferential distance between certain premises and conclusions that follow from them.

---

[3]Cf. Article 23 in this volume.

23

Don't Eliminate Cut

The method of trees, as presented in e.g. Jeffrey's *Formal Logic: Its Scope and Limits*[1] and Smullyan's *First-order Logic*,[2] and standard systems of natural deduction, like the one given in Mates's *Elementary Logic*,[3] are sound and complete methods of logic in the usual sense: they mark an inference as valid if and only if it is valid. There is, however, a significant difference between them in the *manner* in which they can demonstrate validity, a difference that sometimes results in a striking disparity in the *efficiency* or *speed* with which an inference can be shown to be valid. Although a tree that demonstrates the validity of an inference, i.e., a closed tree with the premises and denial of the conclusion at its top, can be transformed into a natural deduction of the conclusion from the premises that it requires approximately the same amount of time to write down, the converse, as we shall see, is emphatically not the case.

It is immediate from the presentation of a standard system of natural deduction such as Mates's that if $(A \rightarrow B)$ and A are derivable in the system, then so is B. On the other hand, it cannot be seen without a considerable amount of work that if there are closed trees for both $(A \rightarrow B)$ and A, then there is also one for B. Thus *modus ponens,* or *cut,* is obviously a valid derived rule of standard natural deduction systems, but *not obviously* a valid derived rule of the method of trees. It is well known, in a general way, that the elimination of cuts from derivations in a system in which cuts are always eliminable can greatly increase the length of derivations. But in view of the efficacy of the method of trees when applied to the usual sorts of examples and exercises found in logic texts, one might think that the danger of encountering a valid inference whose validity cannot feasibly be

From *The Journal of Philosophical Logic* 13 (1984): 373–378. Reprinted with kind permission from Kluwer Academic Publishers. The contents of this paper were first presented at the 1983 AMS Special Session on Proof Theory.

[1](Jeffrey, 1981).
[2](Smullyan, 1968).
[3](Mates, 1972).

demonstrated by the method of trees is rather remote.

Not so. There is a simple inference that can be shown valid by means of a deduction in Mates's system whose every symbol can be written down in one or two pages of normally sized type or handwriting, but for which the smallest closed tree contains more symbols than there are nanoseconds between Big Bangs.

In fact, let H_n ("H" for "heap") be the inference whose premises are

$$\forall x \forall y \forall z + x + yz = + + xyz$$
$$\forall x\, dx = +xx$$
$$L1$$
$$\forall x(Lx \rightarrow L + x1)$$

and whose conclusion is the sentence consisting of L, followed by 2^n consecutive occurrences of d, followed by 1. Thus, for example, the conclusion of H_3 is: $Lddddddddd1$. We shall show that the shortest tree-method proof of the validity of H_n contains $> 2^{2^n}$ characters and that the shortest natural deduction of the conclusion of H_n from the premises contains $< 16(2^n + 8n + 21)$ characters. Thus the smallest closed tree for H_7 contains $> 2^{128} > 10^{38}$ characters, but the smallest natural deduction for H_7 contains < 3280 characters, or, at 5 characters per word and 400 words per page, a bit more than a page and a half.

The extent to which this result provides a reason for favoring natural deduction over trees is an issue that we shall discuss after we have proved our claim about H_n. We first show that any closed tree for H_n contains more than 2^{2^n} symbols.

Let T be a tree with the premises and denial of the conclusion of H_n at its top. Define an interpretation I as follows: The domain of I is the set of positive integers. I assigns one to 1, the function $x \mapsto 2x$ to d, the addition function to $+$ (and an arbitrary n-ary function to any other n-place function sign). Thus the first two premises of H_n are true under I.

Let den(s) be the denotation under I of the (closed) term s. We call a positive integer i *instanced* if for some term s, den(s) $= i$ and the sentence $(Ls \rightarrow L + s1)$ occurs in T.

Finally, I specifies that L applies to a positive integer j iff all positive integers *less than* j are instanced.

Thus the premise $L1$ is (trivially) true under I.

We now want to see that every sentence $(Lt \rightarrow L + t1)$ *that occurs in* T is also true under I. Assume that $(Lt \rightarrow L + t1)$ occurs in T. Let $j = $ den(t). Suppose Lt true under I. Then all $i < j$ are instanced. But since $(Lt \rightarrow L + t1)$ occurs in T, j is also instanced. Thus all $i < j + 1$ are instanced. And since den($+t1$) $= j + 1$, $L + t1$ is also true under I. Thus if $(Lt \to L + t1)$ occurs in T, it is true under I.

Let u be the term in the denial of the conclusion and let $k = \mathrm{den}(u) = 2^{2^n}$. We want to show that if some $j < k$ is not instanced, then T is open. It will then follow that if T is closed, for every $j < k$ there is a term t such that $\mathrm{den}(t) = j$ and $(Lt \to L + t1)$ occurs in T. As it is clear that if $\mathrm{den}(s) = i$, $\mathrm{den}(t) = j$ and $i \neq j$, then $s \neq t$ and $(Ls \to L + s1) \neq (Lt \to L + t1)$, it will also follow that if T is closed, at least $k - 1$ sentences of the form $(Lt \to L + t1)$ occur in T, and therefore T contains more than k symbols.

Accordingly, suppose that some $j < k$ is not instanced, but that T is closed. We shall obtain a contradiction.

Since some $j < k$ is not instanced, the denial $\neg Lu$ of the conclusion is true under I, as are $\forall x \forall y \forall z + x + yz = + + xyz$, $\forall x \, dx = +xx$, and $L1$.

And since T is closed, each of its branches contains some sentence and its denial. Now the only sentences that can occur in any branch of T are the premisses, the denial of the conclusion, and sentences of the forms $\forall y \forall z \, s''(y, z) = t''(y, z)$, $\forall z \, s'(z) = t'(z)$, $s = t$, $(Ls \to Lt)$, Lt, and $\neg Lt$, for the tree rules (i.e., the standard tree rules, which do *not* include the rule XM discussed below) do not lead out of this collection of sentences. Thus as T is closed, each of its branches must contain some sentence Lt and its denial $\neg Lt$.

This is impossible, however, for at every stage of the construction of T, there is at least one branch in which all sentences *other than* $\forall x(Lx \to L + x1)$ are true under I. (Cf. the usual soundness proof for the method of trees.) This is certainly the case at the beginning of the construction of T (when there is only one branch, consisting of the premisses and denial of the conclusion), and, inductively, remains the case throughout the construction of T, since the only tree rules relevant to T are UI, equals for equals, and the rule for the undenied conditional, and these all preserve truth under I. (The statement that the rule for the undenied conditional, which is a branching rule, preserves truth under I means, of course, that if the premiss $(A \to B)$ of the rule is true under I, then either the left conclusion $\neg A$ or the right conclusion B is true under I). The only problematical case in the induction step of the argument is that in which the sentence $\forall x(Lx \to L + x1)$ is a premiss of a rule of inference. But if an identity $s = t$ occurs in a branch, then the term s begins with $+$ or d, as does the term t, and thus neither s nor t is the term 1. Therefore the only rule relevant to $\forall x(Lx \to L + x1)$ is UI, which when applied to this sentence yields a conclusion $(Lt \to L + t1)$. And as we saw earlier, any sentence of this form that occurs in T is automatically true under I.

Therefore if T is closed, T contains more than 2^{2^n} symbols.

The other half of our task is to bound the number of symbols in the shortest deduction in Mates's system of the conclusion of H_n from its premisses.

Let "$d[n]$" denote the sequence consisting of n occurrences of d. Then the conclusion of H_n may be written: $Ld[2^n]1$. And let "$M(y)$" abbreviate the formula: $(Ly \land \forall x(Lx \rightarrow L + xy))$. Thus $M(1)$ is the conjunction of the last two premisses.

One deduction for H_n begins with a subdeduction from the premisses of the two sentences

$$(M(a) \rightarrow M(da))$$
$$\forall y(M(y) \rightarrow M(dy))$$

The cost so far is slightly more than 300 symbols. Then follow n trios of lines

$$(M(d[2^{(i-1)}]a) \rightarrow M(d[2^i]a))$$
$$(M(a) \rightarrow M(d[2^i]a))$$
$$\forall y(M(y) \rightarrow M(d[2^i]y)) \qquad (1 \leq i \leq n),$$

costing in all $14 \cdot 2^n + 114n - 14$ symbols. $(M(1) \rightarrow M(d[2^n]1))$ is then inferred by US (=UI), and the conclusion $Ld[2^n]1$ then follows by T. The total number of symbols in this deduction, according to my count, is $16 \cdot 2^n + 114n + 329$. If $n = 7$, this number is 3175.

There are some annotative comments on the deduction that it may be helpful to make. The associativity of addition is needed to deliver $\forall x(Lx \rightarrow L + xda)$ from $\forall x(Lx \rightarrow L + xa)$; without it we could only obtain $\forall x(Lx \rightarrow L + +xaa)$ (and not $\forall x(Lx \rightarrow L + x + aa)$). Apart from this use of associativity, the subdeduction that ends with the line $(M(a) \rightarrow M(da))$ is a "self-proving" conditionalization on $M(a)$. The first line of the ith trio follows by US from the line $\forall y(M(y) \rightarrow M(d[2^{(i-1)}]y))$ immediately above it. The middle line follows by T from the first line of the trio and the line $(M(a) \rightarrow M(d[2^{(i-1)}]a))$, which is two lines above the first line. Finally, the last line of the trio follows from UG from the middle line.

It should be emphasized that the contrast we have drawn is between *standard formulations* of the method of trees and those of natural deduction. We have been supposing that the method of trees is so formulated that the version of cut appropriate to trees, which Jeffrey calls XM (excluded middle),[4] and which allows one to split any open branch of a tree in two and append any sentence at all to the bottom of one of the new branches and the negation of the sentence to the bottom of the other, is not one of the (underived) rules of the method. Were XM present, one could write down a closed tree for H_n containing approximately the same number of symbols as the natural deduction for H_n that we have just given. It is somewhat ironic that it is the failure of the usual formulations of the tree

[4](Jeffrey, 1981), pp. 34–35.

method to permit a certain sort of *branching* that can be blamed for their inefficiency in treating inferences like H_n.

The most significant feature possessed by natural deduction but not the method of trees, a feature that can easily seem like a virtue, is not so much that natural deduction replicates ordinary reasoning rather more faithfully than the tree method, in which derivations are one and all given the unnatural shape of proofs by contradiction, quantifier stripping and cases, but that it permits the development and utilization within derivations of *subsidiary* conclusions, or, as they would be called in a more informal setting, *lemmas*. In criticizing a certain pair of systems, Feferman once wrote, " ...nothing like sustained ordinary reasoning can be carried on in either logic."[5] Sustained ordinary reasoning cannot be carried on in a tree system unsupplemented by XM, where we are unable to appeal to previously established conclusions. This difficulty with the method of trees, if it is a difficulty, is one for which an obvious remedy exists: add XM. Then from closed trees

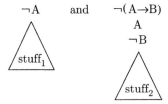

for A and $(A \to B)$ one can use XM to obtain, immediately, a closed tree for B:

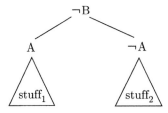

Of course, whether one should favor, adopt, or teach systems in which sustained ordinary reasoning, or rather, a highly idealized version of it, can or cannot be carried out are practical or normative questions on which other features of the systems may bear. The result about feasibility given above hardly decides these issues.

---

[5](Feferman, 1984).

24

The Justification of Mathematical Induction

There is an old (c. 1967) argument due to Dana Scott that is not as well known to philosophers and logicians as it ought to be. I shall come back to it later.

The principle of mathematical induction asserts that every number belongs to any class that contains zero and also contains the successor of any member.

Can the principle of mathematical induction be proved? That is to say, is there a way to show that every number belongs to any class that, etc.?

Like any other statement, the principle of mathematical induction can be derived from itself, in zero lines. This quick and easy derivation is not a proof of mathematical induction: it does not show that induction is true.

The least number principle asserts that if a class contains a natural number whenever it contains all lesser natural numbers, then the class contains every natural number. The principle is so called because of its contrapositive, which asserts that if there is at least one number in a class, then there is a number in the class such that no lesser number is in the class.

With the aid of two other principles we can derive the principle of mathematical induction from the least number principle. These two principles are that every natural number is either zero or the successor of some natural number and that every natural number is less than its successor.

Conversely, with the aid of some other principles, we can derive the least number principle from mathematical induction. This time, the other principles are that no number is less than zero and that a number that is less

From *PSA*, 2 (1984): 469–475. Copyright ©1985 Philosophy of Science Association. Reprinted by kind permission of The University of Chicago Press.

This paper replaces my remarks on Professor Maddy's paper; it was written while I was on a Fellowship for Independent Study and Research from the National Endowment for the Humanities. I am grateful to Scott Weinstein for helpful comments.

than $n + 1$ is either equal to or less than n.

The interderivability of mathematical induction and the least number principle from the four principles just mentioned is exceedingly well known, and the derivation of either from the other hardly counts as a proof of the derived form. What is of interest to us is the question whether induction, say in the form of the least number principle, might be proved from other principles that are noninductive in character and, if possible, appreciably more evident than either mathematical induction or the least number principle. Of course, we cannot hope to prove induction from no assumptions whatsoever. But it is at least conceivable that induction might be obtainable in an interesting manner from some evident truths that do not look too much like induction principles.

The prospect doesn't look too bright. One well-explored avenue of investigation is set-theoretical. One can prove induction in set theory *from suitable definitions*, including one of the natural numbers. This can be done in two ways, neither yielding a derivation of induction from evident, noninductive truths. On both ways, the natural numbers are defined as the objects satisfying some condition or other, and induction is then proved to hold of those objects. But on the more common of the two ways, the condition satisfied by the objects for which induction is proved has a strongly inductive character (e.g., a natural number might be defined to be a member of all classes containing zero and closed under successor or defined to be a set satisfying among other conditions, a "groundedness" condition asserting that any nonempty subset of the set contains a minimal member, that is, a member no member of which is also a member). When this sort of definition is given, the theory in which induction is proved can be quite weak and the principles from which induction is proved are not themselves inductive in character. But this procedure hardly counts as a proof of induction for the objects one was originally interested in, the natural numbers, for if at the outset one doubted whether the natural numbers satisfied induction one will still doubt whether the natural numbers will be (isomorphic to) the objects satisfying the definition, precisely because of the inductive character of the definition. Of course, one can often establish the isomorphism by an appeal to—guess what—induction. Less commonly, the condition defining the natural numbers will not be inductive in character, but the proof of induction for the objects satisfying the definition will appeal to a principle, like the axiom of regularity, possessing the inductive character now missing from the definition of natural number. Thus although one may no longer have to worry how one knows the natural numbers satisfy the definition, one will have to worry how one knows that the inductive principle to which appeal is made in the proof of induction is true.

And not much else appears to be left. We might try to define the natural numbers as certain sorts of reals and then try to derive induction via the least upper bound principle—but this is obviously uninteresting—or, we might follow the intuitionists in defining the natural numbers as certain sorts of proofs;[1] but the well-foundedness of proofs, from which on the intuitionist account induction is supposed to follow, seems no more apparent than does the fact that the natural numbers satisfy induction.

Maybe all isn't lost. In any event, the suggestion I want to make is that induction can be seen to hold because the structure consisting of the natural numbers under *less-than* is embedded in a certain larger structure containing other objects besides the numbers and endowed with other relations in addition to *less-than*. The least number principle can be derived from two principles which describe the numbers, the other objects, *less-than*, and the other relations; the principles do not, either separately or together, have the inductive character possessed by the Frege–Russell definition of the ancestral, the notion of a grounded class, the notion of a well-ordering, the axiom of regularity, the least upper bound principle of analysis, or the notion of a (well-founded) proof; and the principles seem to be obvious truths about the numbers and other objects of which they treat.

The other objects I have in mind are the sets I'll call the natural sets;[2] the other relations are membership (restricted to the natural sets) and a relation between natural sets and numbers I'll call *being formed at stage number*.[3] Here is a (partial) description of the larger structure: Just as the natural numbers are the objects generated from zero by repeated application of the successor operation, so the natural sets are the sets that are formed at stages indexed by natural numbers by repeated application of the following operation: form at each such stage all *possible* sets of natural sets formed at earlier stages, i.e., at stages indexed by lesser natural numbers. Thus at stage number 0, only the null set $\emptyset$ is formed; at stage number 1, only $\emptyset$ and $\{\emptyset\}$; at stage number 2, only $\emptyset, \{\emptyset\}, \{\{\emptyset\}\}$, and $\{\emptyset, \{\emptyset\}\}$; etc. Any natural set is formed repeatedly; indeed, it is formed at every stage that is later than any one at which it is formed. For each natural set x there is a number m, such that x is formed at stage number m, each natural set formed at stage number m is a set whose members are all formed at earlier stages, and, to repeat, at stage number m, all possible sets of natural sets formed at earlier stages are formed.

With the aid of a two-sorted language, L, containing set variables $x, y, \ldots,$

---

[1] Scott Weinstein called this possibility and the obvious objection to it to my attention.

[2] The natural sets are often referred to as the hereditarily finite sets; but because I want to avoid assumptions concerning the term "finite" and also to stress the analogy with the natural numbers, I have chosen to call them the natural sets.

[3] *Being formed at stage number* is more commonly known as *being of rank less than or equal to*.

number variables $m, n, \ldots$, and three binary predicate letters $\in$, $<$, and F, abbreviating "is in," "is less than," and "is formed at stage number," respectively, and perhaps some other predicates or functions of natural numbers and sets, we can conveniently set out the principles concerning natural sets and numbers from which we shall derive the least number principle, as well as other principles of set theoretical interest.

The first principle I'll call Spec:

Spec $\quad \forall m \exists x [x F m \wedge \forall y (y \in x \leftrightarrow [X(y) \wedge \exists n (n < m \wedge y F n)])]$

Spec is an axiom-schema: $X(y)$ is a formula of L not containing free x. Spec expresses the idea that for every m, *all possible sets* of previously formed sets are formed as sets at stage number m. The notion of a "possible set" is expressed through the occurrence of a formula $X(y)$, possibly containing additional free set and number variables, and the universality of "all" is captured (as well as is possible in a first-order language like L) by allowing $X(y)$ to be any formula whatsoever of L. The other principle we'll need is that of the transitivity of *less-than*:

Tra $\quad \forall n \forall m \forall k (n < m \wedge m < k \rightarrow n < k)$.

The discussion may have primed the reader to see the schema Spec as "covertly" inductive in character. But a reader who had been (mis-) informed that the least number principle would be derived not merely from Spec and Tra, but from some other principles as well, would not, I dare say, have been inclined to regard Spec and Tra as at all inductive in character, but would rather have expected some "inductivity" to show up in the additional principles. In fact, it seems to me fair to say that Spec and Tra taken together are significantly less inductive in character than mathematical induction, the least number principle, or any of the other axioms, principles, or definitions mentioned earlier.

How evident is Spec? Recall that for any number m, every possible set of natural sets, each of which is formed at some stage number n, with $n < m$, is a natural set formed at stage m. For any formula $X(y)$, the natural sets y that are formed at stages earlier than stage number m and that satisfy $X(y)$ form a possible set of natural sets, and thus a set x of them all is formed at stage number m. Spec is therefore evident from what we have said about the natural sets, and it seems to me, rather more evident than the least number principle would be had one encountered it or the principle of mathematical induction for the first time already familiar with Spec (a rather remote possibility).

The consistency of Spec and Tra is obvious: They are both true in the trivial structure $\mathcal{M}$ in which the only number is 0, the only set is $\emptyset$, $\emptyset \notin \emptyset$,

$0 \not< 0$, and $\emptyset F0$. The axiom of extensionality and the sentences $\forall m \forall n (m < n \lor m = n \lor n < m)$, $\forall x \exists m \, xFm$ and $\forall x \forall m \forall y (xFm \land y \in x \to \exists n(n < m \land yFn))$ of L, which can also be "read off" the description we have given of the natural numbers and the natural sets, are also true in $\mathcal{M}$. Another sentence of L which can also be so read off, but which is not true in $\mathcal{M}$, is $\forall n \exists m \, n < m$.

The time has come to derive induction. We proceed as in Shoenfield's article in the *Handbook of Mathematical Logic* (Shoenfield, 1967). Call y a *minimal member* of x if $y \in x$ and $\forall z \neg (z \in x \land z \in y)$, and say that y is *grounded* if every set containing y has a minimal member. Then if every member of y is grounded, then y itself is grounded. (*Logic*: Suppose $y \in x$. If for some z, $z \in x$ and $z \in y$, then z is grounded, and x has a minimal member. Otherwise, $\forall z \neg (z \in x \land z \in y)$; but then y is a minimal member of x.)

Definition aRm iff $aFm \land \forall y(y \in a \leftrightarrow y$ is grounded $\land \exists n(n < m \land yFn))$.

Miscellaneous facts:

1. By Spec, for every m, there is an a such that aRm.

2. If aRm, then since all members of a are grounded, a is grounded.

3. Thus if $n < m$, aRm, and bRn, then b is grounded by 2, bFn, and $b \in a$.

We'll derive a theorem-schema from Spec and Tra that expresses the least number principle: $\exists k \, P(k) \to \exists m(P(m) \land \forall n(n < m \to \neg P(n)))$.

Suppose $P(k)$. If for all numbers j such that $j < k$, $\neg P(j)$, then done. So suppose $j < k$ and $P(j)$. By Spec, for some x, $\forall a(a \in x \leftrightarrow [\exists m(m < k \land aRm \land P(m)) \land \exists n(n < k \land aFn)])$. Since aFm if aRm, $\forall a(a \in x \leftrightarrow \exists m(m < k \land aRm \land P(m)))$. Since $j < k$ and $P(j)$, x is nonempty by (1). By (2), all members of x are grounded. Thus x has a minimal member a, and for some m, $m < k$, aRm, and $P(m)$. Now suppose $n < m$. By (1), for some b, bRn. By (3), $b \in a$. By Tra, $n < k$. If $P(n)$, then $b \in x$, a contradiction as a and x are disjoint; thus $\neg P(n)$.

We may remark that not only is the least number principle derivable from Spec and Tra, but the axioms of regularity, together with all the axioms of Zermelo Set Theory except infinity and choice are derivable from Spec, Tra, and the five sentences of L mentioned earlier, viz., the axiom of extensionality, $\forall m \forall n (m < n \lor m = n \lor n < m)$, $\forall x \exists m \, xFm$, $\forall x \forall m \forall y (xFm \land y \in x \to \exists n(n < m \land yFn))$, and $\forall n \exists m \, n < m$. To derive the axiom of infinity, it suffices to adjoin to these a suitable principle asserting the existence of a limit number such as $\exists m (\exists n n < m \land \forall n [n < m \rightarrow \exists k (n < k \land k(m))])$.

Historical note: In 1967 Dana Scott presented a paper[4] to the American Mathematical Society Summer Institute on Axiomatic Set Theory, in which he showed how the axiom-schema of regularity could be derived in a certain elementary theory concerning sets and "partial universes," as he called them. A partial universe, intuitively speaking, is the set of all sets of rank less than some one ordinal. Partial universes are special sorts of sets. and Scott used a two-sorted language with variables $V, V', \ldots$ for partial universes and variables $x, y, \ldots$ for sets. (We omit mention of individuals.) The axioms of the theory are those of extensionality and Aussonderung, an axiom of "accumulation": $\forall V' \forall x (x \in V' \leftrightarrow \exists V \in V' (x \in V \lor x \subseteq V))$, which states that the members of a partial universe are the members or subsets of earlier universes, and an axiom of "restriction": $\forall x \exists V\, x \subseteq V$, stating that every set is a subset of some partial universe. Scott showed that it follows from these axioms alone that all instances of the regularity schema hold, as do all axioms of Zermelo set theory except infinity and choice, and that $\in$ well-orders the partial universes. The most striking mathematical argument in the paper was the utilization of the paradox of the class of grounded classes to demonstrate that $\in$ is well-founded, and it is this argument that we have adapted to deduce the least number principle from Spec and Tra.

I once wrote an article[5] that contained an axiomatic theory of sets and stages, in which no assumption was made to the effect that stages are sets of certain sorts. Axioms of induction for sets and stages were explicitly taken as axioms of this theory. It was on reflecting on the presentation of Scott's argument given in Shoenfield's *Handbook* article[6] that I realized that the least number principle, whose formalized statement is identical to that of the well-foundedness principle for stages, could be formally derived in the weak-looking theory of sets and stages whose axioms are Spec and Tra.

---

[4]Subsequently published as (Scott, 1974).

[5]Reprinted as Article 1 in this volume.

[6](Shoenfield, 1967). Shoenfield does not supply an axiomatic theory concerning the stages at which sets are formed, but infers the axioms of set theory from an informal account of sets and stages.

25

A Curious Inference

The inference is:

I

(1) $\forall n\, fn1 = s1$
(2) $\forall x\, f1sx = ssf1x$
(3) $\forall n\forall x\, fsnsx = fnfsnx$
(4) $D1$
(5) $\forall x(Dx \rightarrow Dsx)$
    $\therefore$
(6) $Dfsssss1sssss1$

I is an inference in the first-order predicate calculus with identity and function signs. (s is a 1-place, f a 2-place function sign.) I is small: it contains 60 symbols or so, fairly evenly distributed among its five premisses and conclusion. And I is logically valid; the Frege–Russell definition of natural number enables us to see that there is a derivation of (6) from (1)–(5) in any standard axiomatic formulation of second-order logic, e.g. the one given in Chapter 5 of Church's *Introduction to Mathematical Logic*.[1] A sketch of a second-order derivation of (6) from (1)–(5) is given in the appendix, and it should be evident from the sketch that there is a derivation of (6) from (1)–(5) in any standard axiomatic system of second-order logic *whose every symbol can easily be written down*.

But it is well beyond the bounds of physical possibility that any actual or conceivable creature or device should ever write down all the symbols of a complete derivation in a standard system of *first*-order logic of (6) from (1)–(5): there are far too many symbols in any such derivation for this to

From *The Journal of Philosophical Logic* 16 (1987): 1–12. Reprinted with kind permission from Kluwer Academic Publishers.

I am grateful to Rohit Parikh, Scott Weinstein, and a referee for the JPL for helpful comments.

[1](Church, 1956).

be possible. Of course in every standard system of first-order logic there is (in the sense in which "there is" is used in and out of mathematics) a derivation of (6) from (1)–(5), for every standard system of first-order logic is complete. But as we shall see, no such derivation could possibly be written down in full detail, in *this* universe.

For definiteness, we shall concentrate our attention on the system M of Mates' book *Elementary Logic*.[2] It is because Mates' book is a standard text and its system M is a perfectly standard system of natural deduction that we have chosen to focus on it. (The rules of M are: premiss introduction, conditionalization, truth-functional consequence, universal instantiation, universal generalization, the usual identity rules, and existential quantification $(\neg\forall\alpha\neg\varphi/\exists\alpha\,\varphi)$.) A result similar to the one we shall obtain for M can be gotten for any other standard formulation of first-order logic, e.g. the axiomatic system of first-order logic contained in Monk's or Shoenfield's *Mathematical Logic*,[3] or any of the systems found in Quine's *Methods of Logic*.[4]

What we shall show is that the number of symbols in any derivation of (6) from (1)–(5) in M is at least the value of an exponential stack

$$2^{2^{\cdot^{\cdot^{2^2}}}} \qquad \text{i.e.} \qquad 2^{(2^{(\cdot^{\cdot^{\cdot(2^2)\dots))}}})}$$

containing 64K, or 65,536, "2"s in all. Do not confuse this number, which we shall call $f(4,4)$, with the number 2^{64K}. The latter number is minuscule in comparison, not even containing as many as 20,000 (decimal) digits. (It is the value of a stack containing only 5 "2"s.) The so-called Skewes' number, which is of interest in prime number theory and has been described as "the largest number found in science," is

$$10^{10^{10^{34}}}$$

Skewes' number is readily seen to be less than the value of a stack of 7 "2"s. It is not hard to show that if "#" denotes Skewes' number, then for some $N < 10$, $f(4,4) >$ the value of a stack of 64K $- N$ "#"s.

In the intended interpretation of I, the variables range over the positive integers. 1 denotes one and s denotes the successor function. There is no particular interpretation intended for D. f denotes an Ackermann-style function $n, x \mapsto f(n,x)$ defined on the positive integers: $f(1,x) = 2x$; $f(n,1) = 2$; and $f(n+1, x+1) = f(n, f(n+1, x))$. Here are some of the early values of f:

[2](Mates, 1972).
[3](Monk, 1976), (Shoenfield, 1967).
[4](Quine, 1972).

n	1	2	3	4	5
x					
1	2	2	2	2	2
2	4	4	$4 = 2^2$	4	4
3	6	8	$16 = 2^{2^2}$	$64\text{K} = 2^{2^{2^2}}$	$2^{2^{\cdot^{\cdot^{\cdot^2}}}}$ 64K "2"s in all
4	8	16	$64\text{K} = 2^{2^{2^2}}$	$2^{2^{\cdot^{\cdot^{\cdot^2}}}}$ 64K "2"s in all	.
5	10	32	$2^{2^{2^{2^2}}}$	.	$f(5,5)$
6	12	64	$2^{2^{2^{2^{2^2}}}}$	.	.
	$2x$	2^x	the value of a stack of x "2"s	.	.

So $f(n,2) = 4$ (all n); $f(2,x) = 2^x$; $f(3,x) =$ the value of an exponential stack containing x 2s; $f(4,3) = f(3,4) = 64\text{K}$; $f(4,4) =$ the value of a stack containing 64K 2s.

Thus by pursuing the obvious strategy of appending a definition of a well-known sort of fast-growing function to a formalization of the premises of the paradox of heap and employing the function to construct a short conclusion to the paradox mentioning a very large number, we obtain an inference which we can see to be valid by means of a simple argument that cannot be replicated in any standard system of first-order logic. I assume that (it is evident that) no formal derivation containing at least $f(4,4)$ symbols can count as *replicating* this argument, or indeed any argument that *we* can comprehend. Indeed, a shorter and even more extravagant conclusion than (6) follows from (1)–(5): $Dffs1ss1ss1 : f(8,3) \gg f(5,5)$. One might wonder whether there is any valid inference interestingly simpler than I whose shortest derivation in some standard system of first-order logic is significantly greater.[5] In brief, I is a simple and natural example of a valid first-order inference the conclusion of which cannot feasibly be derived from the premises in any standard system of first-order logic; but there is a short and simple argument that demonstrates the validity of I, which can be formalized in any standard system of second-order logic.

Of course, it has been known since Gödel's "On the length of proofs"[6] that the use of higher types can drastically reduce the minimum length of derivations in formal systems. In that paper, it will be recalled, Gödel

---

[5]Of course, (1)–(5)/$Dffs1ss1ss1$ is not simpler in an *interesting* way.
[6](Gödel, 1936).

states (without proof) that for any recursive function φ and any i, there are infinitely many arithmetical theorems F of both i^{th} and $(i+1)^{\text{st}}$ order logic[7] such that if k and l are the lengths of the shortest proofs of F in i^{th} and $(i+1)^{\text{th}}$ order logic, respectively, then $k > \varphi(l)$. (The length of a proof is the number of formulae of which it consists.) And it was shown by Statman[8] that there is no function φ provably recursive in second-order arithmetic such that whenever a first-order formula F is derivable in a certain standard system of second-order logic with length $\leq l$, then F is derivable in a certain standard system of first-order logic with length $\leq \varphi(l)$. Noteworthy investigations of speedup have recently been carried out by Harvey Friedman. One of Friedman's theorems is that a certain "finitization" of a combinatorial theorem due to J. Kruskal concerning embeddings of trees can be proved in ZFC in a few pages, but not in the system of second-order arithmetic called ATR (for Arithmetical Transfinite Recursion) in under $f(3, 1000)$ pages.[9]

But our aim is neither to prove a general speedup theorem nor to demonstrate the "practical incompleteness" of first-order logic; rather we are interested in showing that this incompleteness can be demonstrated by means of an inference like I that is remarkably elementary.[10] Indeed. I arises in quite a natural way: the first three premises of I can be taken as defining a kind of function very well known to logicians and computer scientists; the last two premises and conclusion can be used to formalize an ancient and completely familiar logical paradox involving large numbers. Without exaggeration, it may be said that I or a close relative might well be the first inference one would think of if one were *trying* to show first-order logic practically incomplete.

Since Skolem's discovery of non-standard models of arithmetic, it has been well known that there are simple and fundamental logical concepts, e.g., the *ancestral*, that cannot be expressed in the notation of first-order logic. It is also well known that there are notions of a logical character expressible in natural language that cannot be expressed in first-order notation. And it is increasingly well understood that it is neither necessary nor always possible to interpret second-order formalisms as applied first-order set theories in disguise. Thus although the existence of a simple

---

[7]Comparison with Gödel's earlier papers on incompleteness makes it reasonable to suppose that the systems S_i considered in "On the length of proofs" contain the Peano axioms for successor; "i^{th} order arithmetic" might thus be a more apt term for S_i than "i^{th} order logic."

[8](Statman, 1978).

[9](Nerode and Harington, 1984).

[10]An analogy with miniature Universal Turing Machines was suggested to me by Rohit Parikh: miniature UTMs are of interest not in showing the halting problem unsolvable but in showing that unsolvability arises in *such* simple structures.

first-order inference whose validity can be feasibly demonstrated in second-
but not first-order logic cannot by itself be regarded as an overwhelming
consideration for the view that first-order logic ought never to have been
accorded canonical status as *Logic*, it is certainly one further consideration
of some strength for this view.

On the other hand, the fact that we so readily recognize the validity of
I would seem to provide as strong a proof as could be asked for that no
standard first-order logical system can be taken to be a satisfactory ideal-
ization of the psychological mechanisms or processes, whatever they might
be, whereby we recognize (first-order!) logical consequences. "Cognitive
scientists" ought to be suspicious of the view that logic as it appears in
logic texts adequately represents the whole of the science of valid inference.

It may be remarked in passing that the second-order derivation of (6)
from (1)–(5) given in the appendix has a certain foundational interest. If
we interpret the variables in *I* as ranging over a set (species) containing the
positive integers and possibly other objects, 1 as denoting one, *s* as denoting
a one-place function whose restriction to the positive integers is the usual
successor function, *f* as denoting an (unspecified) 2-place function, and
D as denoting the set *N* of positive integers, then (4) and (5) are true.
Consider the following argument, which shows that (6) is true (relative to
the choice of the domain and the denotations of *s* and *f*) if (1)–(3) are.

> We first show by induction on n that for every n in N, for every
> x in N fnx is in N. By (1), $f11 = s1 \in N$. Suppose that
> $x \in N$ and that $f1x \in N$. Then by (2), $f1sx = ssf1x \in N$.
> By induction on x, for every x in N $f1x \in N$. Now suppose
> that $n \in N$ and that for every x in N $fnx \in N$. By (1),
> $fsn1 = s1 \in N$. Suppose that $x \in N$ and $fsnx \in N$. By the
> i.h., $fnfsnx \in N$. By (3), $fsnsx = fnfsnx \in N$. By induction
> on x for every x in N, $fsnx \in N$. By induction on n, for every
> n in N, for every x in N, $fnx \in N$. Since $5 \in N$, $f55 \in N$, and
> (6) is true.

This argument, a simple modification of the derivation of (6) from (1)–(5)
given below, is evidently intuitionistically acceptable. But because of the
presence of the unbounded universal quantifier "for every x in N" in the
induction hypothesis, it cannot be regarded as finitistically acceptable.[11]
Thus the notions of intuitionist and finitist acceptability may readily be
seen to diverge.

The details of the proof that any derivation of (6) from (1)–(5) in *M*
must contain at least $f(4,4)$ symbols are tedious, but an outline of the

---

[11](Tait, 1968) and (Tait, 1981).

reasoning is easily given: We translate derivations in M into derivations in a modification of the system of Schwichtenberg's article in the *Handbook of Mathematical Logic*,[12] a system for which the proof of a cut-elimination theorem is readily available.[13] Any *cut-free* derivation of (6) from (1)–(5) must contain roughly $f(5,5)$ symbols, for in a cut-free system one has to take an instance of premiss (5) for every integer between 1 and $f(5,5)$ to derive the conclusion. Because of the presence of the unanalyzed rule T (tautological inference) in M, translating a derivation in M into one in S may result in an exponential increase in the length of the derivation; and eliminating cuts from a derivation in S may increase its length super-exponentially, of the order of the value of a stack of "2"'s; but such increases are as nothing when compared with the difference between $f(4,4)$ and $f(5,5)$. More of the details of the proof are contained in the appendix to the original version of this article.

Appendix

We present a sketch of a second-order derivation of (6) from (1)–(5) of which a complete formalization in any standard axiomatic formulation of second-order logic can easily be written out:

By the comprehension principle of second-order logic,

$$\exists N \forall z(Nz \leftrightarrow \forall X[X1 \land \forall y(Xy \to Xsy) \to Xz]),$$

and then for some N,

$$\exists E \forall z(Ez \leftrightarrow Nz \land Dz).$$

Lemma 1 $N1$; $\forall y(Ny \to Nsy)$; $Nssss1$; $E1$; $\forall y(Ey \to Esy)$; $Es1$.

Lemma 2 $\forall n(Nn \to \forall x(Nx \to Efnx))$.

Proof. By comprehension, $\exists M \forall n(Mn \leftrightarrow \forall x(Nx \to Efnx))$. We want $\forall n(Nn \to Mn)$. Enough to show $M1$ and $\forall n(Mn \to Msn)$, for then if Nn, Mn.

$M1$: Want $\forall x(Nx \to Ef1x)$. By comprehension, $\exists Q \forall x(Qx \leftrightarrow Ef1x)$. Want $\forall x(Nx \to Qx)$. Enough to show $Q1$ and $\forall x(Qx \to Qsx)$.

$Q1$: Want $Ef11$. But $f11 = s1$ by (1) and $Es1$ by Lemma 1.

$\forall x(Qx \to Qsx)$: Suppose Qx, i.e. $Ef1x$. By (2) $f1sx = ssf1x$; by Lemma 1 twice, $Ef1sx$. Thus Qsx and $M1$.

$\forall n(Mn \to Msn)$: Suppose Mn, i.e. $\forall x(Nx \to Efnx)$. Want Msn, i.e. $\forall x(Nx \to Efsnx)$. By comprehension, $\exists P \forall x(Px \to Efsnx)$. Want $\forall x(Nx \to Px)$. Enough to show $P1$ and $\forall x(Px \to Psx)$.

---

[12](Schwichtenberg, 1977).
[13]The cut-elimination theorem for this system is due to (Tait, 1972).

$P1$: Want $Efsn1$. But $fsn1 = s1$ by (1) and $Es1$ by Lemma 1.

$\forall x(Px \to Psx)$: Suppose Px, i.e. $Efsnx$; thus $Nfsnx$. Want $Efsnsx$. Since $Nfsnx$ and Mn, $Efnfsnx$. But by (3) $fnfsnx = fsnsx$; thus $Efsnsx$. By Lemma 1, $Nsss s1$. By Lemma 2, $Efssss1ssss1$. Thus $Dfssss1ssss1$, as desired. ∎

26

A New Proof of the Gödel Incompleteness Theorem

Many theorems have many proofs. After having given the fundamental theorem of algebra its first rigorous proof, Gauss gave it three more; a number of others have since been found. The Pythagorean theorem, older and easier than the FTA, has hundreds of proofs by now. Is there a great theorem with only one proof?

In this note we shall give an easy new proof[1] of the Gödel incompleteness theorem in the form: *There is no algorithm whose output contains all true statements of arithmetic and no false ones.* Our proof is quite different in character from the usual ones and presupposes only a slight acquaintance with formal mathematical logic. It is perfectly complete, except for a certain technical fact whose demonstration we will outline.

Our proof exploits *Berry's paradox.* In a number of writings, Bertrand Russell attributed to G. G. Berry, a librarian at Oxford University, the paradox of *the least integer not nameable in fewer than nineteen syllables.* The paradox, of course, is that that integer has just been named in eighteen syllables. Of Berry's paradox, Russell once said, "It has the merit of not going outside finite numbers."[2]

Before we begin, we must say a word about algorithms and "statements of arithmetic," and about what "true" and "false" mean in the present context. Let's begin with "statements of arithmetic."

The *language of arithmetic* contains signs + and × for addition and multiplication, a name 0 for zero, and a sign s for successor (plus-one). It also contains the equals sign =, as well as the usual logical signs ¬ (not), ∧

Reprinted with kind permission of the American Mathematical Society from *Notices of the American Mathematical Society* 36 (1989): 388–390 and 676.

[1] Saul Kripke has informed me that he noticed a proof somewhat similar to the present one in the early 1960s.

[2] (Russell, 1973), p. 210.

(and), ∨ (or), → (if ... then ...), ↔ (... if and only if ...), ∀ (for all), and
∃ (for some), and parentheses. The variables of the language of arithmetic
are the expressions x, x', x'', ..., built up from the symbols x and '; they
are assumed to have the natural numbers (0, 1, 2, ...) as their values.
We'll abbreviate variables by single letters: y, z, etc.

We now understand sufficiently well what truth and falsity mean in the
language of arithmetic; for example, $\forall x \exists y\, x = sy$ is a *false* statement,
because it's not the case that every natural number x is the successor of a
natural number y. (Zero is a counterexample: it is not the successor of a
natural number.) On the other hand, $\forall x \exists y(x = (y + y) \lor x = s(y + y))$ is
a true statement: for every natural number x there is a natural number y
such that either $x = 2y$ or $x = 2y + 1$. We also see that many notions can
be expressed in the language of arithmetic, e.g., less-than: $x < y$ can be
defined: $\exists z(sz + x) = y$ (for some natural number z, the successor of z plus
x equals y). And you now see that $\forall x \forall y[(ss0 \times (x \times x)) = (y \times y) \to x = 0]$
is—well, test yourself, is it true or false? (Big hint: $\sqrt{2}$ is irrational.)

For our purposes, it's not really necessary to be more formal than we
have been about the syntax and semantics of the language of arithmetic.

By an *algorithm*, we mean a computational (automatic, effective, mechanical) procedure or routine of the usual sort, e.g., a program in a computer language like C, Basic, Lisp, ..., a Turing machine, register machine,
Markov algorithm, ..., a formal system like Peano or Robinson Arithmetic,
..., or whatever. We assume that an algorithm has an *output*, the set of
things it "prints out" in the course of computation. (Of course an algorithm
might have a *null* output.) If the algorithm is a formal system, then its
output is just the set of statements that are provable in the system.

Although the language of arithmetic contains only the operation symbols
s, $+$, and $\times$, it turns out that many statements of mathematics can be
reformulated as statements in the language of arithmetic, including such
famous unproved propositions as Fermat's last theorem, Goldbach's conjecture, the Riemann hypothesis, and the widely held belief that $P \neq NP$.
Thus if there were an algorithm that printed out all and only the true statements of arithmetic—as Gödel's theorem tells us there is not—we would
have a way of finding out whether each of these as yet unproved propositions is true or not, and indeed a way of finding out whether or not any
statement that can be formulated as a statement S of arithmetic is true:
start the algorithm, and simply wait to see which of S and its negation $\neg S$
the algorithm prints out. (It must eventually print out exactly one of S
and $\neg S$ if it prints out all truths and no falsehoods, for, certainly, exactly
one of S and $\neg S$ is true.) But alas, there is no worry that the algorithm
might take too long to come up with an answer to a question that interests
us, for there is, as we shall now show, no algorithm to do the job, not even

an infeasibly slow one.

To show that there is no algorithm whose output contains all true statements of arithmetic and no false ones, we suppose that M is an algorithm whose output contains no false statements of arithmetic. We shall show how to find a true statement of arithmetic that is not in M's output, which will prove the theorem.

For any natural number n, we let $[n]$ be the expression consisting of 0 preceded by n successor symbols s. For example, $[3]$ is $sss0$. Notice that the expression $[n]$ stands for the number n.

We need one further definition: we say that a formula $F(x)$ *names* the (natural) number n if the following statement is in the output of M: $\forall x(F(x) \leftrightarrow x = [n])$. (Observe that the definition of "names" contains a reference to the algorithm M.) Thus, for example, if $\forall x(x + x = ssss0 \leftrightarrow x = ss0)$ is in the output of M, then the formula $x + x = ssss0$ names the number 2.

No formula can name two different numbers. For if both of $\forall x(F(x) \leftrightarrow x = [n])$ and $\forall x(F(x) \leftrightarrow x = [p])$ are true, then so are $\forall x(x = [n] \leftrightarrow x = [p])$ and $[n] = [p]$, and the number n must equal the number p. Moreover, for each number i, there are only finitely many different formulas that contain i symbols. (Since there are 16 primitive symbols of the language of arithmetic, there are at most 16^i formulas containing i symbols.) Thus for each i, there are only finitely many numbers named by formulas containing i symbols. For every m, then, only finitely many (indeed, $\leq 16^{m-1} + \ldots + 16^1 + 16^0$) numbers are named by formulas containing fewer than m symbols; some number is not named by any formula containing fewer than m symbols; and therefore there is a least number not named by any formula containing fewer than m symbols.

Let $C(x, z)$ be a formula of the language of arithmetic that says that x is a number that is named by some formula containing z symbols. The technical fact mentioned above that we need is that whatever sort of algorithm M may be, there is some such formula $C(x, z)$. We sketch the construction of $C(x, z)$ below, in Comment 3.

Now let $B(x, y)$ be the formula $\exists z(z < y \land C(x, z))$. $B(x, y)$ says that x is named by some formula containing fewer than y symbols.

Let $A(x, y)$ be the formula

$$\neg B(x, y) \land \forall a(a < x \to B(a, y))).$$

$A(x, y)$ says that x is the least number not named by any formula containing fewer than y symbols.

Let k be the number of symbols in $A(x, y)$. $k > 3$.

Finally, let $F(x)$ be the formula $\exists y(y = ([10] \times [k]) \land A(x, y))$. $F(x)$ says that x is the least number not named by any formula containing fewer than

$10k$ symbols.

How many symbols does F contain? Well, [10] contains 11 symbols, [k] contains $k + 1$, $A(x, y)$ contains k, and there are 12 others (since y abbreviates x'): so $2k + 24$ in all. Since $k > 3$, $2k + 24 < 10k$, and $F(x)$ contains fewer than $10k$ symbols.

We saw above that for every m, there is a least number not named by any formula containing fewer than m symbols. Let n be the least such number for $m = 10k$. Then n is not named by $F(x)$; in other words, $\forall x(F(x) \leftrightarrow x = [n])$ is not in the output of M.

But $\forall x(F(x) \leftrightarrow x = [n])$ is a true statement, since n is the least number not named by any formula containing fewer than $10k$ symbols! Thus we have found a true statement that is not in the output of M, namely, $\forall x(F(x) \leftrightarrow x = [n])$. Q.E.D.

Some comments about the proof:

1. In our proof, symbols are the "syllables," and just as "nineteen" contains $2 \ll 19$ syllables, so the term $([10] \times [k])$ contains $k + 15 \ll 10k$ symbols.

2. In his memoir of Kurt Gödel,[3] Georg Kreisel reports that Gödel attributed his success not so much to mathematical invention as to attention to philosophical distinctions. Gregory Chaitin once commented that one of his own incompleteness proofs resembled Berry's paradox rather than Epimenides' paradox of the liar ("What I am now saying is not true").[4] Chaitin's proofs make use of the notion of the *complexity* of a natural number, i.e., the minimum number of instructions in the machine table of any Turing machine that prints out that number, and of various information-theoretic notions. None of these notions are found in our proof, for which the remarks of Kreisel and Chaitin, which the author read at more or less the same time, provided the impetus.

3. Let us now sketch the construction of a formula $C(x, z)$ that says that x is a number named by a formula containing z symbols. The main points are that algorithms like M can be regarded as operating on "expressions," i.e., finite sequences of symbols; that, in a manner reminiscent of ASCII codes, symbols can be assigned code numbers (logicians often call these code numbers *Gödel numbers*); that certain tricks of number theory enable one to code expressions as numbers and operations on expressions as operations on the numbers that code

---

[3](Kreisel, 1980), p. 150.

[4]Cf. (Davis, 1980), pp. 241–267, especially pp. 263–267, for an exposition of Chaitin's proof of incompleteness. Chaitin's observation is found in (Chaitin, 1970).

them; and that these numerical operations can all be defined in terms of addition, multiplication, and the notions of logic. Discussion of symbols, expressions (and finite sequences of expressions, etc.) can therefore be coded in the language of arithmetic as discussion of the natural numbers that code them. To construct a formula saying that n is named by some formula containing i symbols, one writes a formula saying that there is a sequence of operations of the algorithm M (which operates on expressions) that generates the expression consisting of $\forall$, x, (, the i symbols of some formula $F(x)$ of the language of arithmetic, $\leftrightarrow$, x, $=$, n consecutive successor symbols s, 0, and). Gödel numbering and tricks of number theory then allow all such talk of symbols, sequences, and the operations of M to be coded into formulas of arithmetic.

4. Both our proof and the standard one make use of Gödel numbering. Moreover, the unprovable truths in our proof and in the standard one can both be seen as obtained by the substitution of a name for a number in a certain crucial formula. There is, however, an important distinction between the two proofs. In the usual proof, the number whose name is substituted is the code for the formula into which it is substituted; in ours it is the unique number of which the formula is *true*. In view of this distinction, it seems justified to say that our proof, unlike the usual one, does not involve *diagonalization*.

Appendix. A letter from George Boolos[5]

Several readers of my "New Proof of the Gödel Incompleteness Theorem," have commented on its shortness, apparently supposing that the use it makes of Berry's paradox is responsible for that brevity. It would thus seem appropriate to remark that once syntax is arithmetized, an even briefer proof is at hand, essentially the one given by Gödel himself in the introduction to his famous "On Formally Undecidable Propositions ...";

> Say the m applies to n if $F([n])$ is the output of M, where $F(x)$ is the formula with Gödel number m. Let $A(x,y)$ express "applies to," and let n be the Gödel number of $\neg A(x,y)$. If n applies to n, the false statement $\neg A([n],[n])$ is the output of M, impossible; thus n does not apply to n and $\neg A([n],[n])$ is a truth not in the output of M.

What is concealed in this argument is the large amount of work needed to construct a suitable formula $A(x,y)$; proving the existence of the key

[5]First published in *Notices of the American Mathematical Society* 36, 1989, p. 676.

formula $C(x, y)$ in the "New Proof" via Berry's paradox requires at least as much effort. What strikes the author as of interest in the proof via Berry's paradox is not its brevity but that it provides a *different sort of reason* for the incompleteness of algorithms.

27

On "Seeing" the Truth of the Gödel Sentence

In his famous 1931 paper, Gödel showed that for any "sufficiently strong" formal theory T, a sentence S in the language of T equivalent in T to its own T-unprovability cannot be proved in T, *provided that T is consistent*. (In the normal cases, S is equivalent in T to the sentence expressing the consistency of T.) Thus if T proves only true sentences, and is therefore consistent, then S is not provable in T.

Roger Penrose claims that although S is unprovable in T, we can always see that S is true by means of the following argument:[1] If S is provable in T, then S is false, but that is impossible (pp. 107–8: "Our formal system should not be so badly constructed that it actually allows false propositions to be proved!"); thus S is unprovable and therefore true.

There are certain interesting formal theories of which the set of provable sentences can be seen to contain no falsehoods; for the sake of argument we may grant that Peano Arithmetic (PA), say, is one of these. We must then grant that the Gödel sentence for PA, expressing its own PA-unprovability, is true and unprovable in PA.

To concede that we can see the truth of the Gödel sentence for PA, in which only a fragment (albeit non-trivial) of actual mathematical reasoning can be carried out, is not to concede that we can see the truth of Gödel sentences for more powerful theories such as ZF set theory, in which almost the whole of mathematics can be represented. I shall give some reasons for thinking that there is no sense of "see" in which we can see that ZF is consistent; thus we cannot see the truth of the Gödel sentence for ZF either, for that sentence is equivalent (in a much weaker theory than ZF) to the consistency sentence for ZF.

From *Behavioral and Brain Sciences* 13 (1990): 655–656. Reprinted by kind permission of Cambridge University Press.
    [1](Penrose, 1989).

A true story: Once upon a time, distinguished set theorist J sent equally distinguished set theorist M what purported to be a proof that the theory ZFM (ZF+"a measurable cardinal exists"), of which M and many others were fond, is inconsistent. M sat down to work and found the error on p. 39 or so of J's manuscript. As he began to examine J's "proof," M might have been reasonably confident that he would find an error, but by no means did he then know that J's "proof" was fallacious or see the consistency of ZFM. Do we know that some future hotshot will not do to ZF what M feared J had done to ZFM?

I suggest that we do not know that we are not in the same situation vis-à-vis ZF that Frege was in with respect to naive set theory (or, more accurately, the system of his *Basic Laws of Arithmetic*) before receiving, in June 1902, the famous letter from Russell, showing the derivability in his system of Russell's paradox. It is, I believe, a mistake to think that we can see that mathematics as a whole is consistent, a mistake possibly fostered by our ability to see the consistency of certain of its parts.

The verb "should" in the sentence quoted above ought to give us pause. Of course our formal system *should* not be so constructed as to have false theorems. What we may believe or hope to be the case, but cannot "see" to be so, is that the totality of mathematics *is* not badly constructed in that way. Are we really so certain that there isn't some million-page derivation of "0=1" that will be discovered some two hundred years from now? Do we know that we are really better off than Frege in May 1902?

To belabor the point: Penrose has said nothing that shows that we can recognize the truth of the Gödel sentence for ZF or for any other reasonable approximation to whole of the mathematics that we ourselves use. What we can see the truth of is this conditional proposition: the Gödel sentence for ZF is ZF-unprovable (and therefore true) *if* ZF is consistent. We cannot see that the Gödel sentence is true precisely because we cannot see that ZF is consistent. We may hope or believe that it is, but we do not know it, and therefore cannot see it.

Penrose does offer a kind of consideration not advanced in earlier discussions of Gödel's theorem. He states that when a mathematician discovers a proof of some statement, other mathematicians easily and quickly convince one another of its truth.

I don't see that Penrose offers an argument for the conclusion that the ready acceptance of a newly proved proposition shows that mathematicians see that it *is true* rather than that it *follows from the rest of mathematics,* that is, is true *if* the rest of accepted mathematics is. Penrose rightly emphasizes that we must see that each step in an argument can be reduced to something simple and obvious. But such reduction may not be possible: Many regard impredicative comprehension axioms in analysis as neither simple nor obvious; and none of the axioms of set theory forces itself upon us the way "$x + 0 = x$" does.

"When we convince ourselves of the validity of Gödel's theorem we not only 'see' it, but by so doing we reveal the very nonalgorithmic nature of the 'seeing' process itself" (p. 418). Since one of the hypotheses of Gödel's theorem is the consistency of the theories under consideration, Penrose must here mean seeing the truth of the Gödel sentence; but I have argued that we cannot do this if the theory is a reasonable approximation to the whole of mathematics.

The Mandelbrot set has been called the most complex object in all of mathematics, but mathematics itself, of course, outstrips the Mandelbrot set in complexity. Can we really "see" that "$0=1$" is not sitting at the bottom of some lengthy, intricate and ingenious proof perhaps involving concepts and arguments of a kind of which today we are completely unaware?

28

Quotational Ambiguity

According to W. Quine,
Whose views on quotation are fine,
    Boston names Boston,
    And Boston names Boston,
But 9 doesn't designate 9.

Richard Cartwright used to assign to MIT graduate students in philosophy the exercise of supplying quotation marks to that underpunctuated limerick of his so that it says something correct and sensible. One solution is to put pairs of single quotes around the first and fourth occurrences of 'Boston' and a pair of quotes within quotes around the third. Another is to put the single quotes around the second and third occurrences and the quotes within quotes around the first. One of the lessons of this paper is that neither of these solutions is entirely unexceptionable.

It was Quine's *Mathematical Logic*[1] that was responsible for my becoming a philosopher. I came upon a copy of it in the university bookstore during my freshman year; a year later the instructor in "Advanced Logic" counted my having read it as satisfying the course's prerequisite. If I was a bit murky on alphabetic variance and such laws as *Math Logic*'s *159:

If α is not free in $\varphi, \vdash \ulcorner(\alpha)(\varphi \vee \psi) \equiv .\varphi \vee (\alpha)\psi\urcorner$,

I thought I had a pretty good understanding of such arcana as the ancestral and quasi-quotation. I was, I was convinced, an ace on the ordinary kind

First published in Paolo Leonardi and Marco Santambrogio, eds., *On Quine*, Cambridge: Cambridge University Press, 1995, pp. 283–296. Reprinted with kind permission of Cambridge University Press.

I am grateful to David Auerbach, Martin Davis, W. D. Hart, James Higginbotham, David Kaplan, Michael Kremer, Harold Levin, Ruth Marcus, Charles Parsons, W. V. Quine, Nathan Salmon and Göran Sundholm for helpful comments. Research for this paper was carried out under grant no. SES–8808755 from the National Science Foundation.

[1](Quine, 1955).

of quotation.

Less than a decade later I was explaining

'yields a falsehood when appended to its own quotation'
yields a falsehood when appended to its own quotation.

to unfortunates in introductory philosophy who were expecting the meaning
of life. Eventually, though, I wound up teaching courses with titles like
"Paradox and Infinity" or "Logic II," for which the Good Stuff was more
appropriate.

In MIT's "Paradox and Infinity" a few years ago, as I was going over
"yields a falsehood ...," an undergraduate suggested that what I had written on the board, something like:

'blue' appended to the quotation of 'red' = ' 'red' blue'

was ambiguous, and that I needed two kinds of quotation marks, semantic
and syntactic, to say what I wanted to say.

"Groan," I thought. "Another cockamamie undergraduate suggestion.
No undergraduate has anything to teach *me* about quotation." I couldn't
have been more wrong.

The student's name was Michael Ernst ('Michael Ernst'?); he is now a
graduate student in computer science at MIT. And this paper is about his
observation.

When I explained Ernst's observation to Cartwright, he doubled over in
surprise and uttered an oath. I wrote to Quine about it, who replied, "Dear
George, Thanks for Ernst's paradox. I am delighted with it. But I find I am
unable to cope with it, even when I have stopped laughing. Yours, Van."

Here's the problem. *Expressions,* or *strings,* are (or may be identified
with) finite sequences of symbols. We'll use letters of the Greek alphabet
as variables over expressions.

Now where α and β are any expressions at all, β *appended to* α is the
expression obtained by first writing the symbols of α, in the order in which
they occur in α, and then writing immediately after these the symbols of β,
in the order in which they occur in β. Thus, for example, 'apple' appended
to 'pine' is the expression 'pineapple'. The operation of appending is associative: γ appended to (β appended to α) is identical with (γ appended
to β) appended to α. Thus it does not matter how we add parentheses to
such a "term" as: 'VAKIA' appended to 'OSLO' appended to 'CZECH'.
('OSLO', it is well known, is in 'CZECHOSLOVAKIA'.)

So, it would seem, 'b' appended to 'a' is the two-symbol expression 'ab',
consisting of the letters 'a' and 'b' in that order.

The quotation of α is the expression that results when the expression α
is enclosed in a pair of quotation marks, i.e., the result of writing a left

quote, then the symbols constituting α, and then a right quote. Thus the quotation of 'Boston' is ' 'Boston' '.

The quotation of α is supposed to be an expression whose denotation is α; e.g., the denotation of the eight-symbol expression ' 'Boston' ' is the six-symbol expression 'Boston'.

And I hope you are all familiar with the calculation:

'appended to its own quotation' appended to its own quotation

= 'appended to its own quotation' appended to the quotation of 'appended to its own quotation'

= 'appended to its own quotation' appended to ' 'appended to its own quotation' '

= ' 'appended to its own quotation' appended to its own quotation'.

The calculation shows us that there is an expression, viz., 'appended to its own quotation' appended to its own quotation, which denotes itself.

So far all is familiar. But now consider the nonsense string κ:

b' appended to 'a

κ consists of the second letter of the alphabet, a right quote, a space, the eight letters of a certain word, a space, the two letters of a certain other word, a space, a left quote, and the first letter of the alphabet, in that order of course.

There are a lot of things one might want to say about κ. It's ill-formed, it's a nonsense string, it does not contain the letter 'c', it begins with the letter 'b', the number of letters in the English alphabet that alphabetically precede the first letter of κ is one, etc.

Consider now another string, λ:

ab

λ too is not well formed, nor does it contain the letter 'c'. Unlike κ, though, it begins with the letter 'a', and the number of letters of the English alphabet that alphabetically precede its first letter is zero.

Let's write: $N(\alpha)$ as short for: the number of letters of the English alphabet that alphabetically precede the first letter of the expression α. Then $N(\kappa) = 1$ and $N(\lambda) = 0$.

Now consider μ:

'b' appended to 'a'

μ is the quotation of κ; thus the denotation of μ is κ.

But wait! What did we say earlier? μ, note, consists of the quotation of 'b' followed by ' appended to ' followed by the quotation of 'a'. Just as 'apple' appended to 'pine' is 'pineapple', 'b' appended to 'a' is surely 'ab'. Thus μ, which begins with a left quote followed by a 'b' and which is the subject of the second clause of the previous sentence, denotes 'ab', and the denotation of μ is λ, i.e., 'ab'.

So $0 = N(\lambda) = N(\text{the denotation of } \mu) = N(\kappa) = 1$.

What has gone wrong? Obviously, the non-identical κ and λ cannot both be the denotation of μ, and unless they are, our demonstration that $0 = 1$ fails. How did we conclude that they are identical?

In the case of κ we said: μ is the quotation of κ; thus the denotation of μ is κ. We also said: μ ... denotes 'ab', and the denotation of μ is λ. It would thus seem that the inferences:

β is the quotation of α; so α is the denotation of β.

and:

β denotes α; so α is the denotation of β.

are problematic. In any event, μ is ambiguous, for on different parsings it denotes the different expressions κ and λ. And that is Michael Ernst's observation.

Let's treat the matters with which we have been dealing somewhat more formally. A theorem about quotation, appending, and denotation will ensue.

Expressions, as we have said, are finite sequences of symbols, i.e., functions from a finite initial segment of the set of natural numbers to a set of symbols. (Thus, e.g., 'cat' is a function whose domain is $\{0, 1, 2\}$ and whose value at 1 is 'a'.) For any symbol s, $[s]$ is the expression whose sole symbol is s. It is often important to distinguish between s and $[s]$, but it will turn out that we can here safely identify the two.

We define the *length* $\text{lh}(\alpha)$ of the expression α to be the least natural number not in the domain of α. Intuitively, $\text{lh}(\alpha)$ is the number of occurrences of symbols in α. Thus $\text{lh}(\text{'cat'}) = 3$, $\text{lh}(\text{'cattle'}) = 6$ and $\text{lh}(s)$, i.e., $\text{lh}([s])$, $= 1$.

If α and β are expressions of lengths m and n respectively, $\alpha * \beta$ is the expression of length $m + n$ whose first m symbols are those of α in the order in which they occur in α and whose last n symbols are those of β in the order in which they occur in β. (Thus $*$ is associative.) We will sometimes omit asterisks if (we believe that) no confusion will result.

Now assume that L is a language containing two symbols, l and r, and possibly others. Suppose that (1) for any expression α of L, $[l]*\alpha*[r]$ denotes

α. (Thus l and r work like left and right quotation marks.) Suppose further that there is an expression ϑ of L such that (2) for any expressions $\alpha, \beta, \gamma, \delta$ of L, if α denotes γ and β denotes δ, then $\alpha * \vartheta * \beta$ denotes $\gamma * \delta$. (Thus, like, e.g., 'followed by', ϑ "denotes concatenation." Note that 'appended to' switches order, but 'followed by' does not.)

Theorem *Some expression of L denotes two different expressions of L.*

Proof. Let $\alpha = [l] * [l] * [r] * \vartheta * [l] * [r] * [r], \beta = [l] * [r] * \vartheta * [l] * [r]$, and $\gamma = [l] * [r]$. By (1), α denotes β. By (1), $[l] * [l] * [r]$ denotes $[l]$ and $[l] * [r] * [r]$ denotes $[r]$. Then by (2), α denotes γ. But $\beta \neq \gamma$. ∎

An English version of α is:

    ' ' ' followed by ' ' '

which, as we see, ambiguously denotes the two-symbol expression consisting of the left and right quotes in that order and also denotes a longer expression containing oddly placed quotes, spaces, and certain letters of the English alphabet.

I suppose that by now it is hard to resist the suggestion that our puzzles arise from the circumstance that when a left quote is followed by two or more right quotes, it may not be determined which of those right quotes its mate is. Quotation marks differ in this respect from parentheses: any sequence of left and right parentheses is well formed in at most one way. (A left parenthesis and a right parenthesis to its right are mates if (recursively) no unmated parentheses lie between them; a string of parentheses is well formed iff every parenthesis in it has a mate.) In a slogan: "Quotes don't know their mates."

Nevertheless, we should look at the question whether the ambiguity arises from the phrase 'appended to'. Would our problems disappear if we were to abolish 'β appended to α' and to substitute for it 'the result of appending β to α' instead? (Of course our English versions of the semantical paradoxes would thereby become rather more complex—but that is of small concern.) 'The result of appending β to α' resembles function terms such as '$f(x, y)$' in mathematics more closely than does 'β appended to α'; could this deviation from a syntax resembling that of formal languages be responsible for our difficulties?

The suspicion that it's not the use of 'appended to' that has made trouble is easily confirmed if we consider:

    the result of appending 'c' to the result of appending 'b' to 'a'

Does this phrase denote the improbable:

    ac' to the result of appending 'b

or does it denote the more likely:

abc   ?

Would parentheses help? Suppose that instead of writing 'α appended to β' we were always to write 'app(α, β)' instead? But that really wouldn't avail. We would still have:

app('a', 'b', 'c')

to deal with, which may at first appear ill-formed, but need not be considered so. However, if it is not taken as ill-formed, it must be regarded as ambiguous, for on one way of parsing it, it denotes:

ca', 'b

and on another, it denotes:[2]

b', 'ca

I want now to discuss some proposals for dealing with the difficulty that modify or enhance quotation, or replace it with other devices. Eventually I shall offer one that seems to be more or less satisfactory. But first the others.

The first suggestion that comes to mind is to draw a link connecting a left quote with its right mate. We might do this by means of an arch, thus:

‿‿‿‿‿‿‿‿‿‿‿

'a' appended to 'b'

Or we might ⌈ enclose in a box ⌉ the material we wish to quote. Another suggestion is that a pair of quotation marks should be regarded as a discontinuous, two-part, symbol, like the letter 'ы' of the Cyrillic alphabet, or 'i' with its dot. In a language in which a pair of quotation marks is a single symbol, it would be no more possible for an analogue of our nonsense string κ to exist than for 'ɔylophonɔ' to be a word of English.

These proposals have a common defect. Were we to adopt one of them, we could no longer regard expressions as finite sequences of symbols. Were we to introduce arches, boxes, or something similar into the language, or to regard pairs of quotation marks as single but discontinuous symbols, between whose two parts entire expressions could occur, we would violate a requirement of *sequentiality*: expressions must be codable as functions from a finite initial segment of the natural numbers into the set of symbols of the language, i.e., as finite sequences of symbols. In any case we will see how

[2]Michael Kremer pointed out to me that Anil Gupta made a similar observation in (Gupta, 1982), pp. 184–5.

to link the expressions that will turn out to be our substitute for quotation marks without the use of arches, etc.

We might dispense with quotation marks altogether, *italicizing* instead those expressions that are to be referred to. Italicization, of course, has an evident drawback: iterating it is, at best, difficult. What angle with the abscissa shall a capital "I" make after it has been italicized n times? $90/2^n$ degrees?

There is a less noticeable, but more important, difficulty: it would seem that an italicized symbol would have to be regarded as other than the roman version of that symbol, and a doubly italicized symbol as different from both the singly italicized and the roman versions. Were we to allow, then, that the m- and n-fold italicizations of any symbol of a language L are different symbols of L if $m \neq n$, then L would have to be infinite. We desire a solution that does not require the number of symbols in L to be infinite. We may call this requirement the *finiteness* requirement.

Boldfacing obviously satisfies the finiteness requirement no better than italicization.

We have several times displayed expressions, that is, surrounded them with white space, or written them after colons. Although we shall continue to do so when quotation might be impractical or confusing, it is clear that as a general device for quotation, displaying offers difficulties akin to those of italicization, boldfacing, and other ways of changing the style of type.

If underlining is not to fall afoul of the requirements of sequentiality and finiteness, it would appear that we to have to regard the result of underlining an expression α as the result of repeatedly attaching a single symbol, say '_', after (or before) each of the constituent symbols of α. Thus if we wish to regard expressions as consisting of symbols in a linear order, we would seem to have to take the double underlining of 'cat' as the nine-character expression 'c_a_t_'. ('_' contains two expressions; '_', one.) But then, of course, expressions like the three-character '_' are ambiguous as between a non-underlined '_', a '_' whose first '_' is underlined, a '_' whose second '_' is underlined, and a doubly underlined '_'.

In conversation, Ruth Marcus proposed the use of an infinite sequence of ever bigger quotation marks. The suggestion obviously violates the finiteness requirement if each of the infinitely many quotation marks is to be a single symbol; but I confess that it was only after she made it that the finiteness requirement struck me as one that needed to be imposed, and our suggestion will actually turn out to be not so very different from Marcus's.

It seems extremely plausible that a language containing a fixed, finite number of pairs of quotation marks, e.g.:

‘ ’ { } [ ] ( ) ⟦ ⟧

would suffer from ambiguities of the sort we have noticed, but I have no rigorous proof. If we were to order the different pairs of marks and stipulate that no pair earlier in the ordering may surround either member of any later pair, then we would evidently violate another requirement, perhaps the most important one of all, that of *universality*: that for every expression α of the language, no matter how ill-formed, there should exist a (well-formed) expression of the language obtained by enclosing α in a pair of quotation marks. The requirement of universality rules out such devices as Quine's of spelling[3] as cures for our woes, for we wish to insist that the quotation of an expression α actually contain α as a subexpression.

Three familiar features of the ordinary use of quotation marks may be remarked upon.

1) The distinction between left and right quotation marks seems to be of little significance. As those of us who use crude fonts know, one can almost always get by with: ' or: " . In any case our original statement of Ernst's problem mentioned left and right quotes, and abolishing the distinction would certainly not get us out of our jam.

2) According to Collins' *Authors' & Printers' Dictionary*,[4]

> **quotations within quotations** to have only single quotation marks within the double. "The more conspicuous mark to the more inclusive quotation" (Henry Bradley). Quotations within the single quotation, to be double-quoted.

Presumably also, quotations within the double quotation within the single quotation, to be single-quoted, etc. Strict conformity to standard typographical practice, which I take Collins to be describing, is in any case impossible: if we wish to praise X for his adherence to the rule in writing: He asked, "Why not?", then we can't write:

Kudos to X, who wrote, "He asked, "Why not?""

but must, in accordance with Collins' rule, write:

Kudos to X, who wrote, "He asked, 'Why not?' "

But how then are we to criticize Y for violating the rule by writing: He asked, 'Why not?' , since we can't justifiably write:

Fie upon Y, who wrote, "He asked, 'Why not?' "

(for why are we praising X but blaming Y?) and can't adhere to Collins and write either:

[3] (Quine, 1960), p. 143.
[4] (Collins, 1933), p. 314.

Fie upon Y, who wrote, "He asked, "Why not?""

or:

Fie upon Y, who wrote, 'He asked, "Why not?" '

or:

Fie upon Y, who wrote, 'He asked, 'Why not?"

One possibly unfortunate feature of standard usage is that quoting a quoted expression changes it: double quotes become single and single double, as each new pair of outermost double quotes is added.

3) It is commonly thought that the primary function of quotation marks is to produce an expression that refers to another expression, namely the one inside the quotation marks. Perhaps so, but "shudder" quotes, here illustrated, the custom of enclosing titles of works in quotation marks,[5] and the (substandard) use of quotation for emphasis suggest a weaker view: the primary function of quotation marks is to indicate to the reader that quoted expressions are to be treated in some special manner that is inferable from context (it is to be hoped), but is otherwise unspecified.[6]

We come now to our own proposal, according to which to quote an expression is to enclose it in a certain pair of syntactically complex expressions, called the *q-marks* of that expression; the rest of the proposal is a rule for parsing a given expression α to determine which q-marks occurring as subexpressions of α are the mates of which others.

Let $'$ and $^\circ$ be two new symbols. $\{j\}$ is the expression consisting of j consecutive occurrences of $'$; $\mathrm{lh}(\{j\}) = j$.

A q-*mark* is an expression $\{j\}^\circ$, where $j \geq 0$. Thus the q-marks are $^\circ$, $'^\circ$, $''^\circ$, $'''^\circ$, etc.

$\#$ is the null expression; $\mathrm{lh}(\#) = 0$.

For any expression α, $m(\alpha)$, α's q-mark, is the shortest q-mark that is not a subexpression of α. Thus, for example, $m(\#) = m(') = m(s) = {}^\circ$ (s any symbol other than $^\circ$), $m(^\circ) = {}'^\circ, m('^\circ) = {}''^\circ, m(''^\circ) = {}'''^\circ, m(^{\circ'}) = {}'^\circ, m(^{\circ\circ}) = {}'^\circ, m('') = {}^\circ$.

The (revised) quotation of α, $r(\alpha)$, is defined to be $m(\alpha) * \alpha * m(\alpha)$. The first part of our proposal, then, is to revise the notion of quotation so that the quotation of any expression α is $r(\alpha)$. (In an environmentalist spirit, we might wish to recycle the left and right quotes as $'$ and $^\circ$.) It is clear

---

[5] Exercise. Punctuate: Janacek wrote a quartet called the Kreutzer Sonata.

[6] Some years ago an article appeared in *The New Yorker* (I would dearly love the reference) lamenting the overuse of shudder quotes and proposing the introduction of different kinds of quotation marks to indicate different authorial attitudes: raised circles for (wide-eyed) surprise, etc. Gabriel Segal has observed that the carat or circumflex, found on computer keyboards, resembles an eyebrow: skeptics may take note.

that our new sort of quotation respects the requirements of sequentiality, finiteness, and universality.

To complete our proposal we have to specify a procedure for parsing an expression α containing (zero or more) q-marks. It is simply this: Scan α from left to right, looking for an occurrence of $^\circ$. If one is found, locate, by backtracking, the longest q-mark (occurrence) that ends with that occurrence of $^\circ$. Then continue scanning to the right, looking for another q-mark of the same length. If and when one is found, the two q-marks are mates.[7] (The subexpression of α consisting of the two q-marks and the expression β lying between them, which may contain other (shorter) q-marks, may end in a string of occurrences of $'$, or may be null, may be taken to denote β, if desired.) Now apply the procedure to the subexpression of α lying to the right of the right q-mark and repeat in like manner until the end of α is reached. Of course there is no guarantee that α will in any sense be well formed; indeed, after a q-mark has been found, nothing at all ensures that scanning the remainder of α will turn up a mate for it.

We conjecture that the new kind of quotation will avoid the difficulties to which the old is subject, and will offer some evidence for this conjecture.

We shall need to observe that if we define $\langle \alpha, \beta \rangle$ as the expression $r(\alpha) * \beta$, then the ordered pair theorem holds for the operation $\alpha, \beta \mapsto \langle \alpha, \beta \rangle$.

Ordered Pair Theorem *If $\langle \alpha, \beta \rangle = \langle \gamma, \delta \rangle$ then $\alpha = \gamma$ and $\beta = \delta$.*

Proof. Suppose that $\langle \alpha, \beta \rangle = \langle \gamma, \delta \rangle$, i.e., that $\{j\}^\circ \alpha \{j\}^\circ \beta = \{k\}^\circ \gamma \{k\}^\circ \delta$, where $\{j\}^\circ$ and $\{k\}^\circ$ are the shortest q-marks that are not subexpressions of α and γ, respectively. Since the expressions $\{j\}^\circ \alpha \{j\}^\circ \beta$ and $\{k\}^\circ \gamma \{k\}^\circ \delta$ are identical, $j = k$ and thus $\alpha \{j\}^\circ \beta = \gamma \{j\}^\circ \delta$.

Now let $\zeta = \alpha \{j\}^\circ \beta$ and $\eta = \gamma \{j\}^\circ \delta, l = \text{lh}(\alpha)$, and $m = \text{lh}(\gamma)$. Observe that $\zeta(l + j) = {}^\circ, \zeta(l + i) = {}'$ for all $i < j$, and $\eta(m + i) = {}'$ for all $i < j$. Suppose (for *reductio*) that $l < m$. (See Figure 28.) Then $l + j < m + j$, and $\zeta(l + j) \neq \eta(m + i)$ for all $i < j$. Since $\zeta = \eta$, $l + j < m$, and so $l + i < m$ for all $i < j$. Since $\zeta = \eta$, $\eta(l + j) = {}^\circ$ and $\eta(l + i) = {}'$ for all $i < j$, and thus $\{j\}^\circ$ is a subexpression of γ. But since $j = k$, $\{k\}^\circ$ is a subexpression of γ, which contradicts the definition of k.

Similarly, not: $m < l$. Thus $\text{lh}(\alpha) = l = m = \text{lh}(\gamma)$, and since $\zeta = \eta, \alpha = \gamma, \{j\}^\circ \beta = \{j\}^\circ \delta$, and $\beta = \delta$.[8] ■

---

[7]Our scanning procedure will thus correctly parse the expression $^\circ s'^\circ$, which might be taken to be an ill-formed quotation of s, as the (well-formed) quotation of the (ill-formed) expression s'.

[8]The validity of the ordered pair theorem depends upon the paired senses of the definition of ordered pair and of q-mark. Were we to reverse that of the latter, so that the q-marks became $^\circ$, $^{\circ\prime}$, $^{\circ\prime\prime}$, etc., but leave the wording of the definitions of m and r, and ordered pair unchanged, the theorem would fail, as the expression $^{\circ\prime\circ\circ\prime}\gamma$, γ arbitrary, shows.

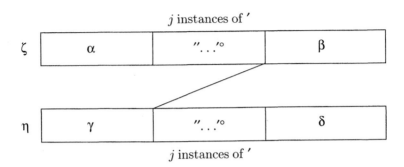

Figure 28.1: Since $\zeta = \eta$, if α is shorter than γ, then the displayed occurrence of $\circ$ in ζ lies in γ, and $''\ldots'^\circ$ is a subexpression of γ, impossible.

Corollary *If $r(\alpha) = r(\beta)$, then $\alpha = \beta$.*

Proof. If $r(\alpha) = r(\beta)$, then $\langle \alpha, \# \rangle = r(\alpha) * \# = r(\alpha) = r(\beta) = r(\beta) * \# = \langle \beta, \# \rangle$, whence by the ordered pair theorem, $\alpha = \beta$. ∎

The ordered pair theorem does not hold for the obvious analogue of ordinary quotation: Let l and r be symbols, $q(\alpha) = [l] * \alpha * [r]$, and $(\alpha, \beta) = q(\alpha) * \beta$. Then, when a and b are any symbols, and $\alpha = a, \beta = lrlbr, \gamma = arl, \delta = lbr, (\alpha, \beta) = q(\alpha) * \beta = larlrlbr = q(\gamma) * \delta = (\gamma, \delta)$, but $\alpha \neq \gamma$ (and $\beta \neq \delta$). Even more simply, let $\alpha = l, \beta = r, \gamma = lr, \delta = \#$; then $q(\alpha) * \beta = q(\gamma) * \delta = llrr$.

We now show that ambiguities like the ones we have been noticing arise if a device like (ordinary) quotation is added in the natural way to logical languages of a familiar sort, but that they can be proved not to arise if we similarly adjoin our new sort of quotation.

The logical languages we have in mind are (first-order) languages written in Polish notation. There are no parentheses in these languages, each operator is written to the left of its operands, and every symbol has a degree: The degree of an n-place predicate or function sign is n, that of a variable or individual constant is 0, and that of the existential quantifier $\exists$, the arrow $\rightarrow$, and the equals sign $=$ is 2. The rules of term and formula composition are the expected ones, e.g.: If φ and ψ are formulas, so is $\rightarrow \varphi\psi$. Following Shoenfield,[9] we call an expression that is either a term or a formula a *designator*.

Now, let L be such a language containing at least one function or predicate

[9](Shoenfield, 1967), esp. pp. 14–16.

symbol f of degree 2. (= will do.) Let l and r be symbols not in L. (l need not be distinct from r.) The natural way to extend L by using l and r as the left and right quote is to say: Let the symbols of L_{bad} be those of L together with l and r; where α is any expression at all of L_{bad} (including expressions containing l and r), the expression $[l] * \alpha * [r]$ is a term of L_{bad} ; the remaining rules of term and formula composition are as they are in L.

L_{bad}, so defined, *is* bad. It lacks unique readability, which requires that no designator of L_{bad} can be parsed in more than one way. More precisely, L_{bad} lacks unique readability if there is some designator σ such that for some symbol s of degree n, and two distinct n-tuples $(\sigma_1, \ldots, \sigma_n)$ and $(\tau_1, \ldots, \tau_n)$ of designators, $\sigma = s\sigma_1 \ldots \sigma_n = s\tau_1 \ldots \tau_n$ (or there are different expressions α, β such that $[l] * \alpha * [r] = [l] * \beta * [r]$—but that is obviously not the case). And there are many such designators in L_{bad}. For let a and b be any symbols of L_{bad} and consider the expression $flarlrlbr$, i.e., $[f] * [l] * [a] * [r] * [l] * [r] * [l] * [b] * [r]$. Since a is an expression (of L_{bad}) (recall that we identify a symbol with the expression of length one whose sole value is that symbol), lar is a term; since rlb is an expression, $lrlbr$ is also a term; and therefore since f is a symbol of degree 2, $flarlrlbr$ is a designator. But arl is also an expression and therefore $larlr$ is a term, and b is an expression and lbr is a term. Thus $flarlrlbr$ has the two distinct parsings:

$f \ lar \ lrlbr$

and:

$f \ larlr \ lbr,$

i.e., there are terms $\alpha, \beta, \gamma, \delta$, viz., $lar, lrlbr, larlr, lbr$, such that $f * \alpha * \beta = f * \gamma * \delta$ even though $\alpha \neq \gamma$ (and $\beta \neq \delta$); thus L_{bad} lacks unique readability.

If L contains the equals sign, interpreted as usual, and $[l] * \alpha * [r]$ is taken as denoting α in L_{bad}, then there are even expressions of L_{bad} true on one parsing and false on another, e.g. $= larlrlarlr$.

It is unique readability, in some clear but as yet undefined sense of the term, that expression μ shows natural languages containing the usual sort of quotation to lack. Expressions κ and μ also show that we would violate universality if we were stipulate that every left quotation mark is to be mated by the nearest right quotation mark to its right.

Let us now show that if we extend L to a language L_{good} by similarly adding to L our new kind of quotation, then L_{good} enjoys unique readability.

Thus let $'$ and $°$ be two symbols not in L. ($'$ and $°$ do not have any degree.)

We now take the symbols of L_{good} to be those of L together with $'$ and $°$. Where α is any expression at all of L_{good} (including expressions containing $'$ and $°$), the expression $r(\alpha)$, defined as above, is a term of L_{good}; the

remaining rules of term and formula composition are as they are in L.
For L_{good} to enjoy unique readability, each designator of L_{good} must be a
designator in exactly one way, that is, that if $r(\alpha) = r(\beta)$, then $\alpha = \beta$,
and if s is a symbol of degree $n, \sigma_1, \ldots, \sigma_n, \tau_1, \ldots, \tau_n$ are designators, σ is a
designator, and $\sigma = s\sigma_1 \ldots \sigma_n = s\tau_1 \ldots \tau_n$, then $\sigma_i = \tau_i$ for all $i, 1 \leq i \leq n$.

Theorem L_{good} *enjoys unique readability.*

Proof. Let σ be a designator. σ either begins with $'$ or $\circ$ or begins with a
symbol of some degree. If the former, then $\sigma = r(\alpha)$ for some expression α;
but then if also $\sigma = r(\beta), \alpha = \beta$, by the corollary to the ordered pair theorem. Thus we may assume that σ begins with a symbol s of degree n and
that for some designators $\sigma_1, \ldots, \sigma_n, \sigma = s\sigma_1 \ldots \sigma_n$. To prove the theorem,
it suffices to suppose that $\sigma = s\tau_1 \ldots \tau_n$, with each τ_i a designator, and show
that $\sigma_i = \tau_i$ for all $i, 1 \leq i \leq n$. Since $\sigma = s\sigma_1 \ldots \sigma_n = s\tau_1 \ldots \tau_n, \sigma_1 \ldots \sigma_n$
(i.e., $\sigma_1 * \ldots * \sigma_n) = \tau_1 \ldots \tau_n$. We conclude the proof by showing that,
more generally, if for all i, $1 \leq i \leq m$, σ_i is a designator, for all i,
$1 \leq i \leq n$, τ_i is a designator, and $\sigma_1 \ldots \sigma_m = \tau_1 \ldots \tau_n$, then $m = n$
and for all $1 \leq i \leq m, \sigma_i = \tau_i$. We proceed by induction on the length of
$\sigma_i \ldots \sigma_m$.

Case 1. σ_1 begins with $\circ$ or $'$. Then τ_1 begins with the same symbol and
for some expressions α and γ , $\sigma_1 = r(\alpha)$ and $\tau_1 = r(\gamma)$. By the ordered
pair theorem, $\sigma_1 = \tau_1$, and $\sigma_2 \ldots \sigma_m = \tau_2 \ldots \tau_n$. Since $\sigma_2 \ldots \sigma_m$ is shorter
than $\sigma_1 \ldots \sigma_m$, by the induction hypothesis, $m - 1 = n - 1$, whence $m = n$,
and for all $2 \leq i \leq m$, $\sigma_i = \tau_i$. Thus for all $1 \leq i \leq m$, $\sigma_i = \tau_i$.

Case 2. The first symbol s of σ_1 is of degree d. Then the first symbol of
τ_1 is also s, and for some designators $\pi_1, \ldots, \pi_d, \rho_1, \ldots, \rho_d, \sigma_1 = s\pi_1 \ldots \pi_d$
and $\tau_1 = s\rho_1 \ldots \rho_d$. Thus $s\pi_1 \ldots \pi_d\sigma_2 \ldots \sigma_m = s\rho_1 \ldots \rho_d\tau_2 \ldots \tau_n$, and
$\pi_1 \ldots \pi_d\sigma_2 \ldots \sigma_m = \rho_1 \ldots \rho_d\tau_2 \ldots \tau_n$. Since $\pi_1 \ldots \pi_d\sigma_2 \ldots \sigma_m$ is shorter
than $\sigma_1 \ldots \sigma_m$, by the induction hypothesis, $d + m - 1 = d + n - 1$, whence
$m = n$, and for all $i, 1 \leq i \leq d$, $\pi_i = \rho_i$, and thus $\sigma_1 = s\pi_1 \ldots \pi_d =$
$s\rho_1 \ldots \rho_d = \tau_1$, and for all $i, 2 \leq i \leq m$, $\sigma_i = \tau_i$. ∎

Returning now to our own language, we want to examine what happens
to the expressions involved in Ernst's paradox on our proposal. κ becomes:

      b° appended to °a,

λ remains:

      ab   ,

and μ becomes either:

      $'$°b° appended to °a$'$°

or:

°b° appended to °a°.

In the former case μ denotes κ; in the latter, λ. In neither case is the denotation of μ ambiguous.

And what of analogues of °app('a','b','c')°? Well, according to our scanning procedure, $'^\circ$app(°a°, °b°, °c°)$'^\circ$ *is* ill-formed, as it contains three operands, instead of the required two.

$$''^\circ\text{app}(°a°,\ '^\circ b°,^\circ c'^\circ)''^\circ$$

and

$$''^\circ\text{app}('^\circ a°,\ °b'^\circ,\ °c°)''^\circ,$$

however, are both well-formed, denoting the oddities $'^\circ b°$, $°ca'^\circ$ and $'^\circ ca°$, $°b'^\circ$, respectively.

It would be a pity if our new kind of quotation prevented us from imitating the calculation involving °'appended to its own quotation' appended to its own quotation° which shows there to be an expression that denotes itself. But it does not. For, taking °quotation° to mean our new sort of quotation, we have that

°appended to its own quotation° appended to its own quotation

= °appended to its own quotation° appended to the quotation of °appended to its own quotation°

= °appended to its own quotation° appended to $'^{\circ\circ}$appended to its own quotation$^{\circ'\circ}$

= $'^{\circ\circ}$appended to its own quotation° appended to its own quotation$'^\circ$.

As usual, the last expression both denotes and is identical with the first, and we have found an expression of the desired kind.

We conclude with the refrain:

According to W. Quine,
Whose views on quotation are fine,
　　°Boston° names Boston,
　　And $'^{\circ\circ}$Boston$^{\circ'\circ}$ names °Boston°,
But 9 doesn't designate 9.

29

The Hardest Logical Puzzle Ever

Some years ago, the logician and puzzle-master Raymond Smullyan devised a logical puzzle that has no challengers I know of for the title of Hardest Logical Puzzle Ever. I'll set out the puzzle here, give the solution, and then briefly discuss one of its more interesting aspects.

The puzzle: Three gods A, B, and C are called, in some order, True, False, and Random. True always speaks truly, False always speaks falsely, but whether Random speaks truly or falsely is a completely *random* matter. Your task is to determine the identities of A, B, and C by asking three yes-no questions; each question must be put to exactly one god. The gods understand English, but will answer all questions in their own language, in which the words for "yes" and "no" are "da" and "ja," in some order. *You do not know which word means which.*[1]

Before I present the somewhat lengthy solution, let me give answers to certain questions about the puzzle that occasionally arise:

- It could be that some god gets asked more than one question (and hence that some god is not asked any question at all).

- What the second question is, and to which god it is put, may depend on the answer to the first question. (And of course similarly for the third question.)

- Whether Random speaks truly or not should be thought of as depending on the flip of a coin hidden in his brain: if the coin comes

Reprinted by kind permission of the editor from *The Harvard Review of Philosophy* 6 (1996): 62–65. An Italian version of this article, translated by Massimo Piattelli-Palmarini, appeared in *La Repubblica* on 16 April 1992 under the title "L'indovinello più difficile del mondo."

[1]The extra twist of your not knowing which the gods' words for "yes" and "no" are is due to the computer scientist John McCarthy.

down heads, he speaks truly; if tails, falsely.

- Random will answer "da" or "ja" when asked any yes-no question.

The Solution: Before solving The Hardest Logic Puzzle ever, we will set out and solve three related, but much easier puzzles. We shall then combine the ideas of their solutions to solve the Hardest Puzzle. The last two puzzles are of a type that may be quite familiar to the reader, but the first one is not well known (in fact the author made it up while thinking about the Hardest Puzzle).

Puzzle 1: Noting their locations, I place two aces and a jack face down on a table, in a row; you do not see which card is placed where. Your problem is to point to one of the three cards and then ask me a single yes-no question, from the answer to which you can, with certainty, identify one of the three cards as an ace. If you have pointed to one of the aces, I will answer your question truthfully. However, if you have pointed to the jack, I will answer your question yes or no, completely at random.

Puzzle 2: Suppose that, somehow, you have learned that you are speaking not to Random but to True or False—you don't know which—and that whichever god you're talking to has condescended to answer you in English. For some reason, you need to know whether Dushanbe is in Kirghizia or not. What one yes-no question can you ask the god from the answer to which you can determine whether or not Dushanbe is in Kirghizia?

Puzzle 3: You are now quite definitely talking to True, but he refuses to answer you in English and will only say "da" or "ja." What one yes-no question can you ask True to determine whether or not Dushanbe is in Kirghizia?

Here's one solution to Puzzle 1: Point to the middle card, and ask "Is the left card an ace?" If I answer "yes," choose the left card; if I answer "no," choose the right card. *Whether the middle card is an ace or not,* you are certain to find an ace by choosing the left card if you hear me say "yes" and choosing the right card if you hear "no." The reason is that if the middle card is an ace, my answer is truthful, and so the left card is an ace if I say "yes," and the right card is an ace if I say "no." But if the middle card is the jack, then *both* of the other cards are aces, and so again the left card is an ace if I say "yes" (so is the right card but that is now irrelevant), and the right card is an ace if I say "no" (as is the left card, again irrelevantly).

To solve Puzzles 2 and 3, we shall use *iff*.

Logicians have introduced the useful abbreviation, "iff," short for "if, and only if." The way "iff" works in Logic is this: when you insert "iff" between two statements that are either both true or both false, you get a statement that is true; but if you insert it between one true and one false statement, you get a false statement. Thus, for example, "The moon is made of Gorgonzola iff Rome is in Russia" is *true,* because "The moon is made of Gorgonzola" and "Rome is in Russia" are both false. But "The moon is made of Gorgonzola iff Rome is in Italy" and "The moon lacks air iff Rome is in Russia" are false. However, "The moon lacks air iff Rome is in Italy" is true. ("Iff" has nothing to do with causes, explanations, or laws of nature.)

To solve Puzzle 2, ask the god, not the simple question "Is Dushanbe in Kirghizia?", but the more complex question "Are you True iff Dushanbe is in Kirghizia?" Then (in the absence of any geographical information) there are four possibilities:

1. The god is True and D. is in K.: then you get the answer "yes."

2. The god is True and D. is not in K.: this time you get "no."

3. The god is False and D. is in K.: you get the answer "yes," because only one statement is true, so the correct answer is "no," and the god, who is False, falsely says "yes."

4. The god is False and D. is not in K.: in this final case you get the answer "no," because both statements are false, the correct answer is "yes," and the god False falsely says "no."

So you get a yes answer to that complex question if D. is in K. and a no answer if it is not, *no matter to which of True and False you are speaking.* By noting the answer to the complex question you can find out whether D. is in K. or not.

The point to notice is that if you ask either True or False "Are you True iff X?" and receive your answer in English, then you get the answer "yes" if X is true and "no" if X is false, regardless of which of the two you are speaking to.

The solution to Puzzle 3 is quite similar: Ask True, not "Is Dushanbe in Kirghizia?" but "Does 'da' mean yes iff D. is in K.?" There are again four possibilities:

1. "Da" means yes and D. is in K.: then True says "da."

2. "Da" means yes and D. is not in K.: then True says "ja" (meaning no).

3. "Da" means no and D. is in K.: then True says "da" (meaning no).

4. "Da" means no and D. is not in K.: then both statements are false, the statement " 'Da' means yes iff D. is in K." is true, the correct answer (in English) to our question is "yes," and therefore True says "ja."

Thus you get the answer "da" if D. is in K. and the answer "ja" if not, regardless of which of "da" and "ja" means yes and which means no. The point this time is that if you ask True "Does 'da' mean yes iff Y?", then you get the answer "da" if Y is true and you get the answer "ja" if Y is false, regardless of which means which.

Combining the two points, we see that if you ask one of True and False (who we again suppose only answer "da" and "ja"), the very complex question "Does 'da' mean yes iff, you are True iff X?" then *you will get the answer "da" if X is true and get the answer "ja" if X is false*, regardless of whether you are addressing the god True or the god False, and regardless of the meanings of "da" and "ja."

We can now solve The Hardest Logic Puzzle Ever.

Your first move is to find a god who you can be certain is not Random, and hence is either True or False.

To do so, turn to A and ask Question 1: *Does "da" mean yes iff, you are True iff B is Random?* If A is True or False and you get the answer "da," then, as we have seen, B is Random, and therefore C is either True or False; but if A is True or False and you get the answer "ja," then B is not Random, therefore B is either True or False.

But what if A is Random?

If A is Random, then neither B nor C is Random!

So if A is Random and you get the answer "da," C is not Random (neither is B, but that's irrelevant), and therefore C is either True or False; and if A is Random and you get the answer "ja," B is not Random (neither is C, irrelevantly), and therefore B is either True or False.

Thus, *no matter whether A is True, False, or Random,* if you get the answer "da" to Question 1, C is either True or False, and if you get the answer "ja," B is either True or False!

Now turn to whichever of B and C you have just discovered is either True or False—let us suppose that it is B; if it is C, just interchange the names B and C in what follows—and ask Question 2: *Does "da" mean yes iff Rome is in Italy?* True will answer "da," and False will answer "ja." Thus, with two questions, you have either identified B as True or identified B as False.

For our third and last question, turn again to B, whom you have now either identified as True or identified as False, and ask Question 3: *Does "da" mean yes iff A is Random?*

Suppose B is True. Then if you get the answer "da," then A is Random, and therefore A is Random, B is True, C is False, and you are done; but if you get the answer "ja," then A is not Random, so A is False, B is true, C is Random, and you are again done.

Suppose B is False. Then if you get the answer "da," then since B speaks falsely, A is not Random, and therefore A is True, B is False, C is Random, and you are done; but if we get "ja," then A is Random, and thus B is False, and C is True, and you are again done. FINIS.

Well, I wasn't speaking falsely or at random when I said that the puzzle was hard, was I?

A brief remark about the significance of the Hardest Puzzle:

There is a law of logic called "the law of excluded middle," according to which either X is true or not-X is true, for any statement X at all. ("The law of non-contradiction" asserts that statements X and not-X aren't *both* true.) Mathematicians and philosophers have occasionally attacked the idea that excluded middle is a logically valid law. We can't hope to settle the debate here, but can observe that our solution to Puzzle 1 made essential use of excluded middle, exactly when we said *"Whether the middle card is an ace or not ... "* It is clear from The Hardest Logic Puzzle Ever, and even more plainly from Puzzle 1, that our ability to reason about alternative possibilities, even in everyday life, would be almost completely paralyzed were we to be denied the use of the law of excluded middle.

By the way, Dushanbe is in Tajikistan, not Kirghizia.

30

Gödel's Second Incompleteness Theorem Explained in Words of One Syllable

First of all, when I say "proved," what I will mean is "proved with the aid of the whole of math." Now then: two plus two is four, as you well know. And, of course, *it can be proved* that two plus two is four (proved, that is, with the aid of the whole of math, as I said, though in the case of two plus two, of course we do not need the *whole* of math to prove that it is four). And, as may not be quite so clear, it can be proved that it can be proved that two plus two is four, as well. And it can be proved that it can be proved that it can be proved that two plus two is four. And so on. In fact, if a claim can be proved, then it can be proved that the claim can be proved. And *that* too can be proved.

Now: two plus two is not five. And it can be proved that two plus two is not five. And it can be proved that it can be proved that two plus two is not five, and so on.

Thus: it can be proved that two plus two is not five. Can it be proved as well that two plus two *is* five? It would be a real blow to math, to say the least, if it could. If it could be proved that two plus two is five, then it could be proved that five is not five, and then there would be *no* claim that could *not* be proved, and math would be a lot of bunk.

So, we now want to ask, can it be *proved* that it can't be proved that two plus two is five? Here's the shock: no, it can't. Or to hedge a bit: *if* it can be proved that it can't be proved that two plus two is five, *then* it can be proved as well that two plus two is five, and math is a lot of bunk. In fact, if math is not a lot of bunk, then no claim of the form "claim X can't be proved" can be proved.

From *Mind* 103 (1994): 1–3. Reprinted by kind permission of Oxford University Press.

So, if math is not a lot of bunk, then, though it can't be proved that two plus two is five, it can't be proved *that* it can't be proved that two plus two is five.

By the way, in case you'd like to know: yes, it *can* be proved that if it can be proved that it can't be proved that two plus two is five, then it can be proved that two plus two is five.

"I wish he would explain his explanation"

If, as we shall assume, the whole of mathematics can be formalized as a formal theory of the usual sort (no small assumption), then there is a formula $\text{Proof}(x, y)$ of the language (of that theory) obtainable from a suitable description of the theory ("as" a formal theory) that meets the following three conditions:

(i)   if $\vdash p$, then $\vdash \Box p$,

(ii) $\vdash (\Box(p \to q) \to (\Box p \to \Box q))$, and

(iii) $\vdash (\Box p \to \Box\Box p)$

for all sentences p, q of the language. We have written: $\Box p$ to abbreviate: $\exists x \text{Proof}(x, \ulcorner p \urcorner)$, where $\ulcorner p \urcorner$ is a standard representation in the language for the sentence p. ($\ulcorner p \urcorner$ might be the numeral for the Gödel number of p.) "$\vdash$" is a preposed verb phrase (of our language) meaning "is provable in the theory." "$\text{Proof}(x, y)$" is a noun phrase (of our language) denoting a formula (of the theory's language) whose construction parallels any standard definition of " ... is a proof of __ in the theory." Thus, for any sentence p of the language, $\Box p$ is another sentence of the language that may be regarded as saying that p is provable in the theory.[1]

Conditions (i), (ii), and (iii) are called the Hilbert–Bernays–Löb derivability conditions; they are satisfied by all reasonable formal theories in which a certain small amount of arithmetic can be proved.

Since the theory is standard, all tautologies in its language are provable in the theory, and all logical consequences in its language of provable statements are provable.

It follows that for all sentences p, q,

(iv)   if $\vdash (p \to q)$, then $\vdash (\Box p \to \Box q)$.

For: if $\vdash (p \to q)$, then by (i) $\vdash \Box(p \to q)$; but by (ii), $\vdash (\Box(p \to q) \to (\Box p \to \Box q))$, and then $\vdash (\Box p \to \Box q)$ by modus ponens.

$\bot$ is the zero-place truth-functional connective that is always evaluated as false. Of course $\bot$ is a contradiction. We shall need to observe later

---

[1] For an extended account of the application of modal logic to the concept of provability in formal theories, see (Boolos, 1993b).

that $(\neg q \rightarrow (q \rightarrow \perp))$ is a tautology. If $\perp$ is not one of the primitive symbols of the language, it may be defined as any refutable sentence, e.g., one expressing that two plus two is five.

With the aid of $\perp$, there is an easy way to say that the theory is consistent: $\not\vdash \perp$, i.e., $\perp$ is not provable in the theory. The sentence of the language stating that the theory is consistent can thus be taken to be $\neg\Box\perp$, which is identical with $\neg\exists x \text{Proof}(x, \ulcorner\perp\urcorner)$.

We may prove Gödel's second incompleteness theorem, which states that if the theory is consistent, then the sentence of the language stating that the theory is consistent is not provable in the theory, as well as the theorem that the second incompleteness theorem is provable in the theory ("the formalized second incompleteness theorem"), as follows.

Via the technique of diagonalization, introduced by Gödel in "On formally undecidable propositions ...,"[2] a sentence p can be found that is equivalent in the theory to the statement that p is unprovable in the theory, i.e. a sentence such that

1.	$\vdash p \leftrightarrow \neg\Box p$	
2.	$\vdash p \rightarrow \neg\Box p$	truth-functionally from 1
3.	$\vdash \Box p \rightarrow \Box\neg\Box p$	by (iv) from 2
4.	$\vdash \Box p \rightarrow \Box\Box p$	by (iii)
5.	$\vdash \neg\Box p \rightarrow (\Box p \rightarrow \perp)$	a tautology
6.	$\vdash \Box\neg\Box p \rightarrow \Box(\Box p \rightarrow \perp)$	by (iv) from 5
7.	$\vdash \Box(\Box p \rightarrow \perp) \rightarrow$ $(\Box\Box p \rightarrow \Box\perp)$	by (ii)
8.	$\vdash \Box p \rightarrow \Box\perp$	truth-functionally from 3, 6, 7, and 4
9.	$\vdash \neg\Box\perp \rightarrow p$	truth-functionally from 8 and 1
10.	$\vdash \Box\neg\Box\perp \rightarrow \Box p$	by (iv) from 9
11.	$\vdash \neg\Box\perp \rightarrow \neg\Box\neg\Box\perp$	truth-functionally from 8 and 10.

(We have omitted outermost parentheses in (1) through (11).)

Thus if $\vdash \neg\Box\perp$, then both $\vdash \neg\Box\neg\Box\perp$, by (11), and $\vdash \Box\neg\Box\perp$, by (i), whence $\vdash \perp$, by the propositional calculus. So if $\not\vdash \perp$, then $\not\vdash \neg\Box\perp$.

[2](Gödel, 1931).

Afterword

Fundamental Theorems of Provability Logic

C. I. Lewis launched modern modal logic when he added to classical sentential logic a two place connective $p \dashv q$ intended to represent "the conclusion that q is logically deducible from the premiss that p." This was later analyzed as $\Box(p \to q)$ or $\neg \Diamond (p \land \neg q)$, where $\Box p$ represents necessity in the specific sense of "it is a logical theorem that p" or "it is logically provable that p" and $\Diamond p$ represents possibility in the specific sense of "it is not logically disprovable that p" or "it is logically consistent that p." Agreement was not achieved as to what are the right laws for box $\Box$ and diamond $\Diamond$, and systems of modal logic proliferated. In formulating modal systems, as primitive connectives may be taken the false $\bot$, the conditional $\to$, and necessity $\Box$. Then negation $\neg p = p \to \bot$, possibility $\Diamond p = \neg \Box \neg p$, and so on may be taken as defined. All modal systems include all axioms and rules of classical logic, so that only their distinctively modal axioms and rules need to be mentioned. Writing $\vdash$ for provability in the system, the *minimal* system K has the following rule A0 of *necessitation* and axiom A1 of *distribution:*

A0   If $\vdash A$, then $\vdash \Box A$
A1 $\vdash \Box(p \to q) \to (\Box p \to \Box q)$

A *normal* modal logic is any having both this rule and axiom, the better-known ones being obtained simply by adding one or a few additional modal axioms to K. Well-known ones include K4, obtained by adding A2 below; and S4, obtained by adding both A2 and A3 below:

A2 $\vdash \Box p \to \Box\Box p$
A3 $\vdash \Box p \to p$

Perhaps the plurality candidate for the correct logic when the box is read as "it is logically provable that" would be S4. The heuristic justification for A0 is the thought that if it something can be logically proved, then exhibiting a logical proof of it should suffice as a logical proof that it can be logically proved. As for A1, the thought is that if a conditional and its antecedent can both be logically proved, logically proving first the one and then the other, and then inferring the consequent from the conditional and antecedent, should suffice as a logical proof of the consequent. As for A2, the thought is much the same as for A0. As for A3, the thought is that any axioms from which a proof begins are true, and the rules of inference involved in a derivation preserve truth, so that anything that is provable is true, and in particular, if it is provable that p, then it is true that p, and it is true that p if and only if p.

Gödel himself raised the question which of the many modal logics is the correct one if we read the box as arithmetical provability or provability in PA^1. To state his question more precisely, recall the notation from the discussion of his work in the Introduction to Part III: $\ulcorner \alpha \urcorner$ is the code numeral of α and $\pi(x)$ is the provability predicate. An assignment $*$ of closed formulas of the language of PA^1 to the sentence letters $p, q, r, \ldots$ of modal logic can be extended to all modal formulas recursively, as follows:

$$\bot^* = \bot$$
$$(A \to B)^* = A^* \to B^*$$
$$(\Box A)^* = \pi(\ulcorner A^* \urcorner)$$

Call such an assignment an *arithmetic realization*. Call a modal formula A *apodictic* or *always-provable* for PA^1 if A^* is provable in PA^1 for any arithmetic realization $*$, and *veracious* or *always-true* if A^* is true for any arithmetic realization $*$. Gödel's question amounts to this: Which of the many modal logics is the one whose theorems are precisely the apodictic formulas? One may ask the same question for veracious formulas. Gödel himself noted that the correct logic would have the rule A0 and the axioms A1 and A2. That is because the provability predicate has the corresponding properties. Writing $\vdash$ for provability in PA^1, for all α and β we have:

B0   If $\vdash \alpha$, then $\vdash \pi(\ulcorner \alpha \urcorner)$

B1 $\vdash \pi(\ulcorner \alpha \to \beta \urcorner) \to (\pi(\ulcorner \alpha \urcorner) \to \pi(\ulcorner \beta \urcorner))$

B2 $\vdash \pi(\ulcorner \alpha \urcorner) \to \pi(\ulcorner \pi(\ulcorner \alpha \urcorner) \urcorner)$

The proofs of B0–B2 are more formal versions of the heuristic arguments given above for A0–A2. But nothing like the heuristic argument for A3, which turned on the notion of truth, for which there is no predicate in the language of PA^1, goes through. In fact the correct modal logic would *not* have the axiom A3, since the provability predicate does *not* have the corresponding property B3, which would be $\vdash \pi(\ulcorner \alpha \urcorner) \to \alpha$

M. H. Löb in the 1950s established a curious additional property of the provability predicate. Gödel had considered a closed formula γ that says of itself that is not provable, in the sense that $\gamma \leftrightarrow \neg \pi(\ulcorner \gamma \urcorner)$ is provable, and he showed that such a γ is in fact unprovable. Leon Henkin considered a closed formula β that says of itself that it *is* provable, in the sense that $\beta \leftrightarrow \pi(\ulcorner \beta \urcorner)$ is provable, and he asked whether such a β is in fact provable.

Löb answered this question in the affirmative, and indeed showed that for any closed formula α, even without the assumption that $\alpha \to \pi(\ulcorner a \urcorner)$ is provable, *just the assumption that* $\pi(\ulcorner \alpha \urcorner) \to \alpha$ *is provable already implies that* α *is provable.* Moreover, this italicized statement is *itself* provable, which is to say that we have:

B4 $\quad \vdash \pi(\ulcorner \pi(\ulcorner \alpha \urcorner) \to \alpha \urcorner) \to \pi(\ulcorner \alpha \urcorner)$

This is something of a brain-twister when considered in isolation from its origin in Henkin's question. Note that Gödel's second incompleteness theorem is an immediate consequence of B4. For we may express the inconsistency of PA^1 by $\pi(\ulcorner \bot \urcorner)$ and hence the consistency of PA^1 by $\neg \pi(\ulcorner \bot \urcorner) = \pi(\ulcorner \bot \urcorner) \to \bot$, and then B4 with $\alpha = \bot$ says that the consistency of PA^1 is provable in PA^1 only if PA^1 is inconsistent. In proving B4, Löb used the *arithmetic fixed-point* theorem, according to which for any formula $\sigma(x)$ in the language of PA^1 there is a closed formula γ in the language of PA^1 such that $\gamma \leftrightarrow \sigma(\ulcorner \gamma \urcorner)$ is provable. (The Gödel sentence and the Henkin sentences are the instances where $\sigma(x)$ is $\neg \pi(x)$ and where $\sigma(x)$ is $\pi(x)$.) Let us give the name GL to the modal logic extending K4 by adding the following axiom A4, the counterpart to B4:

A4 $\quad \vdash \Box(\Box p \to p) \to \Box p$

Then Löb's result gives the following *arithmetic soundness* theorem: Every theorem of GL is apodictic for PA^1. (Actually, as was shown independently by Dick de Jongh and Saul Kripke and Giovanni Sambin, axiom A2 becomes redundant given axiom A4.)

Kripke introduced a model theory for modal logic, as follows. A *frame* consists of a set W, and a dyadic relation R on W. The frame is called finite if the set W is finite, reflexive or symmetric or transitive if the relation R is reflexive or symmetric or transitive, and so on. Heuristically, W is thought of as the set of possible states of the world, and Rxy is thought of as meaning that y is possible *relative* to x—whatever *that* means. A *frame model* consists of these together with a *valuation* or relation V between elements of W and sentence letters. Heuristically, Vxp is thought of as meaning that p is true at x. The notion $w \models A$ of A being true at world w is then defined for compound A by recursion. The recursion clauses read:

$w \not\models \bot$
$w \models A \to B$ iff if $w \models A$, then $w \models B$
$w \models \Box A$ iff for all w' such that wRw', $w' \models A$

Given a class Γ of frames, a sentence is valid for Γ if it comes out true at every world in every model whose frame belongs to Γ. Various modal systems can be characterized as having as theorems precisely the sentences valid for this, that, or the other class of frames, determined by this, that, or the other condition on the relation R. The minimal system K is characterized by the class of arbitrary frames, with no special condition on R. The system K4 is characterized by the class of transitive frames; the system S4 by the class of transitive and reflexive frames. In all these results, there is a pendant to the effect that the system has the *finite model property*. In each case, where the logic has been mentioned as being characterized by some class Γ of frames, it is equally characterized by the class Γ_0 of *finite* members of Γ. This property implies that the system is decidable: There is a "mechanical" procedure for determining, given any modal formula, whether or not it is a theorem of the system, namely, the procedure of just searching through proofs and finite models simultaneously, until one either finds a proof or a finite counter-model. (In some cases these decidability results had been obtained earlier using other methods by J. C. C. McKinsey.)

Krister Segerberg gave a model-theoretic characterization of GL. It can be made to seem more intuitive by considering an alternative *temporal* reading of the box as "it has always been the case that" and of the diamond as "it has sometime been the case that." On such a reading, one doesn't want the axiom A2, which is equivalent to $\neg(\neg p \wedge \neg \diamond \neg p)$, since something may now fail for the *first* time, having previously never failed. And one may, depending on one's conception of the structure of time, want additional axioms not usually considered on the original reading of the box. Consider for instance axiom A4 above, which is equivalent to $\diamond \neg p \rightarrow \diamond(\neg p \wedge \neg \diamond \neg p)$. This says that if something has sometime failed, then it has sometime failed for the first time. One may want this if one conceives of time as so structured that there is no infinite sequence of earlier and earlier times. On the alternative reading of the box, in a model $M = (W, R, V)$ the set W would be thought of as the set of present of past stages of the world, and the relation Rxy would be thought of as meaning that y is past *relative* to x—in other words, that x is *later* than y. Segerberg's result is that the system GL is characterized by the class of transitive and converse-well-founded frames. (According to the author, this result had actually been obtained by Kripke and communicated privately to him, but not published.) What does this last condition mean? It means that there is no infinite sequence of elements $x_1, x_2, x_3, \ldots$ of W such that each stands to the next in the converse of the relation R. And what does *that* mean? It means that if R means "later," so that its converse means "earlier," then there is no infinite sequence of earlier and earlier elements of W. Again the proof gives the finite model property. Note that converse-well-foundedness of R implies *irreflexivity* of R, the non-existence of x such that Rxx. For *finite* transitive frames, converse-well-foundedness simply reduces to irreflexivity.

De Jongh and Sambin highlighted the correspondence between GL and arithmetic provability by establishing a *modal fixed-point* theorem, a kind of parallel to the arithmetic fixed-point theorem. The statement of the full theorem is complicated, and best approached in stages. Let $S(p)$ be a formula in which every occurrence of the sentence letter p is in the scope of a modality $\Box$. First, suppose there are no other sentence letters in $S(p)$. It can be proved that there is a formula G not involving p such that $\vdash G \leftrightarrow S(G)$. For example, for $S(p) = \neg\Box p$, this would be a formula asserting its own unprovability, as in Gödel's first incompleteness theorem. Further, the formula G can be taken to involve no other sentences letters either. In the example, it turns out it can be taken to be $\neg\Box\bot$, the formula asserting consistency, as in Gödel's second incompleteness theorem. Yet further, the formula G is unique in the sense that if G' is any other such formula, then $\vdash G \leftrightarrow G'$. In the example, any formula asserting its own unprovability is equivalent to the formula asserting consistency—a new link between the two incompleteness theorems. Finally, $S(p)$ can be allowed to contain other sentence letters besides p after all (which of course will have to be allowed in G). The conclusion of the full final result may be stated as follows: There is a formula H not involving p or any sentence letter not in $S(p)$ such that $\vdash (p \leftrightarrow S(p)) \leftrightarrow (p \leftrightarrow H)$. For the history of this result, including the work of Claudio Bernardi and Craig Smorynski on preliminary partial results, see the joint paper (Boolos and Sambin, 1991).

Robert Solovay completed the picture by proving the *arithmetical completeness* theorem: Every apodictic modal formula A is a theorem of GL. Or equivalently, if A is *not* a theorem of GL, then A is *not* apodictic for PA^1. Note that to say that A is not a theorem of GL is equivalent to saying that there is a *counter-model* to A, meaning a finite, transitive, irreflexive model $M = (W, R, V)$ in which A comes out false at some x in W. To say that A is not apodictic for PA^1 is the same as saying that there is a *counter-realization* for A, an arithmetic realization $*$ such that A^* is not provable in PA^1. Solovay showed how to construct a counter-realization given any counter-model. Besides thus answering Gödel's original question, he clarified the relationship between the class of apodictic and the class of veracious closed formulas, in such a way as to derive the decidability of the latter from the previous established decidability of the former. In one direction the relationship is clear: α is apodictic or always-provable if and only if $\Box\alpha$ is veracious or always-true. In the other direction, given α, let α_0 be the conjunction of all formulas $\Box\beta \to \beta$ with β a subformula of α. Then α is veracious if and only if $\alpha_0 \to \alpha$ is apodictic.

Further Topics in Provability Logic

None of the author's articles on provability logic has been included in the present volume, partly because of the more technical character of these articles, mainly because of the availability of the author's book *The Logic of Provability.* Throughout the book new proofs, some due to the author and some to others, are given for many of the results reported, whether the author's or others'. The early chapters of the book are mainly devoted to detailed exposition of the results just outlined. The late chapters of the book are mainly devoted to detailed exposition of work of Solovay and of the "ex-Soviet school" (Sergei Artemov, Giorgie Dzhaparidze, Konstantin Ignatiev, Valery Vardanyan, et al.), much of this work having previously been unpublished or published only in Russian. The author's own research results are found especially in the middle chapters. The book supersedes, either by directly including their results or by including more general ones that subsume them, the author's earlier book, *The Unprovability of Consistency,* and the bulk of his research papers, and supersedes as well his various abstracts (Boolos, 1975a) and (Boolos, 1987c), reviews (Boolos, 1981b) and (Boolos, 1988b), and popularization (Boolos, 1984b). (One research paper noted below is not cannibalized in the book; and the co-authored exposition (Boolos and Sambin, 1991) contains historical information still not available elsewhere in print.) The author's work is briefly summarized under seven heads in the paragraphs that follow.

Special classes of formulas in GL. Chapter 7 of the book begins with the author's oldest contribution to the subject, a *normal form theorem,* according to which any letterless formula is provably equivalent in GL to a truth-functional compound of formulas from among the sequence $\perp, \Box\perp$, $\Box\Box\perp$, $\Box\Box\Box\perp$, and so on. (This normal form theorem was subsequently but independently found by Johann van Benthem and by Roberto Magari.) Note that for A letterless, A^* is the same for all arithmetic realizations $*$ Since $(\perp)^* = \perp, (\Box\perp)^* = \pi(\ulcorner\perp\urcorner)$, and so on, are all untrue and unprovable, the normal form theorem gives decidability for the classes of apodictic and veracious letterless formulas (independently of and historically prior to Solovay's theorems). This answered an item from a list of questions Harvey Friedman circulated as a challenge to logicians in the early 1970s. (Friedman's problem was independently solved by Bernardi and Franco Montagna.) The decidability result can be extended to give decidability (still independently of Solovay's theorems) for formulas of the kind whose existence is guaranteed by the basic modal fixed-point theorem, and indeed the proof of the result can be adapted to give a proof of that theorem. The middle of Chapter 7 presents further results of the author on other special classes of formulas, notably on the relationship between *reflection principles* or formulas of the form $\Box\beta \rightarrow \beta$ and *iterated consistency assertions* or formulas of the form $\neg\Box\perp, \neg\Box\Box\perp, \neg\Box\Box\Box\perp$, and so on. Solovay showed that normal forms do not exist even for formulas with just a single sentence letter p, and the author's (simplified) proof is given at the end of Chapter 7.

(The exposition in Chapter 7 generally supersedes articles (Boolos, 1976), (Boolos, 1977), (Boolos, 1979a), and (Boolos, 1982b).)

Global properties of GL. One of the key properties of classical logic is stated in the *Craig interpolation theorem:* If $A \to C$ is valid, then there is a B all of whose non-logical symbols occur both in A and in C such that $A \to B$ is valid and $B \to C$ is valid. The author established that this result holds for GL also, and as expounded in Chapter 8, it yields an alternative proof of the generalized modal fixed-point theorem. Another alternative proof of the generalized modal fixed-point theorem due to the author is also expounded in the same chapter. (The proof of the interpolation theorem in Chapter 8 follows Smorynski, who obtained that result independently.) Another of the key properties of classical logic is decidability, which as already stated holds also for GL in consequence of the finite model property. The author shows in Chapter 10 that there is a decision procedure more practically feasible in examples of interest. For the tree method, familiar from textbooks for classical logic, can be adapted to GL.

Modal systems related to GL. Löb's work was in response to a question of Leon Henkin. The Gödel sentence, which asserts its own unprovability, is indeed unprovable. What, Henkin asked, of the similar sentence that asserts its own provability? Löb showed it is indeed provable. While Löb's work essentially established the proof-theoretic counterpart of the modal axiom A4 above, to answer Henkin's question it would have sufficed to establish the counterpart of the following weaker axiom A4$^-$

A4$^-$ $\vdash \Box(\Box p \leftrightarrow p) \to \Box p$

The author conjectured that, in contrast to the system GL obtained by adding A4 to K, the system GH obtained by adding A4$^-$ does not yield axiom A2 above as a theorem. This conjecture was proved by Magari (whose proof was subsequently simplified by Max Creswell). Since the author and Sambin together proved that A2 is valid for the class of frames for which A4$^-$ is valid, it follows that there is no class of frames such that all and only the formulas that are theorems of GH are valid in that class. (The exposition in the book presents a proof by Lon Berk that A4 and A4$^-$ are valid in exactly the same class of frames.) GH is an *incomplete logic,* and improving on previous examples of such logics, one whose characteristic axiom A4$^-$ involves only a single sentence letter, and nesting of boxes only to depth two. This material is expounded in Chapter 11 (which exposition generally supersedes article (Boolos and Sambin, 1985)).

Modal systems related to GL and alternate notions of realization. Since there is no truth-predicate in the language of PA^1, one cannot literally say in that language that a given closed formula α is both provable and true.

But one can come close, by saying $\pi(\ulcorner \alpha \urcorner) \wedge \alpha$. Suppose we modify the notion of realization by amending the recursion clause for the box to read as follows:

$$(\Box A)^* = \pi(\ulcorner A^* \urcorner) \wedge A^*$$

The question then arises whether there are modal systems whose theorems are precisely the apodictic and precisely the veracious formulas for this alternate notion of realization, and that thus correspond to GL and GLS for the original notion of realization. In fact, one and the same system corresponds both to GL and to GLS, a system known as Grz for Grzegorczyk. It is obtained by adding to K the following axiom:

A5 $\vdash \Box(\Box(p \rightarrow \Box p) \rightarrow p) \rightarrow p$

(It was shown by W. J. Blok and K. E. Pledger that A2 and A3 are then derivable as theorems. It was shown by Segerberg that Grz is characterized by reflexive, transitive, and converse-weakly-well-founded frames, or equivalently by finite reflexive, transitive, and anti-symmetric frames.) Chapter 12 expounds the author's results both about the correspondence with GL (independently obtained by Rob Goldblatt and apparently also by Smorynski) and with GLS. (The exposition in Chapter 12 generally supersedes the articles (Boolos, 1980b), (Boolos, 1980c), and (Boolos, 1980d). Also due to the author, and independently to Arnon Avron, is the Craig interpolation theorem for Grz.)

Alternate notions of realization. The author is one of several who showed that Solovay's theorem holds *uniformly,* meaning that there is a single arithmetic realization $*$ such that for any A that is not apodictic, A^* is not provable in PA^1. This result is expounded just after Solovay's theorem at the end of Chapter 9 (which exposition generally supersedes article (Boolos, 1982a)). Other variations on Solovay's theorem involve alternate interpretations of the box and diamond, for which some background is required. There is a hierarchy of complexity among formulas of the language of PA^1. The hierarchy begins with some very simple formulas called the *limited* formulas $\alpha(x_1, \ldots, x_n)$ Closed formulas of this class can be verified or falsified by computation, and if true can be proved in PA^1, essentially just by exhibiting their computational verification. The Σ^0_0 and Π^0_0 are just the limited formulas, while the Σ^0_{n+1} and Π^0_{n+1} formulas are those of forms $\exists y \, \alpha(x_1, \ldots, x_n, y)$ and $\forall y \, \alpha(x_1, \ldots, x_n, y)$ with α being Π^0_n and with α being Σ^0_n, respectively. A set of natural numbers is called Σ^0_n or Π^0_n if there is a Σ^0_n or Π^0_n formula $\sigma(x)$ or $\pi(x)$ such that the set in question is the set of natural numbers of which that formula is true; and a set is called Δ^0_n if it is both Σ^0_n and Π^0_n. The Δ^0_1 sets are also known as the *recursive* *sets*, and the Σ_1^0 sets as the *recursively enumerable sets*. The *consistency* of a formula β with PA^1, meaning the consistency of the theory $PA^1 + \beta$, is equivalent to the assertion that in $PA^1 + \beta$ all the limited closed formulas that are provable are true. By analogy, one can define n-consistency to mean that all Σ_n^0 closed formulas that are provable are true, and ω-consistency to mean n-consistency for all n. The dual notion to that of β being ω-consistent (the notion of β being such that $\neg\beta$ is not ω-consistent) amounts to β being provable in PA^1 by one application of the ω-*rule*, the infinitary rule that permits $\forall x \alpha(x)$ to be inferred from $\alpha(0), \alpha(\S 0), \alpha(\S\S 0)$, and so on. Provability by unlimited applications of this rule is called ω-*provability*, but for PA^1 this notion simply reduces to that of truth, every true closed formula being ω-provable. For theories with richer languages, beginning with PA^2, ω-provability does not just reduce to truth. Articles (Boolos, 1985a) and (Boolos, 1980a) adapt Solovay's proof to show that GL still characterizes the apodictic formulas if the notion of realization is modified to take 1-consistency or ω-consistency and the dual thereof as the interpretation of the diamond and the box. Solovay himself showed that GL still characterizes the apodictic formulas taking ω-provability and its dual as the box and diamond, provided one considers not arithmetic realizations but analytic realizations, involving formulas of the language of PA^2.

Bimodal logic. Dzhaparidze developed a *bimodal logic* GLB with two styles of box and diamond, and considered arithmetic realizations with one style of diamond interpreted by ordinary consistency, and the other style interpreted by ω-consistency, and showed that GLB yields as theorems all and only the apodictic formulas. (Ignatiev simplified the original very complicated proofs.) Finally, the author showed the techniques of Solovay and Ignatiev could be combined to prove that GLB yields as theorems all and only the apodictic formulas when the two boxes are interpreted as ordinary provability and ω-provability, and analytic rather than arithmetic realizations are considered. The exposition of this material on alternate notions of realization and on bimodal logic, with related results, occupies Chapters 14–16 (the latter part of which largely supersedes article (Boolos, 1993a)).

Predicate provability logic. Quine observed that while we have a rigorously defined notion of what it is for a closed formula α to be provable, we have no rigorously defined notion of what it is for an open formula $\beta(x)$ to be provable *of a thing,* and concluded such combinations of modalities and quantifiers as $\exists x \, \Box Ux$ are meaningless if the box is read as "it is provable that ..." One might hope to make sense of the notion of $\beta(x)$ being provable *of a thing* by defining this to hold just in case $\beta(t)$ is provable for some term t denoting the thing, but this doesn't work unless some reason can be given for privileging one particular such term t over all others, since

$\beta(t)$ may turn out to be provable for some t and not others. The author points out that in arithmetic (in apparent contrast with analysis and set theory) it *does* seem reasonable to privilege certain terms for the things we are concerned with. Namely, the things we are concerned with are the natural numbers, and it seems reasonable to privilege the standard numerals $0, \S 0, \S\S 0$, and so on, as terms for them. And indeed that is what is done in predicate provability logic. That one does *not* get a system of modal logic characterizing the apodictic formulas just by adjoining the apparatus of quantification theory to GL was established by Montagna in 1984. Ultimately, it has turned out that there is *no* system of modal logic that characterizes the apodictic formulas at the predicate level. The reason is that the set of (code numbers of) theorems of any system will be a Σ_1^0 set, since the provability predicate $\pi(x)$ can be written as a Σ_1^0 formula, but the set of (code numbers of) apodictic formulas is not Σ_1^0, by a theorem of Vardanyan, who gave an exact characterization of the complexity of this set. Joint work of Vann McGee and the author, extending results of Artemov, gave an exact characterization (cited in the book as a result of the author, McGee, and Vardanyan) of the complexity of the set of (code numbers of) veracious formulas, which is even farther from being Σ_1^0. The set of (code numbers of) true formulas is not Σ_1^0 or Π_1^0 for any n, since the truth-predicate cannot be written as a formula of the language, and the set of (code numbers of) veracious formulas is even more complicated than that. The exposition of this material occupies Chapter 17 (which generally supersedes article (Boolos and McGee, 1987)).

Bibliography

Barwise, J. 1979. On branching quantifiers in English. *Journal of Philosophical Logic*, *8*, 47–80.

Benacerraf, P. 1960. *Logicism: Some Considerations*. Ph.D. thesis, Department of Philosophy, Princeton University.

Benacerraf, P. 1995. Frege: The last logicist. In (Demopoulos, 1995), pp. 41–67.

Benacerraf, P. and Putnam, H., eds. 1964. *Philosophy of Mathematics: Selected Readings*. First edition. Oxford: Blackwell.

Benacerraf, P. and Putnam, H., eds. 1983. *Philosophy of Mathematics: Selected Readings*. Second edition. Cambridge: Cambridge University Press.

Bolzano, B. 1851. *Paradoxien des Unendlichen*. Leipzig: C. H. Reclam.

Boole, G. 1847. *The Mathematical Analysis of Logic, Being an Essay Toward a Calculus of Deductive Reasoning*. Cambridge: Macmillan, Barclay, and Macmillan.

Boole, G. 1916. *The Laws of Thought* (1854). Chicago: Open Court.

Boolos, G. 1969. Effectiveness and natural languages. In S. Hook, ed., *Language and Philosophy*. New York University Press.

Boolos, G. 1970a. On the semantics of the constructible levels. *Zeitschrift für mathematische Logik und Grundlagen der Mathematik*, *16*, 139–148.

Boolos, G. 1970b. A proof of the Löwenheim–Skolem theorem. *Notre Dame Journal of Formal Logic*, *11*, 76–78.

Boolos, G. 1971. The iterative conception of set. *Journal of Philosophy*, *68*, 215–231. Reprinted in (Benacerraf and Putnam, 1983) and in this volume.

Boolos, G. 1973. A note on Beth's theorem. *Bulletin de l'Academie Polonaise des Sciences*, *21*, 1–2.

Boolos, G. 1974a. Arithmetical functions and minimization. *Zeitschrift für mathematische Logik und Grundlagen der Mathematik*, *20*, 353–354.

Boolos, G. 1974b. Reply to Charles Parsons' "Sets and classes". First published in this volume.

Boolos, G. 1975a. Friedman's 35th problem has an affirmative solution. *Notices of the American Mathematical Society*, *22*, A–646.

Boolos, G. 1975b. On Kalmar's consistency proof and a generalization of the notion of omega-consistency. *Archiv für mathematische Logik und Grundlagenforschung, 17,* 3–7.

Boolos, G. 1975c. On second-order logic. *Journal of Philosophy, 72,* 509–527. Reprinted in this volume.

Boolos, G. 1976. On deciding the truth of certain statements involving the notion of consistency. *Journal of Symbolic Logic, 41,* 779–781.

Boolos, G. 1977. On deciding the provability of certain fixed point statements. *Journal of Symbolic Logic, 42,* 191–193.

Boolos, G. 1979a. Reflection principles and iterated consistency assertions. *Journal of Symbolic Logic, 44,* 33–35.

Boolos, G. 1979b. *The Unprovability of Consistency: An Essay in Modal Logic.* Cambridge: Cambridge University Press.

Boolos, G. 1980a. Omega-consistency and the diamond. *Studia Logica, 39,* 237–243.

Boolos, G. 1980b. On systems of modal logic with provability interpretations. *Theoria, 46,* 7–18.

Boolos, G. 1980c. Provability in arithmetic and a schema of Grzegorczyk. *Fundamenta Mathematicæ, 106,* 41–45.

Boolos, G. 1980d. Provability, truth, and modal logic. *Journal of Philosophical Logic, 9,* 1–7.

Boolos, G. 1980e. Review of Raymond M. Smullyan, *What is the Name of This Book? Philosophical Review, 89,* 467–470.

Boolos, G. 1981a. For every A there is a B. *Linguistic Inquiry, 12,* 465–466.

Boolos, G. 1981b. Review of Robert M. Solovay, *Provability Interpretations of Modal Logic. Journal of Symbolic Logic, 46,* 661–662.

Boolos, G. 1982a. Extremely undecidable sentences. *Journal of Symbolic Logic, 47,* 191–196.

Boolos, G. 1982b. On the nonexistence of certain normal forms in the logic of provability. *Journal of Symbolic Logic, 47,* 638–640.

Boolos, G. 1984a. Don't eliminate cut. *Journal of Philosophical Logic, 13,* 373–378. Reprinted in this volume.

Boolos, G. 1984b. The logic of provability. *American Mathematical Monthly, 91,* 470–480.

Boolos, G. 1984c. Nonfirstorderizability again. *Linguistic Inquiry, 15,* 343.

Boolos, G. 1984d. On "Syllogistic inference". *Cognition, 17,* 181–182.

Boolos, G. 1984e. To be is to be the value of a variable (or some values of some variables). *Journal of Philosophy, 81,* 430–450. Reprinted in this volume.

Boolos, G. 1984f. Trees and finite satisfiability: Proof of a conjecture of Burgess. *Notre Dame Journal of Formal Logic, 25,* 193–197.

Boolos, G. 1985a. 1-consistency and the diamond. *Notre Dame Journal of Formal Logic*, *26*, 341–347.

Boolos, G. 1985b. The justification of mathematical induction. *PSA 1984*, *2*, 469–475. Reprinted in this volume.

Boolos, G. 1985c. Nominalist Platonism. *Philosophical Review*, *94*, 327–344. Reprinted in this volume.

Boolos, G. 1985d. Reading the *Begriffsschrift*. *Mind*, *94*, 331–344. Reprinted in (Demopoulos, 1995) and in this volume.

Boolos, G. 1986. Review of Yuri Manin, *A Course in Mathematical Logic*. *Journal of Symbolic Logic*, *51*, 829–830.

Boolos, G. 1986–87. Saving Frege from contradiction. *Proceedings of the Aristotelian Society*, *87*, 137–151. Reprinted in (Demopoulos, 1995) and in this volume.

Boolos, G. 1987a. The consistency of Frege's *Foundations of Arithmetic*. In J. J. Thomson, ed., *On Being and Saying: Essays for Richard Cartwright*, pp. 3–20. Cambridge, Mass.: The MIT Press. Reprinted in (Demopoulos, 1995) and in this volume.

Boolos, G. 1987b. A curious inference. *Journal of Philosophical Logic*, *16*, 1–12. Reprinted in this volume.

Boolos, G. 1987c. On notions of provability in provability logic. *Abstracts of the 8th International Congress of Logic, Methodology and Philosophy of Science*, *5*, 236–238.

Boolos, G. 1988a. Alphabetical order. *Notre Dame Journal of Formal Logic*, *29*, 214–215.

Boolos, G. 1988b. Review of Craig Smorynski, *Self-Reference and Modal Logic*. *Journal of Symbolic Logic*, *53*, 306–309.

Boolos, G. 1989a. Iteration again. *Philosophical Topics*, *17*, 5–21. Reprinted in this volume.

Boolos, G. 1989b. A new proof of the Gödel incompleteness theorem. *Notices of the American Mathematical Society*, *36*, 388–390. An afterword appeared under the title "A letter from George Boolos," ibid., p. 676. Both are reprinted in this volume.

Boolos, G., ed. 1990a. *Meaning and Method: Essays in Honor of Hilary Putnam*. Cambridge: Cambridge University Press.

Boolos, G. 1990b. On "seeing" the truth of the Gödel sentence. *Behavioral and Brain Sciences*, *13*, 655–656. Reprinted in this volume.

Boolos, G. 1990c. Review of Jon Barwise and John Etchemendy, *Turing's World* and *Tarski's World*. *Journal of Symbolic Logic*, *55*, 370–371.

Boolos, G. 1990d. Review of V. A. Uspensky, *Gödel's Incompleteness Theorem*. *Journal of Symbolic Logic*, *55*, 889–891.

Boolos, G. 1990e. The standard of equality of numbers. In (Boolos, 1990a), pp. 261–278. Reprinted in (Demopoulos, 1995) and in this volume.

Boolos, G. 1991. Zooming down the slippery slope. *NOÛS*, *25*, 695–706. Reprinted in this volume.

Boolos, G. 1993a. The analytical completeness of Dzhaparidze's polymodal logics. *Annals of Pure and Applied Logic*, *61*, 95–111.

Boolos, G. 1993b. *The Logic of Provability*. Cambridge: Cambridge University Press.

Boolos, G. 1993c. Whence the contradiction? *Aristotelian Society Supplementary Volume*, *67*, 213–233. Reprinted in this volume.

Boolos, G. 1994a. 1879? In P. Clark and B. Hale, eds., *Reading Putnam*, pp. 31–48. Oxford: Blackwell. Reprinted in this volume.

Boolos, G. 1994b. The advantages of honest toil over theft. In A. George, ed., *Mathematics and Mind*, pp. 27–44. Oxford: Oxford University Press. Reprinted in this volume.

Boolos, G. 1994c. Gödel's second incompleteness theorem explained in words of one syllable. *Mind*, *103*, 1–3. Reprinted in this volume.

Boolos, G. 1995a. Frege's theorem and the Peano postulates. *Bulletin of Symbolic Logic*, *1*, 317–326. Reprinted in this volume.

Boolos, G. 1995b. Introductory note to **1951*. In (Feferman et al., 1995), pp. 290–304. [Item **1951* is the 1951 Gibbs lecture by Gödel, "Some basic theorems on the foundations of mathematics and their implications."] Reprinted in this volume.

Boolos, G. 1995c. Quotational ambiguity. In (Leonardi and Santambrogio, 1995), pp. 283–296. Reprinted in this volume.

Boolos, G. 1996a. The hardest logical puzzle ever. *Harvard Review of Philosophy*, *6*, 62–65. Reprinted in this volume. An Italian translation by Massimo Piattelli–Palmarini, "L'indovinello piu difficile del mondo," appeared in *La Repubblica* 16 April 1992, pp. 36–37.

Boolos, G. 1996b. On the proof of Frege's theorem. In A. Morton and S. P. Stich, eds., *Benacerraf and his Critics*. Cambridge, Mass.: Blackwell. Reprinted in this volume.

Boolos, G. 1997a. Constructing Cantorian counterexamples. *Journal of Philosophical Logic*, *26*, 237–239. Reprinted in this volume.

Boolos, G. 1997b. Gottlob Frege and the *Foundations of Arithmetic*. First published in this volume.

Boolos, G. 1997c. Is Hume's principle analytic? Written for Richard G. Heck, Jr., ed., *Festschrift for Michael Dummett*, forthcoming. Reprinted in this volume.

Boolos, G. 1997d. Must we believe in set theory? Written for Gila Sher and Richard Tieszen, eds., *Between Logic and Intuition: Essays in Honor of Charles Parsons*, forthcoming from Cambridge University Press. Reprinted in this volume.

Boolos, G. and Heck, Jr., R. G. 1997. *Die Grundlagen der Arithmetik*, §§82–83 In M. Schirn, ed., *Philosophy of Mathematics Today*. Oxford: Clarendon Press. Reprinted in this volume.

Boolos, G. and Jeffrey, R. 1974. *Computability and Logic*. Cambridge: Cambridge University Press.

Boolos, G. and Jeffrey, R. 1985. *Computability and Logic*. Second edition. Cambridge: Cambridge University Press.

Boolos, G. and McGee, V. 1987. The degree of the set of sentences of predicate provability logic that are true under every interpretation. *Journal of Symbolic Logic, 52*, 165–171.

Boolos, G. and Putnam, H. 1968. Degrees of unsolvability of constructible sets of integers. *Journal of Symbolic Logic, 33*, 497–513.

Boolos, G. and Sambin, G. 1985. An incomplete system of modal logic. *Journal of Philosophical Logic, 14*, 351–358.

Boolos, G. and Sambin, G. 1991. Provability: The emergence of a mathematical modality. *Studia Logica, 50*, 1–23.

Burgess, J. P. 1984. Review of Crispin Wright's *Frege's Conception of Numbers as Objects*. *Philosophical Review, 93*, 638–640.

Cantor, G. 1932. *Gesammelte Abhandlungen mathematischen und philosophischen Inhalts*. Berlin: Springer-Verlag. Edited by E. Zermelo.

Carlson, L. 1982. Plural quantifiers and informational independence. *Acta Philosophica Fennica, 35*, 163–174.

Carnap, R. 1931. Die logizistische Grundlegung der Mathematik. *Erkenntnis, 2*. An English translation, "The Logicist Foundations of Mathematics," is contained in (Benacerraf and Putnam, 1983).

Carnap, R. 1937. *The Logical Syntax of Language*. New York: Routledge & Kegan Paul Ltd.

Cartwright, R. L. 1987. *Philosophical Essays*. Cambridge, Mass.: The MIT Press.

Cartwright, R. L. 1994. Speaking of everything. *NOÛS, 28*, 1–20.

Chaitin, G. 1970. Computational complexity and Gödel's incompleteness theorem. *AMS Notices, 17*, 672.

Church, A. 1956. *Introduction to Mathematical Logic, Vol. 1*. Princeton: Princeton University Press.

Collins, F. H. 1933. *Authors' & [sic] Printers' Dictionary*. Seventh edition. London: Humphrey Milford.

Davis, M., ed. 1965. *The Undecidable: Basic Papers on Undecidable Propositions, Unsolvable Problems and Computable Functions*. Hewlett, New York: Raven Press.

Davis, M. 1980. What is a computation? In L. A. Steen, ed., *Mathematics Today*. New York: Vintage books.

Dedekind, R. 1888. *Was sind und was sollen die Zahlen?* Braunschweig: Vieweg. An English translation, "The Nature and Meaning of Numbers," is contained in (Dedekind, 1901).

Dedekind, R. 1901. *Essays on the Theory of Numbers.* Chicago: Open Court. This volume contains translations by Wooster Woodruff Beman of Dedekind's "Stetigkeit und irrationale Zahlen" and "Was sind und was sollen die Zahlen?".

Demopoulos, W., ed. 1995. *Frege's Philosophy of Mathematics.* Cambridge, Mass.: Harvard University Press.

Dugac, P. 1976. *Richard Dedekind et les Fondements des Mathématiques.* Paris: J. Vrin.

Dummett, M. 1991. *Frege: Philosophy of Mathematics.* Cambridge, Mass.: Harvard University Press.

Feferman, S. 1984. Toward useful type-free theories. I. *Journal of Symbolic Logic, 49,* 75–111.

Feferman, S. et al., eds. 1986. *Kurt Gödel, Collected Works, Vol. I: Publications 1929–1936.* Oxford: Clarendon Press.

Feferman, S. et al., eds. 1990. *Kurt Gödel, Collected Works, Vol. II: Publications 1938–1974.* Oxford: Clarendon Press.

Feferman, S. et al., eds. 1995. *Kurt Gödel, Collected Works, Vol. III: Unpublished Essays and Lectures.* Oxford: Clarendon Press.

Felgner, U. 1971. *Models of ZF-Set Theory.* Lecture Notes in Mathematics 223. Berlin: Springer-Verlag.

Field, H. 1989. Platonism for cheap? Crispin Wright on Frege's context principle. In *Realism, Mathematics and Modality,* pp. 147–170. Oxford: Blackwell.

Frege, G. 1879. *Begriffsschrift, eine der arithmetischen nachgebildete Formelsprache des reinen Denkens.* Halle: Louis Nebert. An English translation is contained in (van Heijenoort, 1967).

Frege, G. 1884. *Die Grundlagen der Arithmetik. Eine logisch mathematische Untersuchung über den Begriff der Zahl.* Breslau: Wilhelm Koebner. For an English translation by J. L. Austin see *The Foundations of Arithmetic* (Oxford: Basil Blackwell, 1950).

Frege, G. 1893, 1903. *Grundgesetze der Arithmetik, begriffsschriftlich abgeleitet,* 2 vols. Jena: Pohle. Reprinted by Georg Olms, Hildesheim, 1962.

Frege, G. 1895. [Ausführungen über Sinn und Bedeutung]. Part of Frege's *Nachlaß.* An English translation, "Comments on Sense and Meaning," can be found in (Hermes et al., 1979).

Frege, G. 1897. Logik [1897]. Part of Frege's *Nachlaß.* An English translation, "Logic [1897]," can be found in (Hermes et al., 1979).

Frisch, M. H. et al., eds. 1982. *Writings of Charles S. Pierce: A Chronological Edition.* Bloomington, Ind.: Indiana University Press.

Gabriel, G. et al., eds. 1980. *Gottlob Frege: Philosophical and Mathematical Correspondence.* Chicago: University of Chicago Press. Abridged from the German edition by Brian McGuinness and translated by Hans Kaal.

Gentzen, G. 1936. Die Widerspruchsfreiheit der Stufenlogik. *Mathematische Zeitschrift*, *41*, 357–366. An English translation, "The Consistency of the Simple Theory of Types," is contained in (Szabo, 1969).

Gödel, K. 1931. Über formal unentscheidbare Sätze der *Principia Mathematica* und verwandter Systeme I. *Monatshefte für Mathematik und Physik*, *38*, 173–198. An English translation, "On formally undecidable propositions of *Principia Mathematica* and related systems I," is contained in (Feferman et al., 1986).

Gödel, K. 1936. Über die Länge von Beweisen. *Ergebnisse eines mathematischen Kolloquiums*, *7*, 23–24. An English translation, "On the Length of Proofs," is contained in (Davis, 1965). A new translation by S. Bauer–Mengelberg and J. van Heijenoort is in (Feferman et al., 1986).

Gödel, K. 1944. Russell's mathematical logic. In P. A. Schilpp, ed., *The Philosophy of Bertrand Russell*, vol. 5 of *Library of Living Philosophers*, pp. 123–153. Evanston, Ill.: Northwestern University Press. Reprinted in (Feferman et al., 1990).

Gödel, K. 1947. What is Cantor's continuum problem? *American Mathematical Monthly*, *54*, 515–525. Reprinted in (Feferman et al., 1990).

Gödel, K. 1964. What is Cantor's continuum problem? In (Benacerraf and Putnam, 1964), pp. 470–485. This is a revised version of (Gödel, 1947), and is reprinted in (Feferman et al., 1990).

Gödel, K. 1990. Some remarks on the undecidability results. In (Feferman et al., 1990), pp. 305–306.

Gödel, K. 1995a. Is mathematics syntax of language? In (Feferman et al., 1995), pp. 334–362.

Gödel, K. 1995b. The present situation in the foundations of mathematics. In (Feferman et al., 1995), pp. 45–53.

Gödel, K. 1995c. Some basic theorems on the foundations of mathematics and their implications. In (Feferman et al., 1995), pp. 304–323.

Gödel, K. 1995d. Some observations about the relationship betweeen theory of relativity and Kantian philosophy. In (Feferman et al., 1995), pp. 230–259.

Gödel, K. 1995e. Undecidable diophantine propositions. In (Feferman et al., 1995), pp. 164–175.

Goodman, N. 1972. A world of individuals. In *Problems and Projects*. Indianapolis: Bobbs–Merrill.

Gupta, A. 1982. Truth and paradox. *Journal of Philosophical Logic*, *11*, 1–60. Reprinted in (Martin, 1984). Page references are to the latter.

Hallett, M. 1984. *Cantorian Set Theory and Limitation of Size*. Oxford: Clarendon Press.

Hazen, A. 1985. Review of Crispin Wright's *Frege's Conception of Numbers as Objects*. *Australasian Journal of Philosophy*, *63*, 251–254.

Heck, Jr., R. G. 1992. On the consistency of second-order contextual definitions. *NOÛS*, *26*, 491–495.

Heck, Jr., R. G. 1993. The development of arithmetic in Frege's *Grundgesetze der Arithmetik*. *Journal of Symbolic Logic*, *58*, 579–601.

Heck, Jr., R. G. 1997. The finite and the infinite in Frege's *Grundgesetze der Arithmetik*. In M. Schirn, ed., *Philosophy of Mathematics Today*. Oxford: Clarendon Press.

Henkin, L. 1960. On mathematical induction. *American Mathematical Monthly*, *67*, 323–338.

Hermes, H. et al., eds. 1979. *Gottlob Frege: Posthumous Writings*. University of Chicago Press. Translated by Peter Lang and Roger White.

Hinman, P. G., Kim, J., and Stich, S. P. 1968. Logical truth revisited. *Journal of Philosophy*, *65*, 495–500.

Hintikka, J. 1974. Quantifiers vs. quantification theory. *Linguistic Inquiry*, *5*, 153–177.

Hodes, H. 1984. Logicism and the ontological commitments of arithmetic. *Journal of Philosophy*, *81*, 123–149.

Hodges, W. 1977. *Logic*. New York: Penguin Books.

Hook, S., ed. 1960. *Dimensions of Mind: A Symposium*. New York: New York University Press.

Jech, T. 1973. *The Axiom of Choice*. Amsterdam: North-Holland.

Jeffrey, R. 1967. *Formal Logic: Its Scope and Limits*. New York: McGraw–Hill.

Jeffrey, R. 1981. *Formal Logic: Its Scope and Limits*. Second edition. New York: McGraw–Hill.

Kant, I. 1933. *Critique of Pure Reason*. London: Macmillan. Revised English translation of *Kritik der reinen Vernunft* (1781/1787) by Norman Kemp Smith.

Kelley, J. L. 1955. *General Topology*. Princeton: Van Nostrand.

Kneale, W. and Kneale, M. 1984. *The Development of Logic*. Oxford: Clarendon Press.

Kreisel, G. 1967. Informal rigour and completeness proofs. In I. Lakatos, ed., *Problems in the Philosophy of Mathematics*, pp. 138–171. Amsterdam: North-Holland.

Kreisel, G. 1980. Kurt Gödel: 28 April 1906–14 January 1978. *Biographical Memoirs of Fellows of the Royal Society*, *26*, 149–224.

Leonardi, P. and Santambrogio, M., eds. 1995. *On Quine*. Cambridge: Cambridge University Press.

Lévy, A. 1968. On von Neumann's axiom system for set theory. *American Mathematical Monthly*, *75*, 762–763.

Lévy, A. 1979. *Basic Set Theory*. Berlin: Springer-Verlag.

Lewis, D. 1991. *Parts of Classes*. Oxford: Blackwell.

Littlewood, J. E. 1986. *Littlewood's Miscellany*. Cambridge: Cambridge University Press.

Lucas, J. R. 1961. Minds, machines and Gödel. *Philosophy, 36*, 112–137.

Martin, R. L., ed. 1984. *Recent Essays on Truth and the Liar Paradox*. Oxford: Oxford University Press.

Mates, B. 1972. *Elementary Logic*. Second edition. New York: Oxford University Press.

Mates, B. 1981. *Sceptical Essays*. Chicago: University of Chicago Press.

McGee, V. 1992. Maximal consistent sets of instances of Tarski's Schema (T). *Journal of Philosophical Logic, 21*, 235–241.

Monk, D. 1976. *Mathematical Logic*. New York: Springer-Verlag.

Nagel, E. and Newman, J. 1958. *Gödel's Proof*. New York: New York University Press.

Nerode, A. and Harington, L. A. 1984. The work of Harvey Friedman. *Notices of the American Mathematical Society, 31*, 563–566.

Parikh, R. 1971. Existence and feasibility in arithmetic. *Journal of Symbolic Logic, 36*, 494–508.

Parsons, C. D. 1983a. Frege's theory of number. In (Parsons, 1983b). Reprinted in (Demopoulos, 1995).

Parsons, C. D. 1983b. *Mathematics in Philosophy: Selected Essays*. Ithaca: Cornell University Press.

Parsons, C. D. 1983c. Sets and classes. In (Parsons, 1983b).

Parsons, C. D. 1990. The structuralist view of mathematical objects. *Synthese, 84*, 303–347.

Parsons, C. D. 1995. Quine and Gödel on analyticity. In (Leonardi and Santambrogio, 1995).

Parsons, T. 1987. On the consistency of the first-order portion of Frege's logical system. *Notre Dame Journal of Formal Logic, 28*, 161–168.

Penrose, R. 1989. *The Emperor's New Mind*. New York: Oxford University Press.

Putnam, H. 1960. Minds and machines. In (Hook, 1960), pp. 148–179.

Putnam, H. 1971. *Philosophy of Logic*. New York: Harper & Row.

Putnam, H. 1990. *Realism With a Human Face*. Cambridge, Mass.: Harvard University Press.

Quine, W. V. O. 1951. Two dogmas of empiricism. *Philosophical Review, 60*, 20–43. Reprinted in (Quine, 1953).

Quine, W. V. O. 1953. *From a Logical Point of View*. Cambridge, Mass.: Harvard University Press.

Quine, W. V. O. 1955. *Mathematical Logic.* Cambridge, Mass.: Harvard University Press.

Quine, W. V. O. 1960. *Word and Object.* Cambridge, Mass.: The MIT Press.

Quine, W. V. O. 1969. *Set Theory and Its Logic.* Second edition. Cambridge, Mass.: Harvard University Press.

Quine, W. V. O. 1970. *Philosophy of Logic.* Englewood Cliffs, New Jersey: Prentice–Hall.

Quine, W. V. O. 1972. *Methods of Logic.* Third edition. New York: Holt, Rinehart & Winston.

Quine, W. V. O. 1982. *Methods of Logic.* Fourth edition. Cambridge, Mass.: Harvard University Press.

Raymond, E. S. 1993. *The New Hacker's Dictionary.* Second edition. Cambridge, Mass.: The MIT Press.

Resnik, M. 1980. *Frege and the Philosophy of Mathematics.* Ithaca: Cornell University Press.

Robbin, J. 1969. *Mathematical Logic: A First Course.* New York: W. A. Benjamin.

Russell, B. 1919. *Introduction to Mathematical Philosophy.* London: Allen and Unwin.

Russell, B. 1959. *My Philosophical Development.* New York: Simon and Schuster.

Russell, B. 1973. On "insolubilia" and their solution by symbolic logic. In D. Lackey, ed., *Bertrand Russell, Essays in Analysis.* New York: George Braziller.

Russell, B. 1982. *The Principles of Mathematics.* Second edition. London: Allen & Unwin.

Russinoff, I. S. 1983. *Frege's Problem about Concepts.* Ph.D. thesis, Department of Linguistics and Philosophy, MIT, Cambridge, Mass.

Schwichtenberg, H. 1977. Proof theory: Some applications of cut-elimination. In J. Barwise, ed., *Handbook of Mathematical Logic,* pp. 867–895. Amsterdam: North-Holland.

Scott, D. 1974. Axiomatizing set theory. In T. Jech, ed., *Axiomatic Set Theory (Proceedings of Symposia in Pure Mathematics XIII.2),* pp. 207–214. Providence, Rhode Island: American Mathematical Society.

Shapiro, S. 1984. Principles of logic and principles of reflection. *Journal of Symbolic Logic, 49,* 1446–1147.

Shoenfield, J. R. 1967. *Mathematical Logic.* Reading, Mass.: Addison–Wesley.

Shoenfield, J. 1978. Axioms of set theory. In J. Barwise, ed., *Handbook of Mathematical Logic,* pp. 321–344. Amsterdam: North-Holland.

Skolem, T. 1970. *Selected Works in Logic.* Oslo: Universitetsforlaget.

Smiley, T. J. 1988. Frege's "series of natural numbers." *Mind, 97,* 388–389.

Smullyan, R. 1968. *First-Order Logic.* New York: Springer-Verlag.

Smullyan, R. 1978. *What Is the Name of This Book?* Englewood Cliffs, New Jersey: Prentice–Hall.

Statman, R. 1978. Bounds for proof-search and speed-up in the predicate calculus. *Annals of Mathematical Logic, 15,* 225–287.

Szabo, M., ed. 1969. *The Collected Papers of Gerhard Gentzen.* Amsterdam: North-Holland.

Tait, W. W. 1968. Constructive reasoning. In B. van Rootselaar and J. F. Staal, eds., *Logic, Methodology, and Philosophy of Science III.* Amsterdam: North-Holland.

Tait, W. W. 1972. Normal derivability in classical logic. In J. Barwise, ed., *The Syntax and Semantics of Infinitary Languages,* Lecture Notes in Mathematics 72, pp. 204–236. Berlin: Springer-Verlag.

Tait, W. W. 1981. Finitism. *Journal of Philosophy, 78,* 524–546.

Tarski, A. 1960. On the concept of logical consequence. In A. Tarski, *Logic, Semantics, Metamathematics.* New York: Oxford University Press.

van Heijenoort, J., ed. 1967. *From Frege to Gödel: A Source Book in Mathematical Logic, 1879–1931.* Cambridge, Mass.: Harvard University Press.

Wang, H. 1957. The axiomatization of arithmetic. *Journal of Symbolic Logic, 22,* 145–158.

Wang, H. 1963. *A Survey of Mathematical Logic.* Peking: Science Press.

Wang, H. 1974. *From Mathematics to Philosophy.* New York: Humanities Press.

Wang, H. 1987. *Reflections on Kurt Gödel.* Cambridge, Mass.: The MIT Press.

Whitehead, A. N. and Russell, B. 1910–1913. *Principia Mathematica,* 3 vols. Cambridge: Cambridge University Press.

Whitehead, A. N. and Russell, B. 1927. *Principia Mathematica,* 3 vols. Second edition. Cambridge: Cambridge University Press.

Wright, C. 1983. *Frege's Conception of Numbers as Objects.* Scots Philosophical Monographs, vol. 2. Aberdeen: Aberdeen University Press.

Zermelo, E. 1908. Untersuchungen über die Grundlagen der Mengenlehre, I. *Mathematische Annalen, 65,* 261–281. An English translation, "Investigations in the Foundations of Set Theory, I," is contained in (van Heijenoort, 1967).

Index